ENGINEERING CHEMISTY

CONCEPTS IN CHEMISTRY FOR ENGINEERING

REVISED EDITION

DR. MANISHA AGRAWAL
Dean, Basic Sciences,
Chhattisgarh Swami Vivekananda Technical University, Bhilai,
Professor and Head, Department of Chemistry
Rungta College of Engineering & Technology, Bhilai (C.G.)

DR. SATYA PRAKASH
Professor
Department of Applied Sciences
Guru Teg Bahadur Institute of Technology
(GGSIP University)

KHANNA BOOK PUBLISHING CO. (P) LTD.
Publisher of Engineering and Computer Books
4C/4344, Ansari Road, Darya Ganj, New Delhi-110002
Phone : 011-23244447-48 **Mobile:** +91-9910909320
E-mail : contact@khannabooks.com
Website : www.khannabooks.com

Dear Readers,

To prevent the piracy, this book is secured with HIGH SECURITY HOLOGRAM on the front title cover. In case you don't find the hologram on the front cover title, please write us to at contact@khannabooks.com or whatsapp us at +91-99109 09320 and avail special gift voucher for yourself.

Specimen of Hologram on front Cover title:

Moreover, there is a SPECIAL DISCOUNT COUPON for you with EVERY HOLOGRAM.

How to avail this SPECIAL DISCOUNT:

Step 1: Scratch the hologram
Step 2: Under the scratch area, your "coupon code" is available
Step 3: Logon to www.khannabooks.com
Step 4: Use your "coupon code" in the shopping cart and get your copy at a special discount
Step 5: Enjoy your reading!

Price: ₹ 399.00

ISBN: 978-93-80016-75-7

Engineering Chemistry
by Dr. Manisha Agrawal, Dr. Satya Prakash

First Edition: 2010
Second Edition: 2022
Reprint: 2023, **2024**

To view complete list of books, Please scan the QR Code:

KPH

Published by:
Khanna Book Publishing Co. (P) Ltd.
CIN: U22110DL1998PTC095547

Visit us at: **www.khannabooks.com**
Write us at: **contact@khannabooks.com**

Printed in India.

PREFACE

This book titled– **"Engineering Chemistry: Concepts in Chemistry for Engineering"** covers the syllabus of Chemistry – I, as per AICTE Model Curriculum for First Year Undergraduate degree courses in Engineering & Technology 2018. The Curriculum has been designed where the students can understand the industry requirements and have hands-on experience. The students will develop a problem - solving approach and will be able to meet the challenges of future.

To fulfil the aims and objectives of the above, this book is written for undergraduate students of Engineering. The Book contains six chapters, which strictly follows the model syllabus revised and framed by honourable team members. Each chapter contains the objectives, diagram, tables, flowchart, equations, examples, summary and related practice questions to make readers friendly. The separate reference section is cited at the end of this book and authors are thankful to all. We hope that this book will be advantageous for the students to enhance their skills and knowledge.

Valuable suggestions, corrections, and feedbacks of students, teachers, and readers are appreciated for improvement in to next edition.

Authors

CONTENTS

1 Atomic and Molecular Structure

Schrodinger equation. Particle in a box solutions and their applications for conjugated molecules and nanoparticles. Forms of the hydrogen atom wave functions and the plots of these functions to explore their spatial variations. Molecular orbital of diatomic molecules and plots of the multicenter orbitals. Equations for atomic and molecular orbitals. Energy level diagrams of diatomic. Pi-molecular orbitals of butadiene and benzene and aromaticity. Crystal field theory and the energy level diagrams for transition metal ions and their magnetic properties. Band structure of solids and the role of doping on band structures.

Introduction

In 1911 Lord Rutherford proposed nuclear theory of atomic structure. In 1913 Bohr modified this theory by applying quantum theory of M. Planck. According the Bohr Model, electron of an atom move in close circular path called orbits. Each orbit is associated with a definite amount of energy. Electrons moving in orbits neither absorb or emit energy. This Bohr model could explain successfully the hydrogen atom spectrum. However, it could not explain fine structure and atomic spectrum of atoms having more than one electron. Nowadays nature of electrons in atom is very well defined by wave mechanical concept of atom which is based on de Broglie theory, Heisenberg's uncertainty principle and Schrodinger Wave Equation.

SCHRODINGER EQUATION

Schrodinger was the first person to write down such a wave equation. The schrodinger equation stands for "quantum mechanics" it basically used to find the energy level in a mechanical system and the wave function gives the probability of finding a particle. The eigen values of the wave equation were shown to be equal to the energy levels of the quantum mechanical system, and the best test of the equation was when it was used to solve for the energy levels of the Hydrogen atom, and the energy levels were found to be in accord with Rydberg's Law.

Debroglie 1924 proposed dual behaviour of electrons and gave the equation

$$E = mc^2 \qquad ...(1.1)$$

Where,

E = energy, m = mass and c = speed of light

Using Planck's theory which states every quantum of a wave has a discrete amount of energy given by Planck's equation :

$$E = h\nu \qquad ...(1.2)$$

with

E = energy, h = Plank's constant (6.62607×10^{-34} J s) and ν = frequency

Since de Broglie believed particles and wave have the same traits, he hypothesized that the two energies would be equal :

$$mc^2 = h\nu \quad ...(1.3)$$

Because real particles do not travel at the speed of light, De Broglie submitted velocity (v) for the speed of light (c).

$$mv^2 = h\nu \quad ...(1.4)$$

Where, m = mass and v = velocity

For assigning sense and meaning of probability approach, Schrodinger derived an equation known after his name as Schrodinger's wave equation calculation of the probability of finding the electrons at various points in an atom was the main challenges before Schrodinger. His equation is the key note of wave mechanics and is based upon the idea of electrons "standing wave" around the nucleus. The equation for the standing wave , comparable with that of a stretched string is

$$\Psi = A \sin \frac{2\pi x}{\lambda} \quad ...(1.5)$$

Where,

Ψ = amplitude of the wave

λ = wavelength of wave

X = distance from the origin

A = constant

On the double differentiation w.r.t. x, we get

$$\frac{d\psi}{dx} = A . \frac{2\pi}{\lambda} . \cos \frac{2\pi x}{\lambda} \quad ...(1.6)$$

$$\frac{d^2\psi}{dx} = A . \frac{2\pi}{\lambda} . \frac{2\pi}{\lambda} \left[-\sin \frac{2\pi x}{\lambda} \right] \quad ...(1.7)$$

$$= \frac{4\pi^2}{\lambda^2} A \sin \frac{2\pi x}{\lambda} \quad ...(1.8)$$

Substituting $\psi = A \sin 2\pi x/\lambda$ from equation (1.5) in to Eq. (1.8) we get

$$\frac{d^2\Psi}{dx} = \frac{4\pi^2}{\lambda^2}\Psi \quad ...(1.9)$$

$$\frac{d^2\Psi}{dx} + \frac{4\pi^2}{\lambda^2}\Psi = 0 \quad ...(1.10)$$

or

This is the equation for one dimensional standing wave which can be extended to describe motion of an electron in three dimensional by a second order partial differential equation (1.11)

$$\frac{d^2\Psi}{dx^2} + \frac{d^2\Psi}{dy^2} + \frac{d^2\Psi}{dz^2} + \frac{4\pi^2}{\lambda^2}\Psi = 0 \qquad ...(1.11)$$

Where, ψ is a function of Cartesian coordinates x, y, z.

This may be written more precisely as

$$\nabla^2\psi + \frac{4\pi^2}{\lambda^2}\psi = 0 \qquad ...(1.12)$$

Where, $\nabla^2\Psi$ stands for $\frac{d^2\psi}{dx^2} + \frac{d^2\psi}{dy^2} + \frac{d^2\psi}{dz^2}$

On combining eq. (1.11) and (1.12) we get

$$\nabla^2\psi + \frac{4\pi^2 m^2 v^2}{h^2}\psi = 0 \qquad ...(1.13)$$

Considering the electron as a particle , the total energy (E) of the system is sum of kinetic ($\frac{1}{2}$ mv²) and potential energy (V) energies, i.e.,

$$E = \frac{1}{2}mv^2 + V$$

$$V^2 = \frac{2(E-V)}{m} \qquad ...(1.14)$$

Or

In order to calculate values for the energy state value of V² (eq. 1.14)is substituted into (eq. 1.13) to obtained a fundamental form of Schrodinger's wave equation for a single particle in three dimensions.

$$\nabla^2\psi + \frac{8\pi^2 m}{h^2}(E-V)\psi = 0 \qquad ...(1.15)$$

Or

$$\frac{d^2\psi}{dx^2} + \frac{d^2\psi}{dy^2} + \frac{d^2\psi}{dz^2} + \frac{8\pi^2 m}{h^2}\psi = 0 \qquad ...(1.16)$$

Where,

ψ = amplitude of the wave function associated with the electron

x, y, z = Cartesian coordinates

m = mass of electron

h = Planck's constant

E = Total energy of the system

V = Potential energy of the system

The potential energy term V is Schrodinger wave equation is given by following equation:

$$V = \frac{-ze^2}{r} \qquad ...(1.17)$$

Where, + Z_e = Charge of nucleus

Z = atomic number

– e = Charge of electron

Consequently, the equation (1.16) for one electron atomic system in three dimensions may be written as

$$\frac{d^2\psi}{dx^2}+\frac{d^2\psi}{dy^2}+\frac{d^2\psi}{dz^2}+\frac{8\pi^2 m}{h^2}\left[E+\frac{ze^2}{r}\right]\psi = 0 \qquad ...(1.18)$$

The Schrodinger wave equation, being a second order differential equation, has innumerable solutions, many of them have no significance at all. Out of these, only those solutions are significant which give certain definite and acceptable values of the total energy, E. The acceptable value of total energy E given by a solution called eigen values and the corresponding function which gives eigen value of energy is called eigen function. Thus , every function is not an eigen (or acceptable) function.

In terms of polar co-ordinates, above equation can be written as follows:

$$\frac{1}{r^2}\frac{\partial}{\partial r}\left[r^2\frac{\partial\psi}{\partial r}\right]+\frac{1}{r^2\sin\theta\partial\theta}\left[\sin\theta\frac{\partial\psi}{\partial\theta}\right]+\frac{1}{r^2\sin^2\theta}\frac{\partial^2\psi}{\partial\theta^2}+\frac{8\pi^2 m}{h^2}\left[E+\frac{ze^2}{r}\right]\psi = 0 \qquad ...(1.19)$$

Schrodinger Wave equation may be written as:

$$\frac{\partial^2\psi}{\partial X^2}+\frac{8\pi^2 m}{h^{2-}}(E-V)\psi = 0 \qquad ...(1.20)$$

Where,

$\frac{\partial^2\psi}{\partial X^2}$ = second derivation with respect to X

X^2 = Position

ψ = Schrodinger Wavefunction

E = Energy

V = Potential Energy

Application of Schrodinger wave equation

- To explain behaviour of micro particles such as electrons, protons and atoms.
- To determine total energy E and wave function ψ if potential energy is known.
- To determine translational energy of a microparticle in space.
- To explain atomic and molecular structure.
- To analyse molecular motions.

THE HYDROGEN ATOM; ATOMIC ORBITALS

Atomic Spectra

When gaseous hydrogen in a glass tube is excited by a 5000–volt electrical discharge, four lines are observed in the visible part of the emission spectrum red at 656.3 nm, blue-green at 486.1 nm, blue violet at 434.1 nm and violet at 410.2 nm:

Other series of lines have been observed in the ultraviolet and infrared regions. Rydberg (1890) found that all the lines of the atomic hydrogen spectrum could be fitted to a single formula

$$1/\lambda = RZ^2(1/n_1^2 - 1/n_2^2) \qquad ...(1.21)$$

where

λ is the wavelength of the photon (wavenumber = 1/wavelength)

R = Rydberg's constant (1.0973731568539 (55) × 10^7 m^{-1})

Z = atomic number of the atom

n_1 and n_2 are integers where $n_2 > n_1$. where n_1 = 1, 2, 3 ..., $n_2 > n_1$

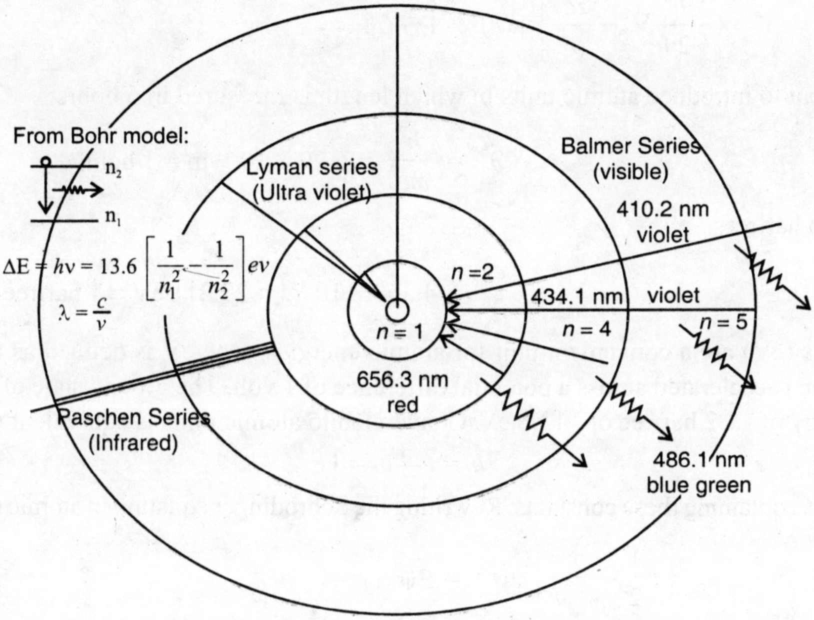

Fig.1.1: Visible spectrum of atomic hydrogen

where R, known as the Rydberg constant, has the value 109,677 cm^{-1} for hydrogen. The Balmer series of spectral lines in the visible region, shown in Fig. 1.1, correspond to the values $n_1 = 2$, $n_2 = 3, 4, 5$ and 6.

The lines with $n_1 = 1$ in the ultraviolet make up the Lyman series.

The line with $n_2 = 2$, designated the Lyman alpha, has the longest wavelength (lowest wavenumber) in this series, with $1/\lambda = 82.258$ cm^{-1} or $\lambda = 121.57$ nm.

Other atomic species have line spectra, which can be used as a "fingerprint" to identify the element. However, no atom other than hydrogen has a simple relation analogous to (1) for its spectral frequencies.

Bohr in 1913 proposed that all atomic spectral lines arise from transitions between discrete energy levels, giving a photon such that

$$\Delta E = h\nu = hc / \lambda \qquad ...(1.22)$$

This is called the Bohr frequency condition. We now understand that the atomic transition energy ΔE is equal to the energy of a photon, as proposed earlier by Planck and Einstein.

Various models and theories are proposed by scientists to know the structure of Hydrogen atom. These are Rutherford model, Bohr's model, Sommerfield-Wilson quantum conditions model and Quantum Mechanics of Hydrogen Atom.

Quantum Mechanics of Hydrogen like Atoms

In contrast to the particle in a box and the harmonic oscillator, the hydrogen atom is a real physical system that can be treated exactly by quantum mechanics. In addition to their inherent significance, these solutions suggest prototypes for atomic orbital's used in approximate treatments of complex atoms and molecules. For an electron in the field of a nucleus of charge $+Z_e$, the Schrodinger equation can be written

$$\left\{-\frac{h^2}{2m}\nabla^2 - \frac{ze^2}{r}\right\}\psi(r) = E\psi(r) \qquad ...(1.23)$$

It is convenient to introduce atomic units in which length is measured in 6 bohrs:

$$\partial^0 = \frac{h^2}{me^2} = 5.29 \times 10^{-11}\ \text{m} \equiv 1\ \text{bohr}$$

and energy in hartrees:

$$\frac{e^2}{\partial^0} = 4.358 \times 10^{-18}\text{J} = 27.211\ \text{eV} \equiv 1\ \text{hartree}$$

Electron volts (eV) are a convenient unit for atomic energies. One eV is defined as the energy an electron gains when accelerated across a potential difference of 1 volt. The ground state of the hydrogen atom has an energy of –1/2 hartree or –13.6 eV. Conversion to atomic units is equivalent to setting

$$h = e = m = 1$$

in all formulas containing these constants. Rewriting the Schrodinger equation in atomic units, we have

$$\left\{-\frac{1}{2}\nabla^2 - \frac{z}{r}\right\}\psi(r) = E\psi(\text{r}) \qquad ...(1.24)$$

Since the potential energy is spherically symmetrical (a function of r alone), it is obviously advantageous to treat this problem in spherical polar coordinates r, θ, φ. Expressing the Laplacian operator in these coordinates

$$-\frac{1}{2}\left\{\frac{1}{r^2}\frac{\partial}{\partial r}\partial^2\frac{\partial}{\partial r} + \frac{1}{r^2 \sin\theta}\frac{\partial}{\partial\theta}\sin\theta\frac{\partial}{\partial\theta} + \frac{1}{r^2\sin^2\theta}\frac{\partial^2}{\partial\varphi^2}\right\}$$

$$\text{X}\,\psi(r, \theta, \varphi) - \psi(r, \theta, \varphi) = \text{E}\,\psi(r, \theta, \varphi) \qquad ...(1.25)$$

Eq 1.25 will have separable solutions of the form

$$\psi(r, \theta, \varphi) = \text{R}(r)\,\text{Y}\,m(\theta, \varphi) \qquad ...(1.26)$$

Substituting (1.25) into (1.26) and using the angular momentum eigenvalues equation , we obtain an ordinary differential equation for the radial function R (r):

$$\left\{-\frac{1}{2r^2}\frac{d}{dr}r^2\frac{d}{dr} + \frac{1(1+1)}{2r^2} - \frac{2}{r}\right\}\text{R}(r) = \text{ER}(r) \qquad ...(1.27)$$

Note that in the domain of the variable r, the angular momentum contribution acts as an effective addition to the potential energy. It can be identified with centrifugal force, which pulls the electron outward, in opposition to the Coulomb attraction. Carrying out the successive differentiations in (1.28) and simplifying, we obtain

$$\frac{1}{2}R''(r)+\frac{1}{r}R'(r)\left[\frac{z}{r}-\frac{l(1+l)}{2r^2}+E\right]R(r) = 0 \qquad ...(1.28)$$

another second-order linear differential equation with non-constant coefficients. It is again useful to explore the asymptotic solutions to (1.28), as $r \to \infty$. In the asymptotic approximation,

$$R''(r) - 2|E|R(r) \approx 0 \qquad ...(1.29)$$

having noted that the energy E is negative for bound states. Solutions to (1.29) are

$$R(r) \approx \text{const } e^{\pm\sqrt{2|E|}\,r} \qquad ...(1.30)$$

We reject the positive exponential on physical grounds, since $R(r) \to \infty$ as $r \to \infty$, in violation of the requirement that the wave function must be finite everywhere. Choosing the negative exponental and setting $E = -Z^2/2$, the ground state energy in the Bohr theory (in atomic units), we obtain

$$R(r) \approx \text{const } e^{-Zr} \qquad ...(1.31)$$

It turns out, very fortunately, that this asymptotic approximation is also an exact solution of the Schrodinger equation (1.28) with = 0, .The solutions , designated $R_{nl}(r)$, are labelled by n, known as the principal 8 quantum number, as well as by the angular momentum, which is a parameter in the radial equation. The solution (1.31) corresponds to R10 (r). This should be normalized according to the condition

$$\int_0^\infty \left[R10(r^2)r^2dr = 1\right] \qquad ...(1.32)$$

A useful definite integral is

$$\int_0^\infty r^n e^{-\alpha r}\,dr = \frac{n!}{\alpha^{n-1}} \qquad ...(1.33)$$

The normalized radial function is thereby given by

$$R_{10}(r) = 2Z^{3/2}\,e^{-Zr} \qquad ...(1.34)$$

Since this function is nodeless, we identify it with the ground state of the hydrogen like atom. Multipyling (1.34) by the spherical harmonic $Y_{00} = 1/\sqrt{4\pi}$,

we obtain the total wavefunction (1.35)

$$\psi 100\,(r) = \left(\frac{z^3}{\pi}\right)^{1/2} e^{-Zr} \qquad ...(1.35)$$

This is conventionally designated as the 1s function $\psi 1s(r)$. Integrals in spherical-polar coordinates over a spherically-symmetrical integrand can be significantly simplified. We can do the reduction

$$\int_0^\infty\int_0^\pi\int_0^{2\pi} f(r)\,r^2\sin\theta\,dr\,d\theta\,d\varphi = \int_0^\infty f(r)\,4\pi r^2 dr \qquad ...(1.36)$$

since integration over θ and φ gives 4π, the total solid angle of a sphere. The normalization of the 1s wavefunction can thus be written as

$$\int_0^{\infty} [\psi 1s(r)]^2 \, 4\pi r^2 \, dr \qquad ...(1.37)$$

INTRODUCTION TO MOLECULAR ORBITAL THEORY

Valence Bond Theory fails to answer certain questions like Why He_2 molecule does not exist and why O_2 is paramagnetic? Therefore in 1932 F. Hood and R.S. Mulliken came up with theory known as Molecular Orbital Theory. According to Molecular Orbital Theory individual atoms combine to form molecular orbitals, as the electrons of an atom are present in various atomic orbitals and are associated with several nuclei.

LINEAR COMBINATION OF ATOMIC ORBITALS (LCAO)

Molecular Orbitals

Homonuclear Diatomics and Heteronuclear Diatomics

Molecules consisting of two identical atoms ,which are said to be homonuclear diatomic, such as: H_2, N_2, O_2, and F_2.The diatomic molecules are linear and bond is non-polar due to the electronegativity difference of zero.

Molecules consisting of two non-identical atoms are said to be heteronuclear diatomic, such as: CO, NO, HF, and LiF.

Phases and Nodes

Phases are designated either (+) or (–) relative to their wave "up" or wave "down" displacements. A node occurs if the phase signs change from (+) to (–) or vice versa. It is important to notice that the phase signs do NOT symbolize charges. Nodes are regions where the probability of finding an electron is ZERO.

Sigma and Pi Bonds

A sigma-bond (σ) is an "end-to-end" bond formed from symmetric atomic orbitals. A pi-bond (π) is formed from a "sideways" overlap.

Sigma bond exist in the region directly between two bonded atoms and pi bond exist in the region above and below a line drawn between two bonded atoms.

Table 1.1: Sigma and pi bonds

Single bond (_)	1 σ bond
Double bond (=)	1 σ bond and 1 π bonds
Triple bond (≡)	1 σ bond and 2 π bonds

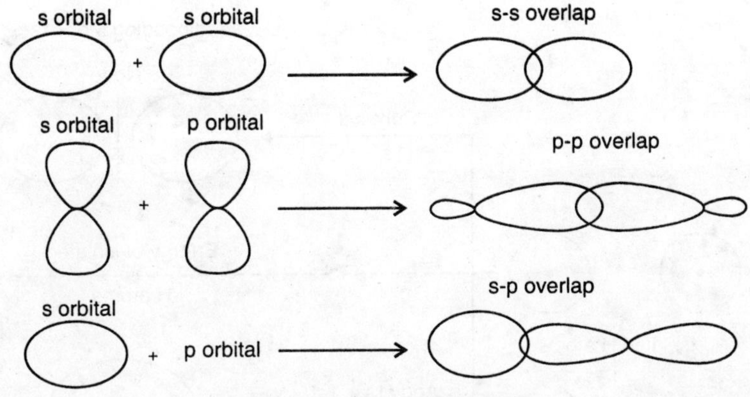

Fig.1.2: Sigma and pi bonds

Linear Combination of Atomic Orbitals (LCAO)

As per this method the formation of orbitals is because of Linear Combination (addition or subtraction) of atomic orbitals which combine to form molecule. Consider two atoms A and B which have atomic orbitals described by the wave functions ψ_A and ψ_B .If electron cloud of these two atoms overlap, then the wave function for the molecule can be obtained by a linear combination of the atomic orbitals ψ_A and ψ_B i.e. by subtraction or addition of wave functions of atomic orbitals

$$\psi_{MO} = \psi_A \pm \psi_B \qquad ...(1.38)$$

The above equation forms two molecular orbitals

a. Bonding Molecular Orbitals

When addition of wave function takes place, the type of molecular orbitals formed are called Bonding Molecular orbitals and is represented by

$$\psi_{MO} = \psi_A + \psi_{B.} \qquad ...(1.39)$$

They have lower energy than atomic orbitals involved. It is similar to constructive interference occurring in phase because of which electron probability density increases resulting in formation of bonding orbital. Molecular orbital formed by addition of overlapping of two s orbitals shown in figure no. 1.3. It is represented by s.

Anti-Bonding Molecular Orbitals

When molecular orbital is formed by subtraction of wave function, the type of molecular orbitals formed are called Antibonding Molecular Orbitals and is represented by

$$\psi_{MO} = \psi_A - \psi_{B.} \qquad ...(1.40)$$

They have higher energy than atomic orbitals. It is similar to destructive interference occurring out of phase resulting in formation of antibonding orbitals. Molecular Orbital formed by subtraction of overlapping of two s orbitals are shown in figure no. 1.3. It is represented by σ^* (* is used to represent antibonding molecular orbital) called Sigma Antibonding.

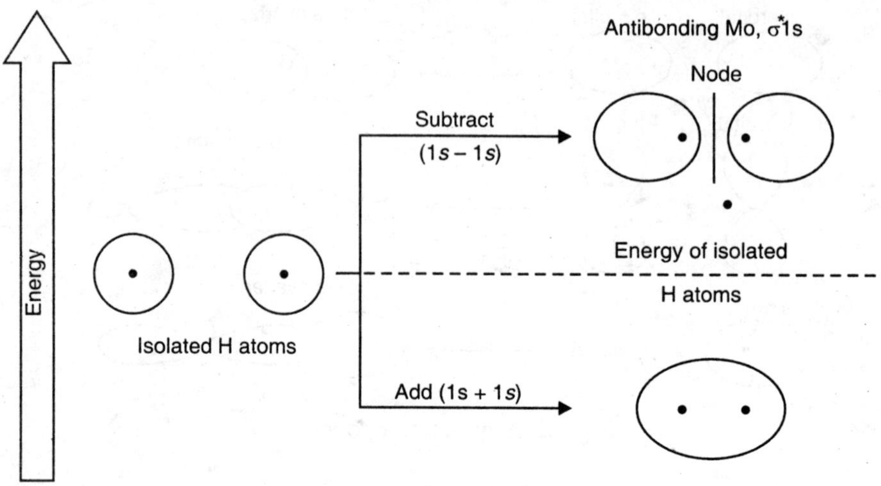

Fig.1.3: Formation of Bonding and Anti-Bonding Orbital

Therefore, Combination of two atomic orbitals results in formation of two molecular orbitals, bonding molecular orbital (BMO) whereas other is anti-bonding molecular orbital (ABMO).

BMO has lower energy and hence greater stability than ABMO. First BMO are filled then ABMO starts filling because BMO has lower energy than that of ABMO.

Formation of molecular orbitals occurs by the combination of atomic orbitals of proportional symmetry and comparable energy. Therefore, a molecular orbital is polycentric and atomic orbital is monocentric. Number of molecular orbitals formed is equal to the number of atomic orbitals.

Table 1.2: Differences between Molecular Orbital and Atomic Orbital

	Molecular Orbital		Atomic Orbital
1.	An electron Molecular orbital is under the influence of two or more nuclei depending upon the number of atoms present in the molecule.	1.	An electron in atomic orbital is under the influence of only one positive nucleus of the atom.
2.	Molecular orbitals are formed by combination of atomic orbitals	2.	Atomic orbitals are inherent property of an atom.
3.	They have complex shapes.	3.	They have simple shapes.

Molecular orbital diagrams

For a diatomic molecule, an MO diagram effectively shows the energetics of the bond between the two atoms, whose AO unbonded energies are shown on the sides.

For simple polyatomic molecules with a "central atom" such as methane (CH_4) or carbon dioxide (CO_2), a MO diagram may show one of the identical bonds to the central atom.

Overlapping of atomic orbitals – In MO theory molecular orbitals form by the overlap of atomic orbitals. Because σ bonds feature greater overlap than π bonds, σ and σ* bonding and antibonding orbitals feature greater energy splitting (separation) than π and π* orbitals. Two atomic orbitals can overlap in two ways depending on their phase relationship. The phase of an orbital is a direct consequence of the wave-like properties of electrons.

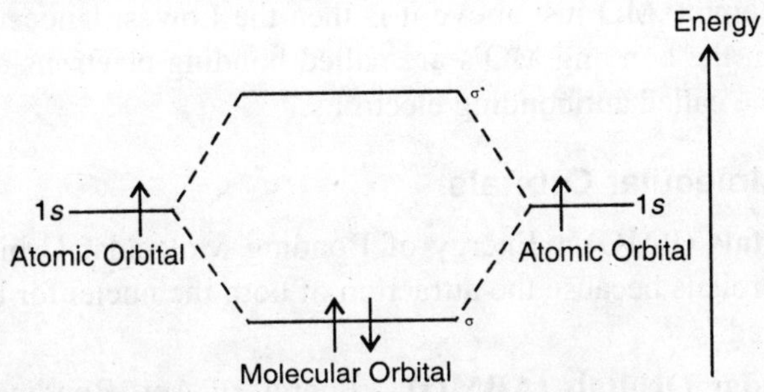

Fig.1.4: Molecular orbital diagrams

Constructive Overlap (Bonding Orbitals) – Two same-sign orbitals have a constructive overlap forming a molecular orbital with the bulk of the electron density located between the two nuclei. This MO is called the bonding orbital and its energy is lower than that of the original atomic orbitals.

Symmetry labels are further defined by whether the orbital maintains its original character after an inversion about its centre; if it does, it is defined gerade, *g*. If the orbital does not maintain its original character, it is ungerade, *u*.

Destructive Cancellation (Anti-Bonding Orbitals) – Atomic orbitals can also interact with each other out-of-phase which leads to destructive cancellation and no electron density between the two nuclei at the so-called nodal plane depicted as a perpendicular dashed line.

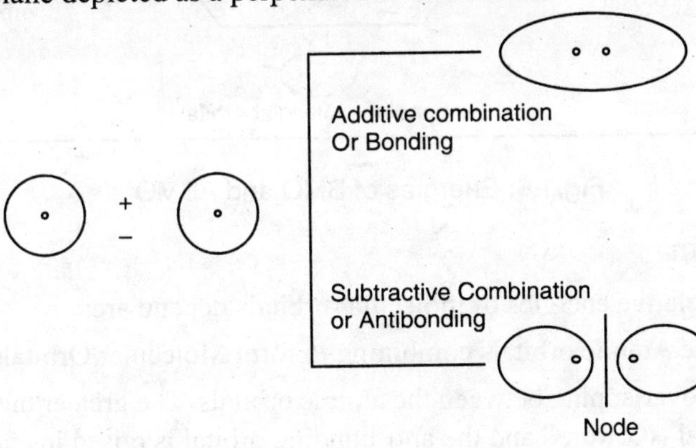

Fig.1.5: Bonding and anti-bonding molecular orbital

Filling of Electrons

The next step in constructing an MO diagram is filling the newly formed molecular orbitals with electrons. Three general rules apply:

- The Aufbau principle states that orbitals are filled starting with the lowest energy
- The Pauli exclusion principle states that the maximum number of electrons occupying an orbital is two, with opposite spins

- Hund's rule states that when there are several MO's with equal energy, the electrons occupy the MO's one at a time before two electrons occupy the same MO.

HOMO and LUMO – The filled MO highest in energy is called the Highest Occupied Molecular Orbital or HOMO and the empty MO just above it is then the Lowest Unoccupied Molecular Orbital or LUMO. The electrons in the bonding MO's are called bonding electrons and any electrons in the antibonding orbital would be called antibonding electrons.

Relative Energies of Molecular Orbitals

Bonding Molecular Orbitals (BMO) – Energy of Bonding Molecular Orbitals is less than that of Anti Bonding Molecular Orbitals because the attraction of both the nuclei for both the electron (of the combining atom) is increased.

Anti-Bonding Molecular Orbitals (ABMO) – Energy of Anti Bonding Molecular Orbitals is higher than Bonding Molecular Orbitals because the electron try to move away from the nuclei and are in repulsive state.

The Energies of Bonding Molecular Orbitals and Anti-Bonding Molecular Orbitals are shown in figure below:

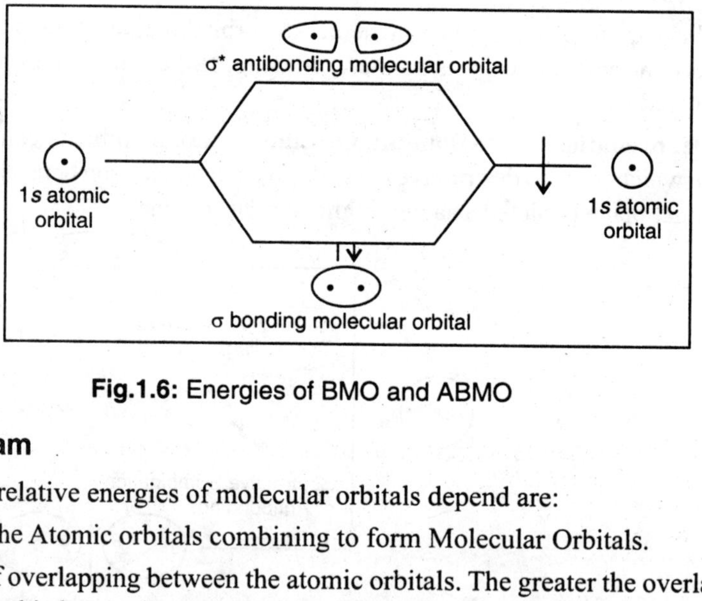

Fig.1.6: Energies of BMO and ABMO

Energy Level Diagram

The factors upon which relative energies of molecular orbitals depend are:

(*i*) Energies of the Atomic orbitals combining to form Molecular Orbitals.

(*ii*) The extent of overlapping between the atomic orbitals. The greater the overlap, the more the bonding orbital is lowered and the anti-bonding orbital is raised in energy relative to AOs.

1s Atomic Orbitals (AOs) of two atoms form two Molecular Orbitals (MOs) designated as s1s and s *1s.The 2s and 2p orbitals (eight AOs of two atoms) form four bonding MOs and four anti-bonding MOs as:

Bonding MOs: σ 2s, σ $2p_z$, π $2p_x$, π $2p_y$

Anti – Bonding MOσ: σ *2s, σ $*2p_z$, π $*2p_x$, π $*2p_y$

Using Spectroscopy, the energy levels of these molecular orbitals are determined experimentally. The order of increasing energy of molecular orbitals obtained by combination of 1s, 2s and 2p orbitals of two atoms is →

σ1s, σ *1s, σ 2s, σ *2s, σ $2p_z$, π $2p_x$ = π $2p_y$, π *$2p_x$ = π *$2p_y$, σ *$2p_z$

(Energy Increases from left to right)

The molecular orbital diagram representing this order of energy levels is shown in fig 1.7.

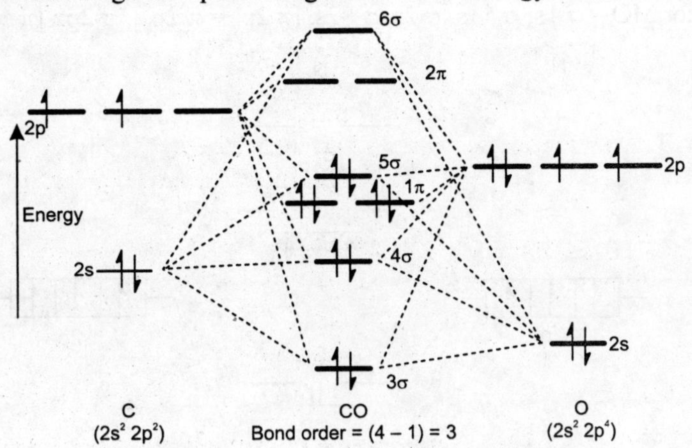

Fig.1.7: Order of Energy level

But experimental evidence for some diatomic molecules have shown that the above sequence of energy levels of MOs is not correct for all the molecules. For example, homonuclear diatomic molecules of second row elements like Li_2, Be_2, B_2, C_2, N_2, the σ $2p_z$ MOs is higher in energy than π $2p_x$ and π $2p_y$ MOs

For these atoms, the order is: →

σ1s, σ *1s, σ 2s, σ *2s, [π $2p_x$ = π $2p_y$], σ $2p_z$ [π *$2p_x$ = π *$2p_y$], σ *$2p_z$

ENERGY LEVEL DIAGRAM OF DIATOMIC MOLECULES

1. Hydrogen Molecule

Atomic number of hydrogen atom is 1

Electronic configuration – $1s^1$

Order of energy for MO – σ1s, σ *1s

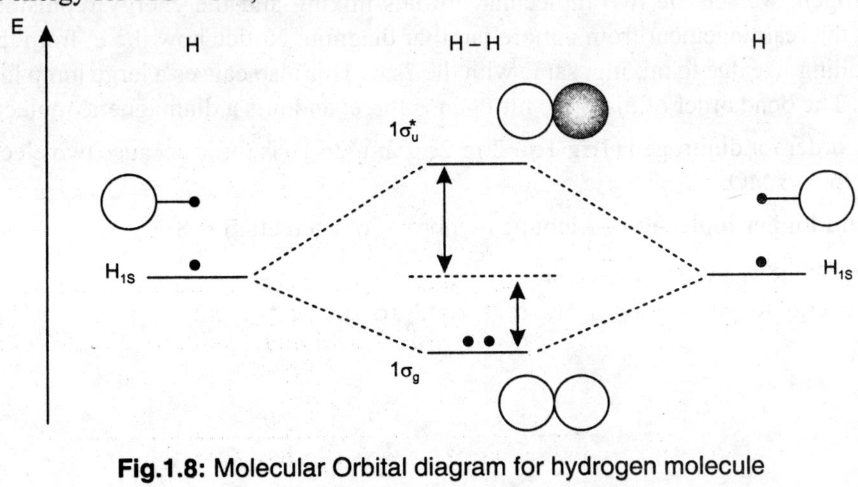

Fig.1.8: Molecular Orbital diagram for hydrogen molecule

For N_2 and lower molecules →

Molecular weight of nitrogen atom will ne – 7

Electronic configuration – $1s^2, 2s^2, 2p^3$

Order of energy for MO – σ 1s, σ *1s, σ 2s, σ *2s, [π $2p_x$ = π $2p_y$], σ 2px [π *$2p_x$= π *$2p_y$], σ*$2p_z$

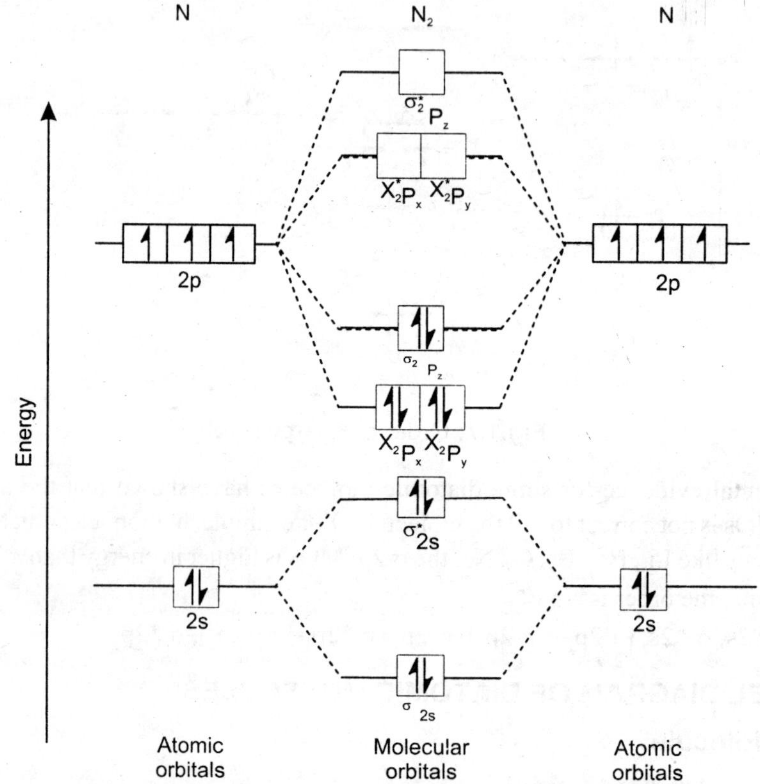

Fig.1.9: Order of Energy for N_2 and lower molecules

With nitrogen, we see the two molecular orbitals mixing and the energy repulsion. This is the reasoning for the rearrangement from a more familiar diagram. Notice how the e^- from the 2p behaves more non-bonding like due to mixing, same with the 2s σ. This also causes a large jump in energy in the 2p σ* orbital. The bond order of diatomic nitrogen is three, and it is a diamagnetic molecule.

The bond order for dinitrogen ($1\sigma g^2 1\sigma u^2 2\sigma g^2 2\sigma u^2 2\pi u^4 2\sigma g^2$) is three because two electrons are now also added in the 3σ MO.

For O_2 and higher molecules → atomic number of oxygen atom is 8

Electronic configuration – $1s^2, 2s^2, 2p^4$

Order of energy for MO – σ1s, σ *1s, σ 2s, σ *2s, σ $2p_z$, [π$2p_x$ = π$2p_y$], [π*$2p_x$= π*$2p_y$], σ *$2p_z$

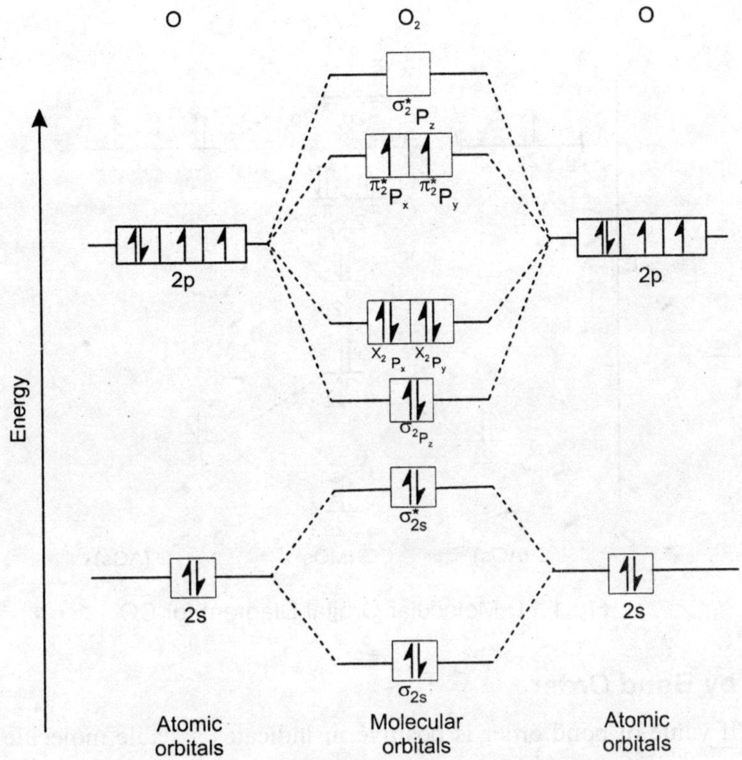

Fig.1.10: Order of Energy for O_2 and Higher molecules

Oxygen has a similar setup to H_2, but now we consider 2s and 2p orbitals. When creating the molecular orbitals from the p orbitals, notice the three atomic orbitals split into three molecular orbitals, a singly degenerate σ and a doubly degenerate π orbital. Another property we can observe by examining molecular orbital diagrams is the magnetic property of diamagnetic or paramagnetic. If all the electrons are paired, there is a slight repulsion and it is classified as diamagnetic. If unpaired electrons are present, it is attracted to a magnetic field, and therefore paramagnetic. Oxygen is an example of a paramagnetic diatomic. Also notice the bond order of diatomic oxygen is two.

Bond Order

It may be defined as the half of difference between the number of electrons present in the bonding orbitals and the antibonding orbitals that is,

$$\text{Bond order (B.O.)} = \frac{(\text{No.of electrons in BMO} - \text{No.of electrons in ABMO})}{2} \quad ...(1.41)$$

Those with positive bond order are considered stable molecule while those with negative bond order or zero bond order are unstable molecule.

Magnetic Behavior: If all the molecular orbitals in species are spin paired, the substance is diamagneti. But if one or more molecular orbitals are singly occupied it is paramagnetic. For Example, if we look at CO Molecule, it is diamagnetic as all the electron in CO are paired as in the figure below:

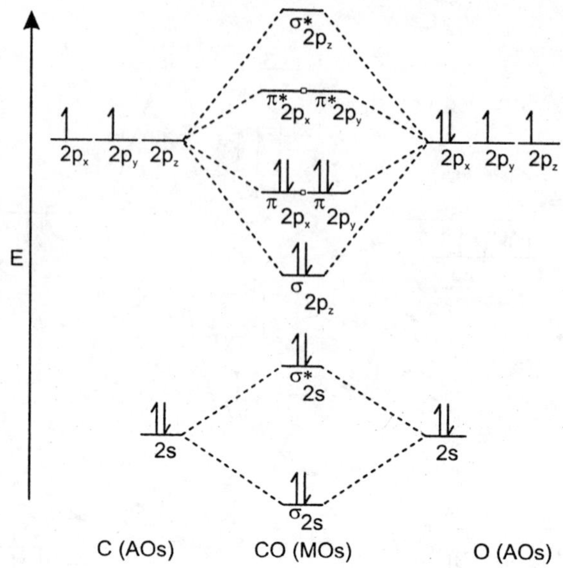

Fig.1.11: Molecular Orbital Diagram for CO

Analysis done by Bond Order

1. **Stability:** If value of bond order is positive, it indicates a stable molecule and if the value is negative or zero, it means that the molecule is unstable.
2. **Bond Dissociation Energy:** The stability of a molecule is measured by its bond dissociation energy. But the bond dissociation energy is directly proportional to the bond order. Greater the bond order, greater is the bond dissociation energy.
3. **Bond Length:** Bond order is inversely proportional to the bond length. The higher the bond order value, smaller is the bond length. For Example the bond length in nitrogen molecule is shorter than in oxygen molecule
4. **Magnetic Character:** If all the electrons in the molecule of a substance are paired, the substance is diamagnetic (repelled by the magnetic field). On the other hand, if there are unpaired electrons in the molecule, the substance is paramagnetic (attracted by the magnetic field).

pi – MOLECULAR ORBITALS OF BUTADIENE

The π-molecular orbitals of butadiene is a result of combining the π-molecular orbitals of two ethene molecules. This shows the effect of conjugation.The overall energy of the two bonding butadiene molecular orbitals is lower than that of the two molecular orbitals for ethene. This means that butadiene is thermodynamically more stable than we might expect if its structure were just two isolated double bonds.

Explaination

- The HOMO for butadiene is higher in energy relative to the HOMO for ethene. This means butadiene should be more reactive than ethene towards nucleophiles.
- The LUMO for butadiene is lower in energy than the LUMO for ethene. Consequently, butadiene would be expected to be more reactive towards nucleophiles than ethane.

- So whilst butadiene is more stable than two isolated double bonds, it is also more reactive.
- Together, these are the frontier orbitals of the allylcation, and they control the chemistry of the species.
- Nucleophiles will interact with the LUMO, and electrophiles with the HOMO, creating new pairs of MOs.

Finally, here are the MOs for a system built from four p-orbitals: butadiene, $CH_2 = CH - CH = CH_2$.

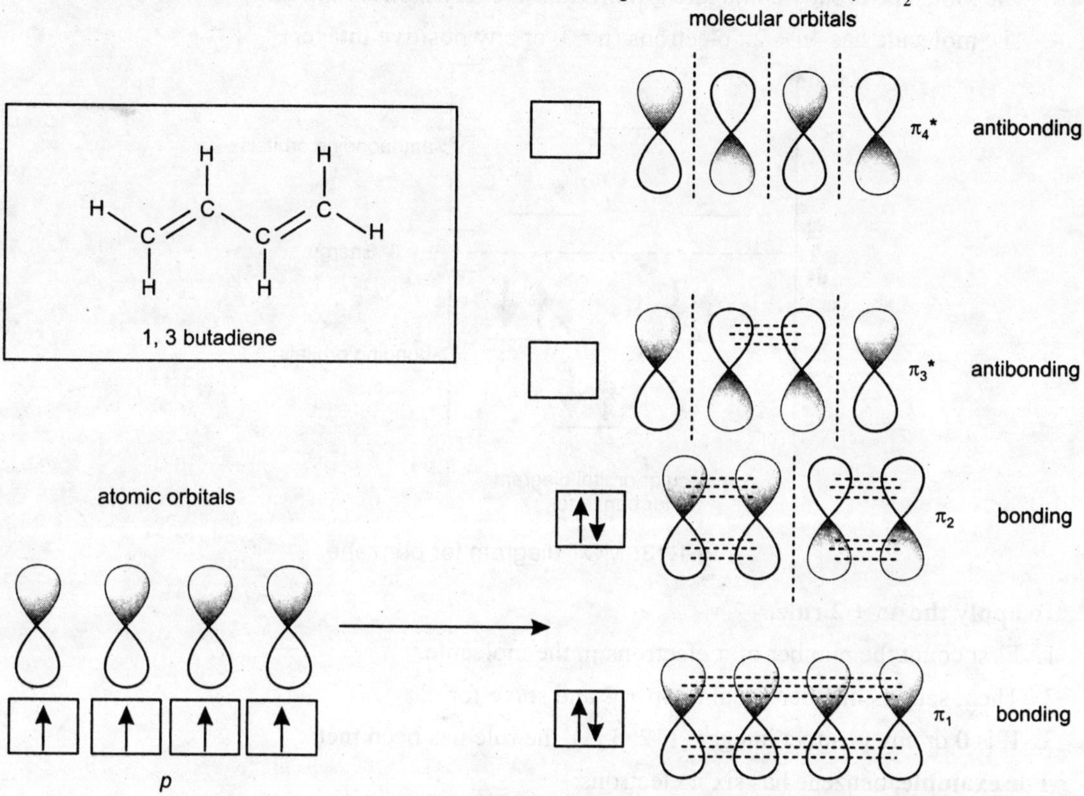

Fig.1.12: Bonding and antibonding orbitals in 1,3-butadiene molecule

The lowest energy molecular orbital, pi_1, has only constructive interaction and zero nodes. Higher in energy, but still lower than the isolated p orbitals, the pi_2 orbital has one node but two constructive interactions - thus it is still a bonding orbital overall. Looking at the two antibonding orbitals, π_3* has two nodes and one constructive interaction, while π_4* has three nodes and zero constructive interactions.

Butadiene has four p electrons. Two go in the lowest MO, and two in the next. All electrons thus occupy orbitals lower in energy than individual p orbitals, and thus the molecule is stabilized by the interaction of the two double bonds.

AROMATICITY

Benzene and other organic compounds display certain characteristic property that differentiate them from aliphatic compounds. These characteristic properties constitute which is called aromatic character or aromaticity.

Criteria for Aromaticity

When deciding if a compound is aromatic, go through the following checklist. If the compound does not meet all the following criteria, it is likely not aromatic in nature.

1. The molecule is cyclic (a ring of atoms)
2. The molecule is planar (all atoms in the molecule lie in the same plane)
3. The molecule is fully conjugated (p orbitals at every atom in the ring)
4. The molecule has $4n + 2\pi$ electrons (n = 0 or any positive integer)

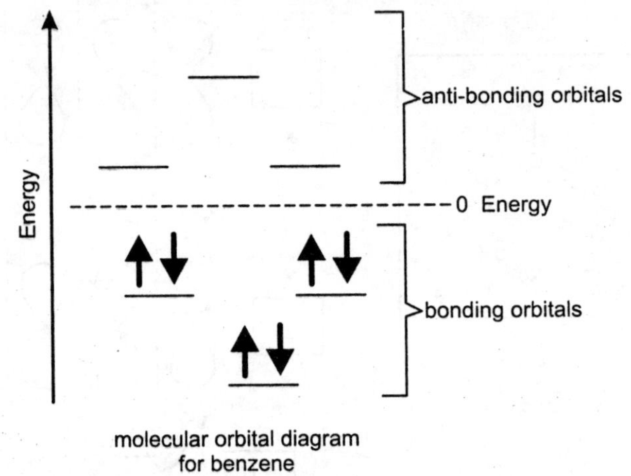

Fig.1.13: MOT diagram for benzene

To apply the 4n + 2 rule,

1. First count the number of π electrons in the molecule.
2. Then, set this number equal to 4n + 2 and solve for n.
3. If is 0 or any positive integer (1, 2, 3,...), the rule has been met.

For example, benzene has six π electrons:

$$4n + 2 = 6\pi$$
$$4n = 4$$
$$n = 1$$

For benzene, we find that n = 1, which is a positive integer, so the rule is met.

HUCKEL'S RULES

On the basis of Quantum mechanics E Hucken in 1931 proposed a new law related to aromaticity which is called Huckel's rule, "A cyclic conjugated compound will be aromatic, if its ring contains $(4n + 2)\pi$ electrons".

where n = an integer i.e. 0, 1, 2, 3.... Thus all those monocyclic compounds whose ring contains 2, 6, 10, 14 etc, π electrons should be aromatic i.e., extra stable and exhibits electrophilic substitution reactions easily. Thus benzene is aromatic whereas cyclobutadiene and cyclooctatetraene, though, monocyclic conjugated polymethin, are aromatic.

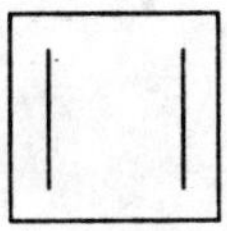

Cyclobutadiene has 4 pi electrons
(There are 2 double bonds in the right each double bond is counted as 2 pi electrons)

4n + 2 = 4
Find n:

4n + 2 = 4
4n = 4 – 2
4n = 2
n = ½

An aromatic compound follows Huckel's rule if n in equal to zero or a positive whole number.   Cyclobutadiere is not aromatic Because n is not a whole number

Fig.1.14: Explanation of aromaticity in Cyclobutadiene

Benzene

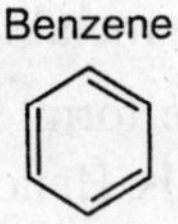

Since benzene has 6 pi electrons:

4n + 2 = 6
Find in:

4n + 2 = 6
4n = 6 – 2
4n = 4
n = 1 ←

An aromatic compound follows Huckel's rule if n is equal to zero or a positive whole number.   Benzene is aromatic

Fig.1.15: Explanation of aromaticity in Benzene

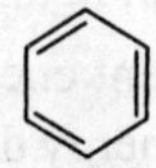

$4n + 2 = 6 : 4n = 4 : n = \frac{4}{4} = 1$

Benzene 6 electrons
n = 1

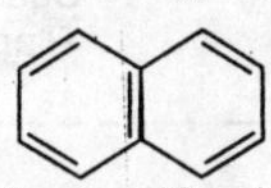

Naphthalene 10π electrons
n = 2

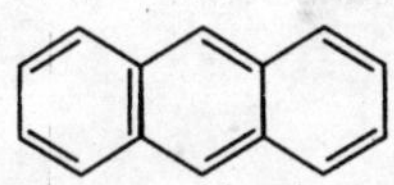

Anthracene 14π electrons
n = 3

Fig.1.16: Calculation of aromaticity in various organic compounds

CRYSTAL FIELD THEORY

CFT was proposed by the physicist Hans Bethe in 1929. Subsequent modifications were proposed by J.H. Van Vleck in 1935 to allow for some co-valency in the interactions. Crystal Field Theory was developed to describe important properties of complexes (magnetism, absorption spectra, oxidation states, coordination).

Basic Concept

In Crystal Field Theory, it is assumed that the ions are simple point charges (a simplification). When applied to alkali metal ions containing a symmetric sphere of charge, calculations of bond energies are generally quite successful. The approach taken uses classical potential energy equations that take into account the attractive and repulsive interactions between charged particles (that is, Coulomb's Law interactions).

$$E \propto \frac{q_1 q_2}{r^2} \quad ...(1.42)$$

Where,

E = bond energy between the charges

q_1 and q_2 = charges of the interacting ions

r = separating distance

Postulates of CFT

- Ligand used in the complex formation are either negative ions e.g., F^-, Cl^-, Br^-, CN^-, etc. or neutral molecule consisting dipole e.g. NH_3, H_2O etc.
- Complex is a group of central metal ion and ligand surrounding the central metal ion.
- At the time of formation of complex by metal ion and ligands, ligends as point charge move towards the metal ions with their negative charge facing the metal ion.
- Attraction between the central metal ion and ligands in the form of point charges is perfectly electrostatic. Overlapping does not take place between orbitals. Hence , the bond formed is purely ionic.
- Energy of d-orbitals of the metal ion increases when ligands move towards central metal ions.
- Increase in the energy of d- orbitals is not uniform. Energy of orbitals whivh are found direct in the way of of ligand increases to a greater extent as compared to the energy of the orbitals.
- Nature of ligand, nature of metal ions and number of ligands affect the approaching way of ligands.
- Metal ion orbitals get splitted in the set of different energy due to non- uniform increase in the energy of orbitals of metal ion. This splitting of energy depends on the geometry of the complex.

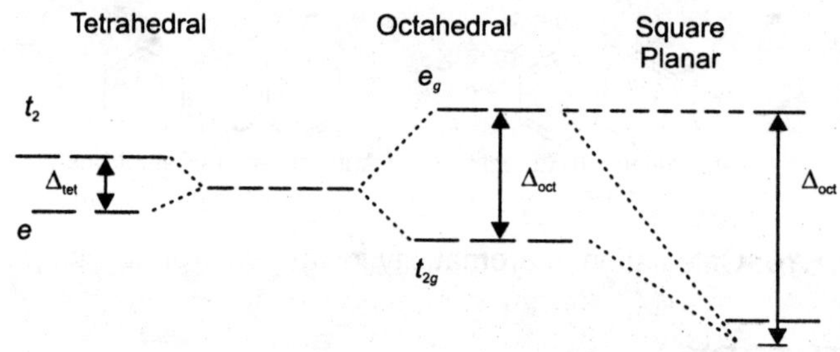

Fig.1.17: CFT of central metal ion in different geometries of the complex

When examining a single transition metal ion, the five d-orbitals have the same energy. When ligands approach the metal ion, some experience more opposition from the d-orbital electrons than others based on the geometric structure of the molecule. Since ligands approach from different directions, not all d-orbitals interact directly. These interactions, however, create a splitting due to the electrostatic environment. Orientation of orbitals have two types related to their the axis

Axial orientation – In this orientation the lobes of orbitals lies along the axis are dz^2 and $dx^2 - y^2$ orbitals. According to group theory this group of orbital is called e_g group and these orbital are called e_g orbital.

Non axial orientation – In this orientation the lobes of orbitals are found between the axis. dxy, dxz, and dyz orbitals are this type of orbitals. According to group theory this group of orbital is called t_{2g} group and these orbital are called t_{2g} orbitals.

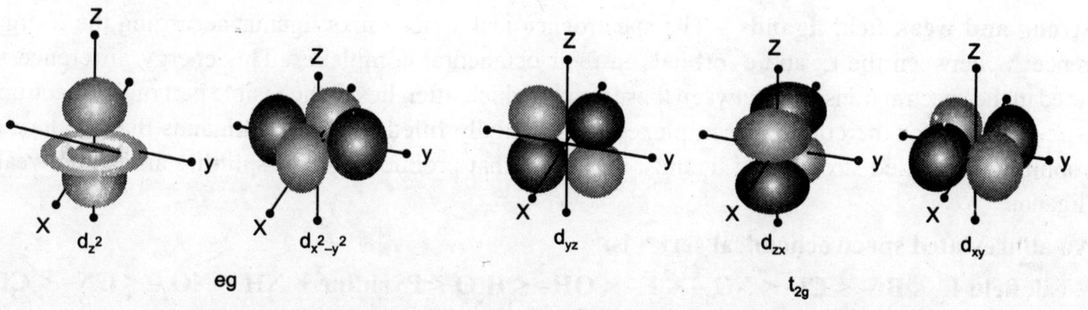

Fig.1.18: Angular dependence functions of d-orbitals t_2g and e_g

Orbital Splitting

The five d-orbitals are given the symbols d_{xy}, d_{zx}, d_{yz}, d_{x2-y2} and d_{z2}. In a complex they are all differently aligned relative to the incoming charge. Depending on the geometry of the complex, some of the d-orbitals will point directly towards the ligands, while some will point between them. The orbitals pointing at the ligands will be less stable and higher in energy. Now all the d-orbitals are no longer equivalent, giving rise to the phenomenon of orbital splitting, and the difference in energy between the more and less repelled orbitals is called the crystal field splitting parameter.

This splitting is affected by the following factors

- The nature of the ligands. The stronger the ligand, the greater is the splitting.
- Oxidation state of the central metal ion. A higher oxidation state leads to larger splitting.
- Size of d orbitals (i.e., transition series).
- Geometry of the complex.

The Spectrochemical Series

The variation of the magnitude of the crystal field splitting (Δ) with the nature of the ligand follows a regular order, known as spectrochemical series. This series is given below in the order in which they produce increasing value of Δ.

I– < Br– < S^2–< SCN– <Cl– < N_3^- < F^- < urea, OH^- < ox, O^{2-} < H_2O < NCS^- < py, NH_3 < en < bipy, phen < NO_2^- < CH_3^-, C_6H_5–< CN^-

- Weak field ligands have small Δ and will form high spin complexes.
- Strong field ligands have large Δ and will form low spin complexes.

The electrons can go into either a high spin or low spin arrangement depending on the magnitude of the crystal field splitting energy.

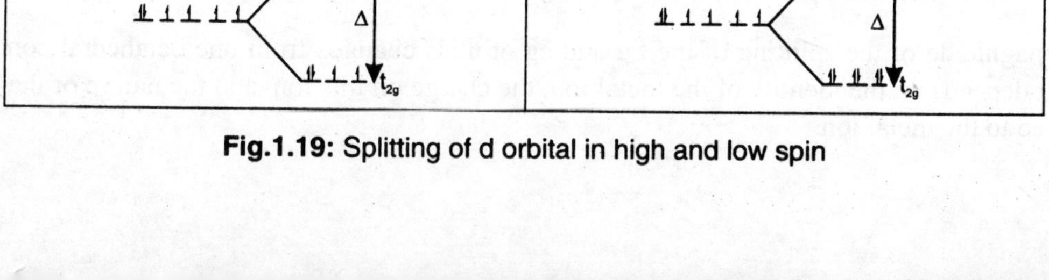

Fig.1.19: Splitting of d orbital in high and low spin

Strong and weak field ligands – The spectrochemical series ranks ligands according the energy difference Δ_O between the t_{2g} and e_g orbitals in their octahedral complexes. This energy difference is measured in the spectral transition between these levels, which often lies in the visible part of the spectrum and is responsible for the colors of complexes with partially filled d-orbitals. Ligands that produce a large splitting are called strong field ligands, and those that produce a small splitting are called weak field ligands.

An abbreviated spectrochemical series is:

Weak field I^- < Br^- < Cl^- < NO_3^- < F^- < OH^- < H_2O < Pyridine < NH_3 < NO_2^- < CN^- < CO Strong field

OCTAHEDRAL COMPLEXES

The octahedral arrangement of six ligands surrounding the central metal ion is as shown in the below figure.

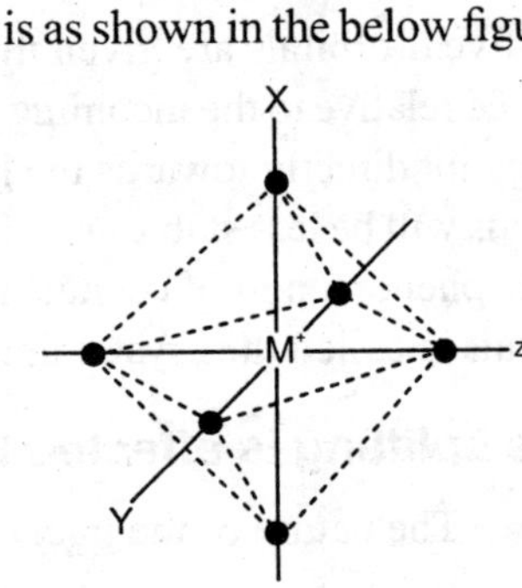

In an octahedral complex, the metal ion is at the centre and the ligands are at the six corners. In the figure, the directions x, y and z point to the three adjacent corners of the octahedran. The lobes of the e_g orbitals ($d_{x^2-y^2}$ and d_{z^2}) point along the x, y and z axis while the lobes of the t_{2g} orbitals (d_{xy}, d_{zx} and d_{yz})point in between the axes. As a result, the approach of six ligands along the x, y z, –x, –y and –z directions will increase the energy of $dx^2 - y^2$ and dz^2 orbitals (which point towards the ligands) much more than that it increases the energy of d_{xy}, d_{zx} and d_{yz} orbitals (which point in between the metal-legand bond axis).

Thus, under the influence of an octahedral field, the d orbitals split into triply degenerate orbitals with less energy and another as doubly degenerate orbitals with higher energy. The main energy level between these two sets of orbitals is taken as zero, which is called bari centre. The splitting between these two orbitals is called crystal field splitting. The magnitude of stabilization will be 0.4 Δ_0 and the magnitude of destabilization will be 0.6 Δ_0.

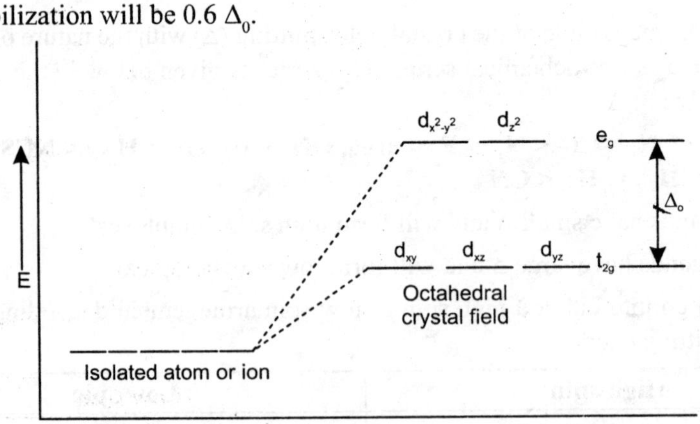

Fig.1.20: The magnitude of splitting stabilization in Octahedral complex

The magnitude of the splitting of the t_2g and eg orbitals changes from one octahedral complex to another. It depends on the identity of the metal ion, the charge on this ion, and the nature of the ligands coordinated to the metal ion.

TETRAHEDRAL COMPLEX

The tetrahedral arrangement of four ligands surrounding the metal ions is as shown in the figure.

A regular tetrahedron is a cube. One atom is at the centre of the cube and four of the eight corners of the cube are occupied by ligands. The directions x, y and z point to the face centres. The $d_{x^2-y^2}$ and d_{z^2} orbitals point along the x, y and z directions and d_{xy}, d_{zx} and d_{yz} orbitals point in between x, y and z directions.

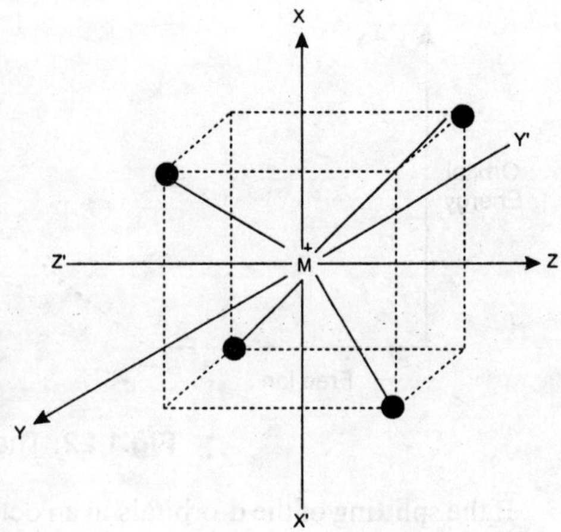

The direction of approach of ligands does not coincide exactly with either the e or t_2 orbitals. The t_2 orbitals are pointing close to the direction in which the e orbitals are lying in between the ligands. As a result, the energy of t_2 orbitals increases compared to the energy of e orbitals. Thus, d orbitals again split into two sets- triply degenerate t_2 of higher energy and doubly degenerate e orbitals of lower energy. That is, t_2 orbitals are raised by 0.4 Δ_t in energy and the e orbitals are stabilized by 0.6 Δ_t in energy.

The energy difference between the two sets of orbitals (Δ_t) will be about half the magnitude of that in an octahedral complex (Δ_o).

The d-orbitals will thus split as shown below:

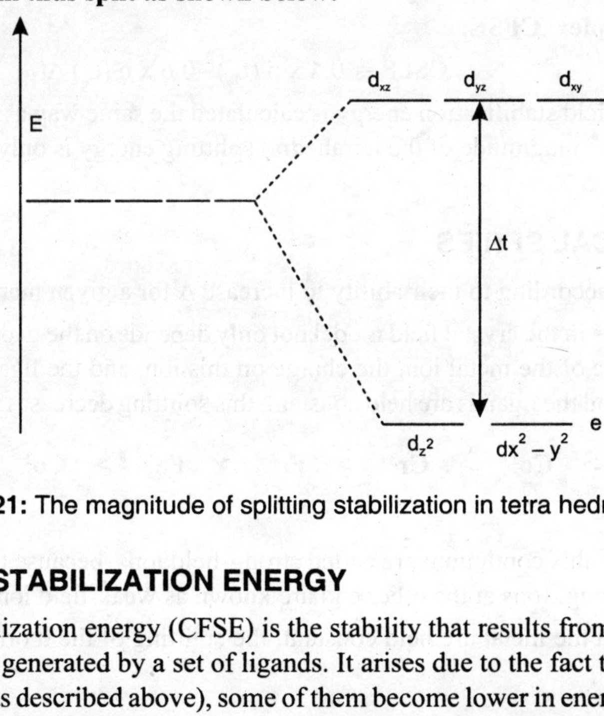

Fig.1.21: The magnitude of splitting stabilization in tetra hedral complex

CRYSTAL FIELD STABILIZATION ENERGY

The crystal field stabilization energy (CFSE) is the stability that results from placing a transition metal ion in the crystal field generated by a set of ligands. It arises due to the fact that when the d-orbitals are split in a ligand field (as described above), some of them become lower in energy than before with respect to a spherical field known as the bari centre in which all five d-orbitals are degenerate. For example, in an octahedral case, the t_2g set becomes lower in energy than the orbitals in the bari centre. Owing to the splitting of the d orbitals in a complex, the system gains an extra stability due to the rearrangement of the d electrons filling in the d levels of lower energy. The consequent gain in bonding energy is known as crystal field stabilization energy (CFSE).

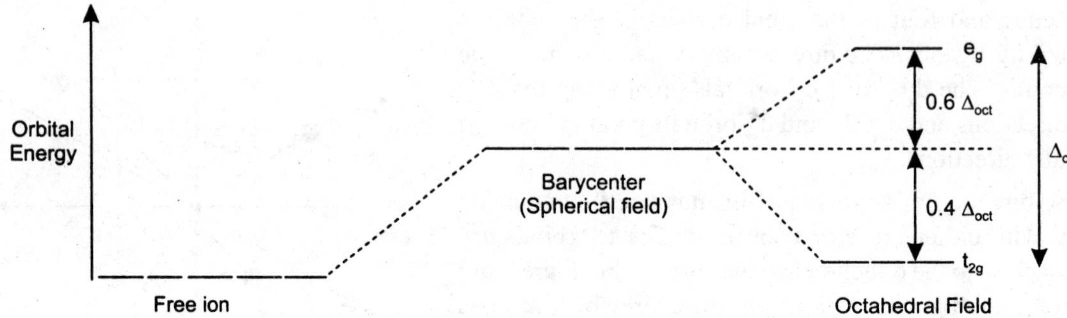

Fig.1.22: The CFSE in octahedral complex

If the splitting of the d-orbitals in an octahedral field is Δ_{oct}, the three t_{2g} orbitals are stabilized relative to the bari centre by 2/5 Δ_{oct}, and the e_g orbitals are destabilized by 3/5 Δ_o.

For an octahedral complex, CFSE:

$$CSFE = -0.4 \times n(t_{2g}) + 0.6 \times n(e_g)\ \Delta_0$$

Where, n (t_{2g}) and n (e_g) are the no. of electrons occupying the respective levels.

For a tetrahedral complex, CFSE:

$$CSFE = 0.4 \times n(t_{2g}) - 0.6 \times n(e_g)\ \Delta t$$

The tetrahedral crystal field stabilization energy is calculated the same way as the octahedral crystal field stabilization energy. The magnitude of the tetrahedral splitting energy is only 4/9 of the octahedral splitting energy, or $\Delta_t = 4/9\ \Delta_0$.

THE SPECTROCHEMICAL SERIES

An arrangement of ligands according to their ability to increase Δ for a given metal center.

The splitting of d orbitals in the crystal field model not only depends on the geometry of the complex, it also depends on the nature of the metal ion, the charge on this ion, and the ligands that surround the metal. When the geometry and the ligands are held constant, this splitting decreases in the following order.

$Pt^{4+} > Ir^{3+} > Rh^{3+} > Co^{3+} > Cr^{3+} > Fe^{3+} > Fe^{2+} > Co^{2+} > Ni^{2+} > Mn^{2+}$

strong-field ligands **weak-field ligands**

Metal ions at one end of this continuum are called strong-field ions, because the splitting due to the crystal field is unusually strong. Ions at the other end are known as weak-field ions.

When the geometry and the metal are held constant, the splitting of the d orbitals decreases in the following order.

$O > CN^- > NO_2^- > NH_3 > -NCS^- > H_2O > OH^- > F^- > -SCN^- > Cl^- > Br^-$

strong-field ligands **weak-field ligands**

HIGH-SPIN VERSUS LOW-SPIN OCTAHEDRAL COMPLEXES

Once we know the relative energies of the d orbitals in a transition-metal complex, we have to worry about how these orbitals are filled. Degenerate orbitals are filled according to Hund's rules.

- One electron is added to each of the degenerate orbitals in a subshell before a second electron is added to any orbital in the subshell.
- Electrons are added to a subshell with the same value of the spin quantum number until each orbital in the subshell has at least one electron.

Octahedral transition–

1. Metal ions with d^1, d^2, or d^3 configurations for low and high spins are same. It can therefore be described by the following diagrams.
2. Metal ions with d^4, d^5, d^6 or d^7 configurations When we try to add a fourth electron, we are faced with a problem. This electron could be used to pair one of the electrons in the lower energy (t_{2g}) set of orbitals or it could be placed in one of the higher energy (e_g) orbitals. One of these configurations is called high-spin because it contains four unpaired electrons with the same spin. The other is called low-spin because it contains only two unpaired electrons. The same problem occurs with octahedral d^5, d^6, and d^7 complexes.
3. Metal ions with d^8, d^9, or d^{10} configurations For octahedral d^8, d^9, and d^{10} complexes, there is only one way to write satisfactory configurations

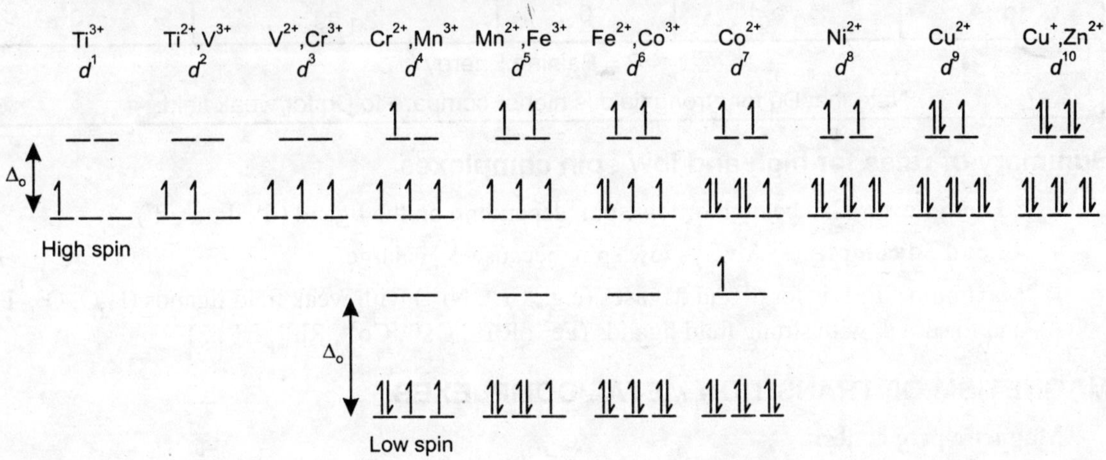

Fig.1.23: Filling of electron in d orbital in octahedral complex in effect of low spin and high spin complex

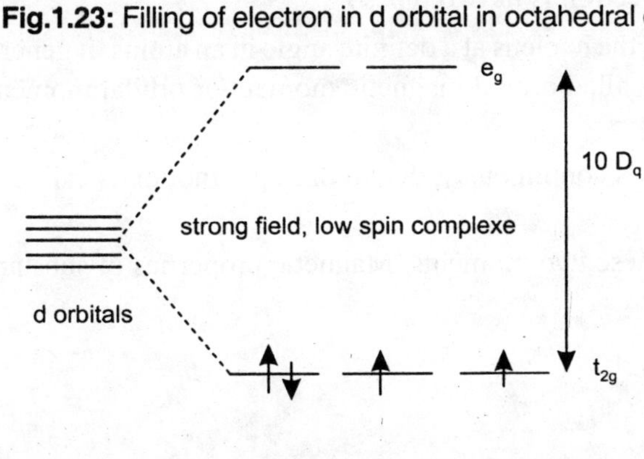

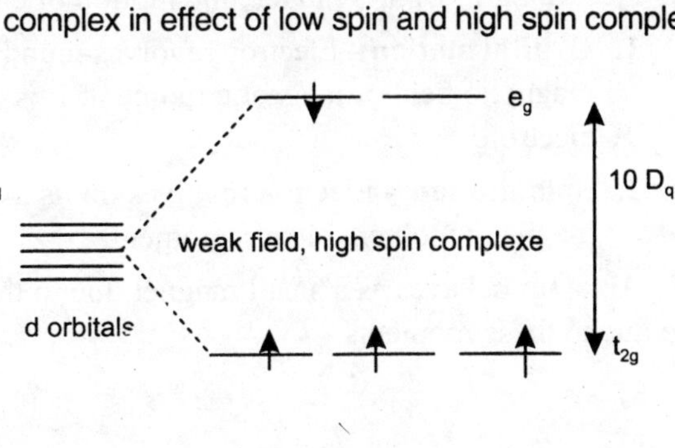

Table 1.3: CFSE for d^1 to d^{10} electrons for low spin and high spin complexes

The Electron Arrangement and CFSE for Octahedral Complexes				
Configuration	**Strong field (low spin)**	**CFSE**	**Weak filed (high spin)**	**CFSE**
d^1	$t_{2g}^1e_g^0$	4Dq	$t_{2g}^1eg^0$	4Dq
d^2	$t_{2g}^2e_g^0$	8Dq	$t_{2g}^2eg^0$	8Dq
d^3	$t_{2g}^3e_g^0$	12Dq	$t_{2g}^3eg^0$	12Dq
d^4	$t_{2g}^4e_g^0$	16Dq – p	$t_{2g}^3eg^1$	6Dq
d^5	$t_{2g}^5e_g^0$	20Dq – 2p	$t_{2g}^3eg^2$	0Dq
d^6	$t_{2g}^6e_g^0$	24Dq – 2p	$t_{2g}^4eg^2$	4Dq
d^7	$t_{2g}^6e_g^1$	18Dq – p	$t_{2g}^5eg^2$	8Dq
d^8	$t_{2g}^6e_g^2$	12Dq	$t_{2g}^6eg^2$	16Dq
d^9	$t_{2g}^6e_g^3$	6Dq	$t_{2g}^6eg^3$	6Dq
d^{10}	$t_{2g}^6e_g^4$	0	$t_{2g}^6eg^4$	0
P = Paining Energy Note that Dq for strong field is higher compare to Dq for weak field.				

Summary of rules for high and low spin complexes

- **3d complexes:** Can be high or low spin, depending on the ligand (d^4, d^5, d^6, d^7)
- **4d and 5d complexes:** Always low spin, because Δ_o is large
- Maximum CFSE is for d^3 and d^8 cases (e.g., Cr^{3+}, Ni^{2+}) with weak field ligands (H_2O, O^{2-}, F^-,...) and for d^3 - d^6 with strong field ligands (Fe^{2+}, Ru^{2+}, Os^{2+}, Co^{3+}, Rh^{3+}, Ir^{3+},...)

MAGNETISM OF TRANSITION METAL COMPLEXES

Magnetic properties:

According to Bohr's model , the motion of electron is of two types–

1. **Orbital motion** – Electron revolves round the nucleus at a definite angle in an atoms. It generates magnetic field or magnetic moment. It is called orbital magnetic moment or orbital moment of electron.
2. **Spin motion** – Electron revolves on its axis continuously, due to this spin motion is called spin magnetic moment or spin moment.

An atom behaves as a small magnet due to these two moments. Magnetic properties of substances are due to these moments.

Fully, empty and partially filled d orbital of transition metal shows various magnetic properties. A range of magnetic properties can be expected, depending on the oxidation state, electronic configuration and coordination number of the central metal.The unpaired electrons in (n – 1) d orbitals are responsible for the magnetic properties.

Characteristics for Magnetism

- Unpaired electrons exist when the complex has an odd number of electrons or because electron pairing is destabilized.
- The more unpaired electrons, the stronger the magnetic property.
- Tetrahedral complexes have weaker splitting because none of the ligands lie within the plane of the orbitals.
- Square planar compounds are always low-spin and therefore are weakly magnetic.
- In bi-and polymetallic complexes, the electrons may couple through the ligands, resulting in a weak magnet, or they may enhance each other.

Classification of substance according to magnetic properties

Diamagnetism

When any substance is placed in an external magnetic field, there is an induced circulation of electrons producing a net magnetic moment aligned in opposition to the applied field. This is the diamagnetic effect and it arises from paired electrons within a sample, since all compounds contain some paired electrons, diamagnetism is a universal property of matter. If a substance has only paired electrons, this effect will dominate, the material will be classified as diamagnetic, and it will be slightly repelled by a magnetic field.

Paramagnetism

It is produced by unpaired electrons in a sample. The spins and orbital motions of these electrons give rise to permanent molecular magnetic moments that tend to align themselves with an applied field, Because it is much larger than the diamagnetic effect, the paramagnetic effect cancels any repulsions between an applied field and paired electrons in a sample, Thus even substances having only one unpaired electron per molecule will show a net attraction into a magnetic field. The paramagnetic effect is observed only in the presence of an external field.

Para magnetism is due to motion and spin of electron. Contribution of orbital moment is negligible. Para magnetism depends on temperature and decreases with increase in temperature. Paramagnetic substances are attracted in magnetic field. If number of unpaired electrons is n, magnetic moment,

$$\mu = \sqrt{n(n+2)} \qquad ...(1.43)$$

Number of unpaired electrons is calculated by magnetic measurement. Cuprous ion with d^{10} configuration forms diamagnetic salts. Cupric salts are paramagnetic. It means it has d^9 configuration for Cu^{2+} i.e. one unpaired electron. Following table represents magnetic moment of important ions of first transition series elements.

Table 1.4: Magnetic moments of ions of first transition series

Ion	Outer electronic configuration	Number of unpaired electrons	Magnetic moments	
			Calculated	Experimental
Sc^{3+}	$3d^0$	0	0	0
Ti^{3+}	$3d^1$	1	1.73	1.75
Ti^{2+}	$3d^2$	2	2.84	2.86
V^{2+}	$3d^3$	3	3.87	3.86
Cr^{2+}	$3d^4$	4	4.90	4.80
Mn^{2+}	$3d^5$	5	5.92	5.95
Fe^{2+}	$3d^6$	4	4.90	5.0 – 5.5
Co^{2+}	$3d^7$	3	3.87	4.4 – 5.2
Ni^{2+}	$3d^8$	2	2.84	2.9 – 3.4
Cu^{2+}	$3d^9$	1	1.73	1.4 – 2.2
Zn^{2+}	$3d^{10}$	0	0	0

Ferromagnetism – A substance is said to be ferromagnetic if it is very high paramagnetic character. This character is found in some compounds and alloys of Fe, Co. Ni, Mn. Ferromagnetic compounds retain magnetism ever after removing from magnetic field. Such compounds consists of a number of tiny magnets arranged randomly. In magnetic field, these magnets align themselves in one direction. Due to this, such compounds exhibit high degree of magnetism.

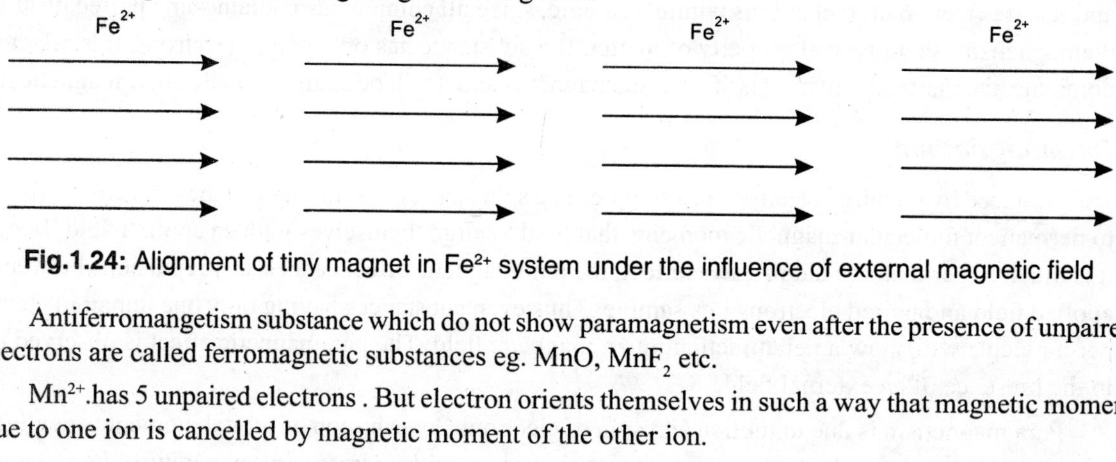

Fig.1.24: Alignment of tiny magnet in Fe^{2+} system under the influence of external magnetic field

Antiferromangetism substance which do not show paramagnetism even after the presence of unpaired electrons are called ferromagnetic substances eg. MnO, MnF_2 etc.

Mn^{2+}.has 5 unpaired electrons . But electron orients themselves in such a way that magnetic moment due to one ion is cancelled by magnetic moment of the other ion.

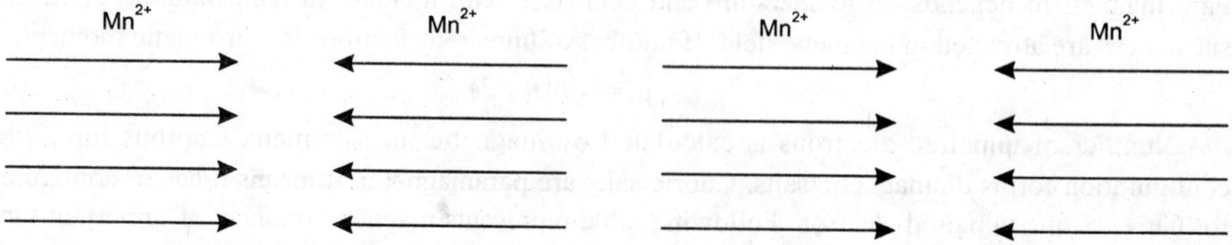

Fig.1.25: Effect of external magnetic field on unpaired electrons of Mn^{2+}

Table 1.5: On the basis of the magnetic properties solids can be classified as follows:

Properties	Description	Alignment of magnetic dipoles	Examples	Application
Diamagnetic	They are weakly repelled by the magnetic fields	All the electrons in the orbitals are paired and are completely filled.	NaCl, Benzene	Behaves like an insulator.
Paramagnetic	They are weakly attracted by the magnetic fields.	Contains at least one unpaired electron in the orbital.	O_2, Cu^{2+}etc.	Electronic appliances
Ferromagnetic	Strongly attracted by the magnetic field.The can be magnetised permanently	Consists of unpaired electrons, all having the same direction	Cobalt, nickel, CrO_2etc.	CrO_2 is commonly used in making cassette recorder.
Antiferromagnetic	Net magnetic moment is zero.	Dipole moments are arranged in a compensatory way	NiO, MnO, V_2O_3 etc.	–
Ferrimagnetic	Possess small net magnetic moments	Unequal number of parallel and antiparallel arrangement of magnetic moments	Fe_3O_4	–

BAND STRUCTURE OF SOLIDS AND THE ROLE OF DOPING ON BAND STRUCTURE

According to the Bohr atomic model, in an isolated atom the energy of any of its electrons is decided by the orbit in which it revolves. But when the atoms come together to form a solid they are close to each other.Each electron will have a different energy level are called energy bands. These are of two types–

1. **Valence band** – The energy band which includes the energy levels of the valence electrons is called the valence band.
2. **Conduction band** – The energy band above the valence band is called the conduction band. Normally the conduction band is empty.

Solids may be classified in three types on the basis of their electrical property conductors, insulators and semiconductors.

A. Conductors – In conductors, there is no band gap between their valence and conduction bands, since they overlap. There is a continuous availability of electrons in these closely spaced orbitals. Through these materials electricity can pass. Examples – Metals.

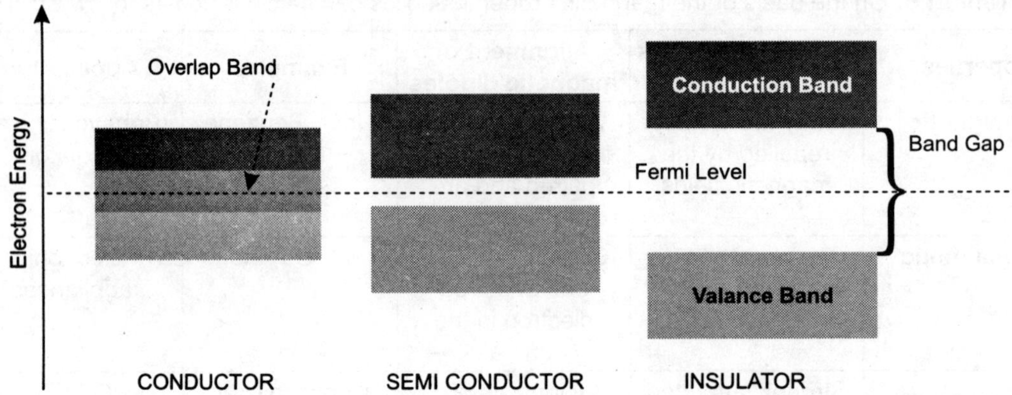

Fig.1.26: Conductors, Insulators and Semiconductors

B. Insulators – In insulators, the band gap between the valence band the conduction band is so large that electrons cannot make the energy jump from the valence band to the conduction band. Through these materials electricity cannot pass. Example – wood, non metals, plastics.

C. Semiconductors – Semiconductors have a small energy gap between the valence band and the conduction band. Electrons can make the jump up to the conduction band, but not with the same ease as they do in conductors. Semiconductors are those materials whose electrical conductivity is between conductors and insulators. Examples – silicon and germanium.

There are two different kinds of semiconductors: intrinsic and extrinsic.

i. Intrinsic Semiconductors

An intrinsic semiconductor is a semiconductor in its pure state. For every electron that jumps into the conduction band, the missing electron will generate a hole that can move freely in the valence band. The number of holes will equal the number of electrons that have jumped.

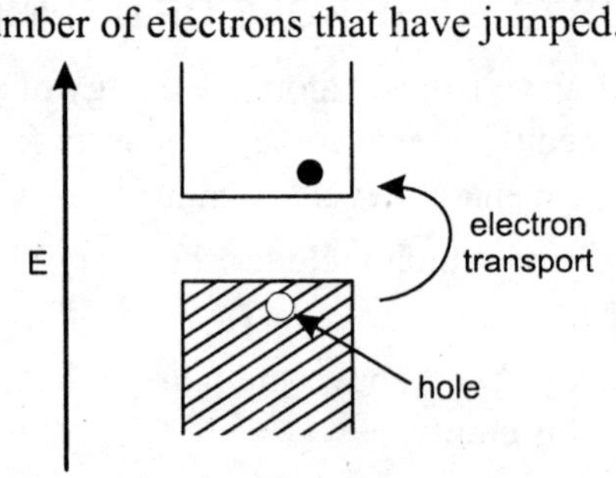

Fig.1.27: Intrinsic Semiconductors

ii. Extrinsic Semiconductors

In extrinsic semiconductors, the band gap is controlled by purposefully adding small impurities to the material. This process is called doping. Doping, or adding impurities to the lattice can change the electrical conductivity of the lattice and therefore vary the efficiency of the semiconductor. In extrinsic semiconductors, the number of holes will not equal the number of electrons jumped. There are two different kinds of extrinsic semiconductors, p-type (positive charge doped) and *n*-type (negative charge doped).

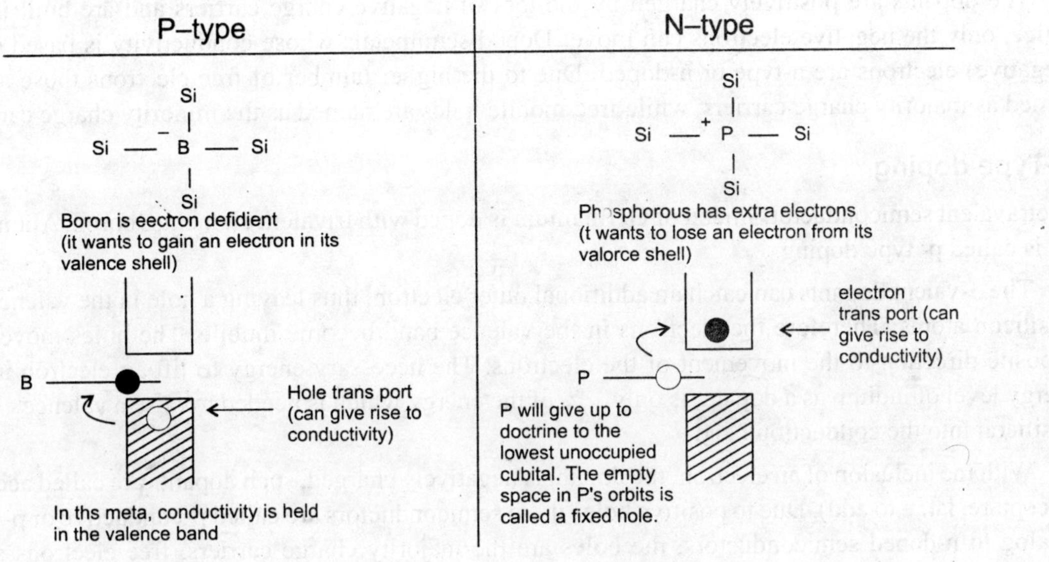

Fig.1.28: Extrinsic Semiconductors

Doping

Doping means the introduction of impurities into a semiconductor crystal to the defined modification of conductivity.

Doping may be performed two types–

n–Type doping

If tetravalent semiconductor Silicon or Germanium doped with penta valent elements (Phosphorous, Arsenic) it is called n- type doping.

The 5-valent dopant has an outer electron more than the silicon atoms. Four outer electrons combine with ever one silicon atom, while the fifth electron is free to move and serves as charge carrier. This free electron requires much less energy to be lifted from the valence band into the conduction band, than the electrons which cause the intrinsic conductivity of silicon. The dopant, which emits an electron, is known as an electron donor (donare, lat. = to give).

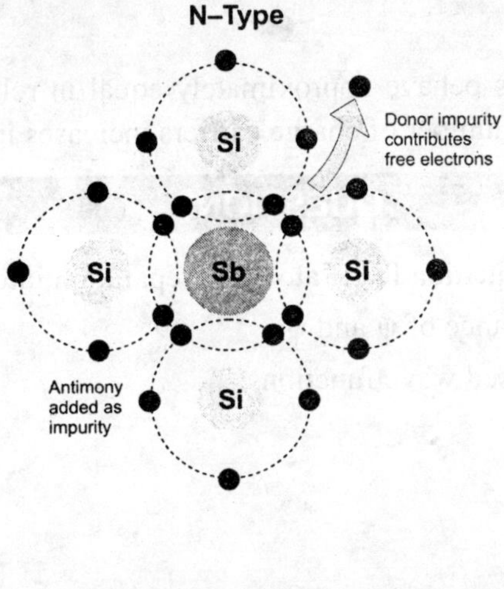

The dopants are positively charged by the loss of negative charge carriers and are built into the lattice, only the negative electrons can move. Doped semimetals whose conductivity is based on free (negative) electrons are n-type or n-doped. Due to the higher number of free electrons those are also named as majority charge carriers, while free mobile holes are named as the minority charge carriers.

p–Type doping

If tetravalent semiconductor Silicon or Germanium is doped with trivalent elements (boron, Aluminium,) it is called p- type doping.

The 3-valent dopants can catch an additional outer electron, thus leaving a hole in the valence band of silicon atoms. Therefore the electrons in the valence band become mobile. The holes move in the opposite direction to the movement of the electrons. The necessary energy to lift an electron into the energy level of indium as a dopant, is only 1 % of the energy which is needed to raise a valence electron of silicon into the conduction band.

With the inclusion of an electron, the dopant is negatively charged, such dopants are called acceptors (acceptare, lat. = to add). Due to positive holes these semiconductors are called p-conductive or p-doped. Analog to n-doped semiconductors, the holes are the majority charge carriers, free electrons are the minority charge carriers.

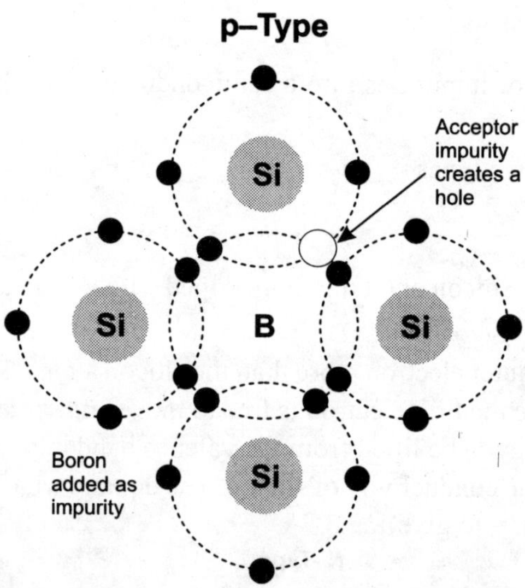

n-and p-doped semiconductors behave approximately equal in relation to the current flow. With increasing amount of dopants, the number of charge carriers increases in the semiconductor crystal.

QUESTIONS

1. Solve schrodinger wave equation for H atom, its separation into three equations.
2. Write the physical significance of ψ and ψ^2.
3. What is meant by normalised wave functions.

4. With the help of MO diagram, arrange the following molecules/ ion in order of their increasing bond length.
   O_2, O_2- and $O_2^{2}-$.
5. Draw Molecular orbital diagram of nitrogen, helium and CO molecules.
6. Explain why the Co $(NH_3)_6^{3+}$ ion is a diamagnetic, low-spin complex, whereas the CoF_6^{3-} ion is a paramagnetic, high-spin complex.

SHORT ANSWER QUESTIONS

1. How do you distinguish between a valence band and a conduction band?
   Ans. The valence band is the highest band with electrons in it, and the conduction band is the highest band with no electrons in it.
2. Is the energy gap between an insulator smaller or larger than the energy gap between a semiconductor?
   Ans. Larger
3. What two methods bring conductivity to semiconductors?
   Ans. Electron transport and hole transport
4. You are more likely to find electrons in a conduction band if the energy gap is smaller/larger?
   Ans. Smaller
5. The property of being able to be drawn into a wire is called.
   Ans. Ductility
6. To be aromatic, a molecule must be planar conjugated, and obey the 4n + 2 rule. Predict , is the following molecule will be aromatic?

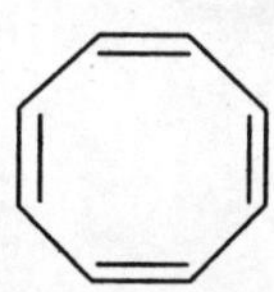

   Ans. No, it is not. It does not obey the 4n + 2 rule. Also it is not planar.
7. Using the criteria for aromaticity, determine if the following molecules are aromatic:

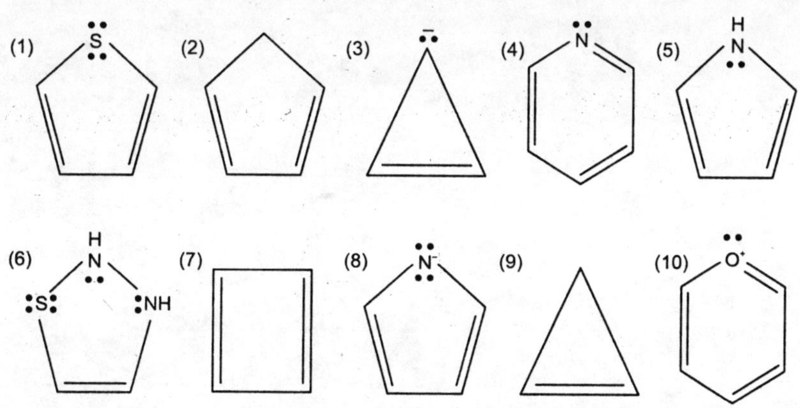

Ans.

1. Aromatic – only 1 of S's lone pairs counts as π electrons, so there are 6 π electrons, n = 1
2. Not aromatic – not fully conjugated, top C is sp^3 hybridized
3. Not aromatic – top C is sp^2 hybridized, but there are 4 π electrons, n = 1/2
4. Aromatic – N is using its 1 p orbital for the electrons in the double bond, so its lone pair of electrons are not π electrons, there are 6 π electrons, n = 1
5. Aromatic – there are 6 π electrons, n = 1
6. Not aromatic – all atoms are sp^2 hybridized, but only 1 of S's lone pairs counts as π electrons, so there 8 π electrons, n = 1.5
7. Not aromatic – there are 4 π electrons, n = 1/2
8. Aromatic – only 1 of N's lone pairs counts as π electrons, so there are 6 π electrons, n = 1
9. Not aromatic – not fully conjugated, top C is sp^3 hybridized
10. Aromatic – O is using its 1 p orbital for the elections in the double bond, so its lone pair of electrons are not π electrons, there are 6 π electrons, n = 1.

2 Spectroscopic Techniques and Applications

Principles of spectroscopy and selection rules. Electronic spectroscopy. Fluorescence and its applications in medicine. Vibrational and rotational spectroscopy of diatomic molecules. Applications. Nuclear magnetic resonance and magnetic resonance imaging, surface characterization techniques. Diffraction and scattering.

Introduction

At the end of the nineteenth century, spectroscopy was limited to the absorption, emission, and scattering of visible, ultraviolet, and infrared electromagnetic radiation. Since its introduction, spectroscopy has expanded to include other forms of electromagnetic radiation, such as X-rays, microwaves, and radio waves and other energetic particles, such as electrons and ions.

Electromagnetic waves : Electromagnetic waves or EM waves are waves that are created as a result of vibrations between an electric field and a magnetic field. In other words, EM waves are composed of oscillating magnetic and electric fields.

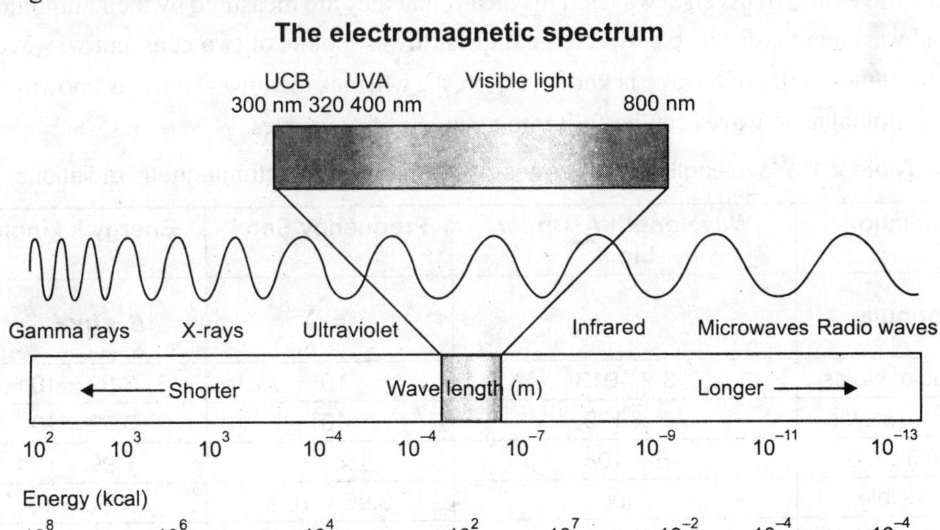

Fig.2.1: Region of Electromagnetic radiations

Description: Electromagnetic waves are formed when an electric field comes in contact with a magnetic field. They are hence known as 'electromagnetic' waves. The electric field and magnetic field of an electromagnetic wave are perpendicular (at right angles) to each other. They are also perpendicular to the direction of the EM wave.

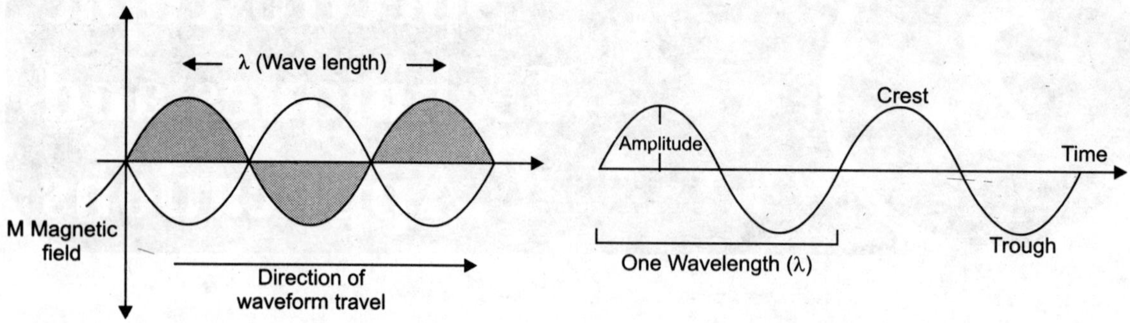

Fig.2.2: Electromagnetic wave indicating parameters of radiations.

Properties of EM

1. EM waves travel with a constant velocity of 3.00×10^8 ms^{-1} in vacuum.
2. They are deflected neither by the electric field, nor by the magnetic field.
3. However, they are capable of showing interference or diffraction.
4. An electromagnetic wave can travel through anything in space or in air, a solid material or vacuum.
5. It does not need a medium to propagate or travel from one place to another.
6. Mechanical waves (like sound waves or water waves), on the other hand, need a medium to travel.
7. EM waves are 'transverse' waves. This means that they are measured by their amplitude (height) and wavelength (distance between the highest/lowest points of two consecutive waves). The highest point of a wave is known as 'crest', whereas the lowest point is known as 'trough'.
8. Electromagnetic waves can be split into a range of frequencies.

Table 2.1: Wavelengths, frequencies and energies of electromagnetic radiations

Radiation↓	Wavelength A⁰ Upper Limit	Frequency Sec^{-1}	Energy kJ mole^{-1}
Formula →	λ	$\nu = \frac{c}{\lambda}$	$E = h\nu = \frac{hc}{\lambda}$
Radio waves	3 x 1013	10^5	3.99×10^{-8}
Microwaves	3 x 109	10^9	3.99×10^{-4}
Infra red	6 x 106	5×10^{11}	1.99×10^{-4}
Visible	7600	3.95×10^{14}	1.58×10^2
Ultra violet	3800	7.9×10^{14}	3.16×10^2
X- Rays	150	2×10^{16}	8.89×10^3
Y- Rays	1	3×10^{18}	1.20×10^6
Cosmic rays	0.001	3×10^{21}	1.20×10^9

Spectroscopy is a technique that uses the interaction of energy with a sample to perform an analysis.

The optical system which allows production and viewing of the spectrum is called a spectroscope. There are many other forms of light which are not visible to the human eye and spectroscopy is extended to cover all these. Theories and application of various spectroscopic regions are mentioned below in table 2.2.

Table 2.2: Electromagnetic radiations name of spectroscopy its theory and applications

Electro magnetic Region/source	Name of Spectroscopy	Theory	Application
Ultraviolet/ Visible	**Electronic**	Electron Transitions between Orbitals	Analysis of UV/Vis Spectra
Infrared	**Vibrational**	(i) Simple Stretching Vibrations (ii) Higher Order Vibrations	Analysis of Vibrational Banding Patterns
Between IR and microwave	**Raman**	Change of frequency between incident and transmitted radiations due to scattering	To identify molecules and Study chemical bonding.
Microwave	**Rotational**	Measurement of the energies of transitions between quantized rotational states of molecules in the gas phase	Identifying stable reaction products. For purity analysis and reaction efficiency.
Radio waves	**Nuclear Magnetic Resonance**	(i) Nuclei with Spin 1/2 (ii) The Chemical Shift Scale (iii) Scalar Coupling between Nuclei (iv) Analysis of 1D NMR Spectra (v) Nuclear Overhauser Effect	Introduction to analysis of 2D NMR Spectra
Ionization of atom	**Mass**	(i) Ionization Techniques (ii) Mass Analyzers (iii) Molecular Ions and High Resolution MS	Analysis of Mass Spectral Fragmentation Patterns

BASIC PRINCIPLES OF SPECTROSCOPY

Born Oppenheimer Approximation

The atomic spectra arises from the transition of an electron between the atomic energy levels, whereas the molecular spectra arise from three types of energy changes (molecular rotation, molecular vibration and electronic transition).According to Born Oppenheimer approximation, the total energy of a molecule is given by ,

$$E = E_{translational} + E_{rotational} + E_{vibrational} + E_{electronic} \quad ...(2.1)$$

The equation indicates that the total energy E of the molecule is some of translational energy, rotational energy, vibrational energy and electronic energy. All of these energies are quantized except translational energy. It is negligibly small and so the Born Oppenheimer approximation can be represented as

$$E = E_{rotational} + E_{vibrational} + E_{electronic}$$

It should also be noted that

$$E_{electronic} >> E_{vibrational} >> E_{rotational} >> E_{translational}$$

In case of diatomic molecule, the molecule possesses translational energy when its centre of gravity changes as a consequence of motion. **Rotational energy** arises when a molecule undergoes rotation about an axis perpendicular to the interrnuclear axis and passing through the centre of gravity of the molecule. **Vibrational energy** arises because of to and fro motion of the nuclei of the molecule in such a way that centre of gravity does not change. **Electronic energy** is associated with the transition of an electron from the ground state energy level to an excited state energy level of the molecule because of the absorption of a photon of a suitable frequency.

It should be noted that:

- The vibrational spectra of a molecule is observed in the infra red region of the electromagnetic spectrum.
- The rotational spectra of a molecule is observed in the microwave region of the electromagnetic spectrum.
- The electronic spectra of a molecule is observed in the UV and visible region of the electromagnetic spectrum.

LIGHT ABSORPTION AND FATE OF EXCITATION ENERGY: FRANCK–CONDON PRINCIPLE

When light is absorbed by an atom or molecule in form of photon, each quantum is taken up and the whole energy of the quantum is communicated to it. The absorbing atom or molecule is thus excited, that is, lifted from its normal state of lowest energy (highest stability) to an excited, energy-rich state. According to Niel Bohr's theory of atomic and molecular structure, an atom or molecule can exist only in a series of discrete states of electronic energy. This is represented below.

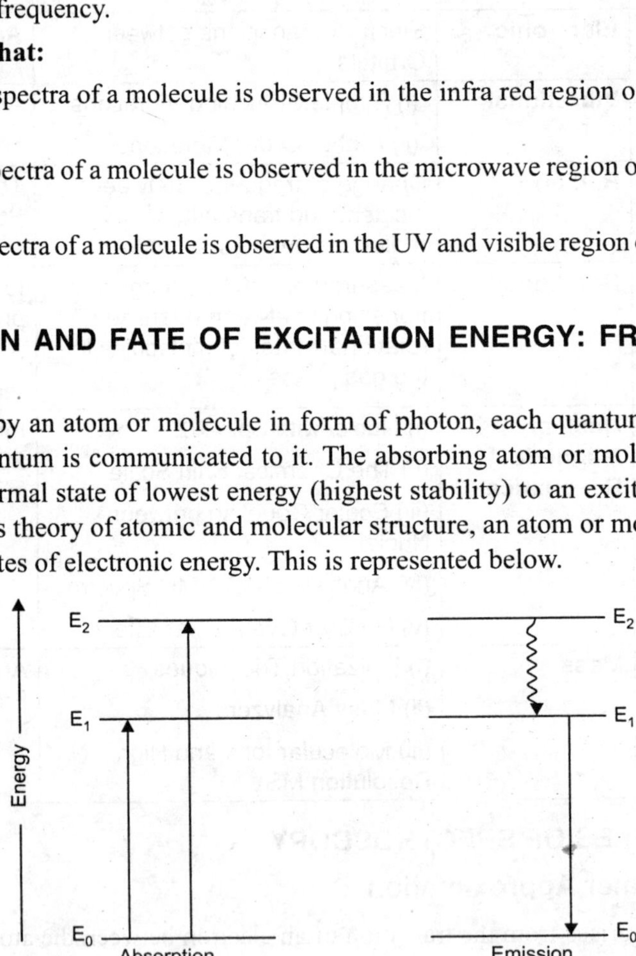

Fig.2.3: Ground state (E_0) and two excited states (E_1, E_2) of a molecule

Explanation

- The lowest line represents the ground state, E_0, of the atom or molecule, means without absorption of photon.
- The higher lines (E_1 and E_2) represent excited electronic states, the only kind of excited states possible in an atom.

The Franck–Condon principle is a rule in spectroscopy and quantum chemistry that explains the intensity of vibronic transitions. Vibronic transitions are the simultaneous changes in electronic and vibrational energy levels of a molecule due to the absorption or emission of a photon of the appropriate energy.

Franck–Condon principle is the approximation that an electronic transition is most likely to occur without changes in the positions of the nuclei in the molecular entity and its environment. The resulting state is called a Franck–Condon state, and the transition involved, a vertical transition. The quantum mechanical formulation of this principle is that the intensity of a vibronic transition is proportional to the square of the overlap integral between the vibrational wave functions of the two states that are involved in the transition.

Franck-Condon principle (vertical transitions):

- Electrons respond much faster than nuclear motion, therefore an excitation proceeds without a change to the nuclear geometry.
- Light will be resonant with electronic energy gap at equilibrium nuclear geometry.

The two potential curves show the potential energy of the molecule as a function of this distance for two electronic states, a ground state and an excited state. Excitation is represented, according to the Franck-Condon principle, by a vertical arrow (A). This arrow hits the upper curve, except for very special cases, not in its lowest point, corresponding to a non vibrating state, but somewhere higher. This means that the molecule finds itself, after the absorption act, in a non equilibrium state and begins to vibrate like a spring. This vibration is described, in the figure, by the molecule running down, up, down again, etc., along the upper potential curve, like a pendulum. The periods of these vibrations are of the order of 10^{-13}, or 10^{-12} seconds. Since the usual lifetimes of excited electronic states are of the order of 10^{-9} s, there is enough time during the excitation period for many thousands of vibrations.

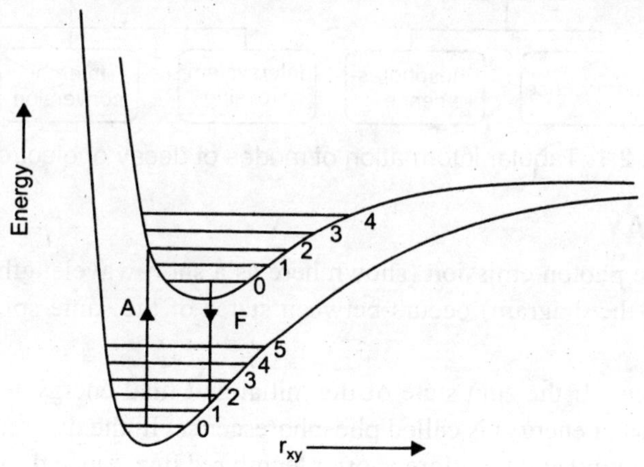

Fig.2.4: Frank-Condon Principle

JABLONSKI DIAGRAM : RELAXATION MECHANISM FOR EXCITED STATE MOLECULES

Once a molecule has absorbed energy in the form of electromagnetic radiation, there are a number of routes by which it can return to ground state (the statistically most common energy state for room temperature chemical species). The following graphic, termed a Jablonski diagram, shows a few of these processes.

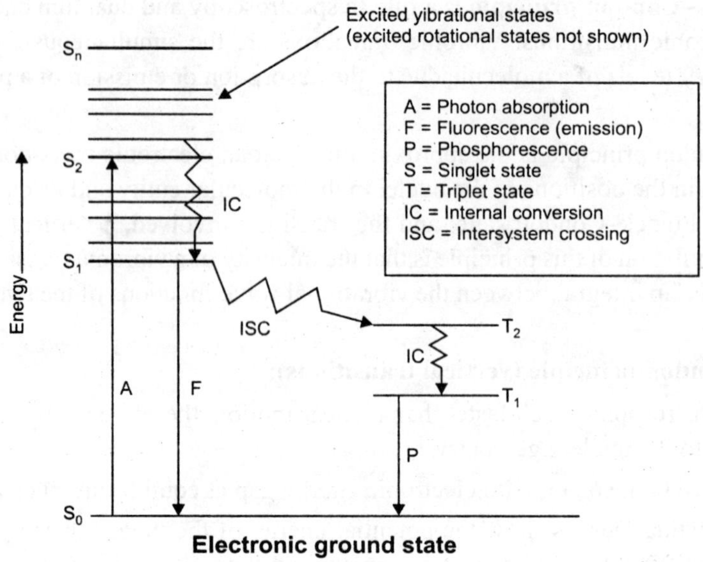

Fig.2.5: Jablonski diagram

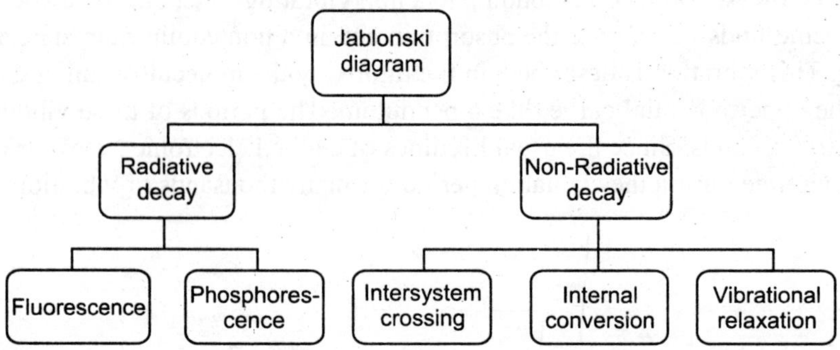

Flow Chart. 2.1: Tabular information of modes of decay of electromagnetic radiations

RADIATIVE DECAY

Fluorescence : If the photon emission (shown here as a short wavelength, that is downward pointing, "long" green line in the diagram) occurs between states of the same spin state (e.g. $S_1 \rightarrow S_0$) this is termed fluorescence.

Phosphorescence : If the spin state of the initial and final energy levels are different (e.g. $T_1 \rightarrow S_0$), the emission (loss of energy) is called phosphorescence. In the diagram this is depicted by a longer wavelength (lower energy) and therefore shorter length red line. Since fluorescence is statistically much more likely than phosphorescence for most molecules, the lifetimes of fluorescent states are very short (1×10^{-5} to 10^{-8} seconds) and phosphorescence somewhat longer (1×10^{-4} seconds to minutes or even hours; think about glow-in-the-dark flying disks).

Non-radiative Deacy

1. **Internal conversion (IC) :** Internal conversion is the **radiation less** transition between energy states of the same spin state (compare with fluorescence-a radiative process).

2. **Intersystem crossing (ISC) :** Intersystem crossing is a radiation less transition between different spin states (compare to phosphorescence),
3. **Vibrational relaxation (VR) :** Vibrational relaxation, the most common of the three--for most molecules, occurs very quickly ($<1 \times 10^{-12}$ seconds) and is enhanced by physical contact of an excited molecule with other particles with which energy, in the form of vibrations and rotations, can be transferred through collisions. This means that most excited state molecules never emit any energy because in liquid samples the solvent or, in gas phase samples, other gas phase molecules that are present "steal" the energy before other deactivation processes can occur.

Points to Remember

- Electronic transitions involve much larger energies than vibrational and rotational transitions.
- Each electronic transition is accompanied by vibrational and rotational transitions.
- The combination of vibrational transitions with electronic transitions is called "vibronic" transitions.
- There is no selection rule for vibrational transitions between two electronic states.
- The Franck-Condon principle governs the intensity distribution of vibronic transitions.
- The R branches in the rotational-electronic spectra show bandheads.

Electronic Spectroscopy

The energy difference between the electronic states in a molecule are much larger than those between the successive vibrational levels. It has been observed that a smallest change in electronic energy will produce a large change in the vibrational energy. A change in the vibrational energy produces change in vibrational quantum number giving bands in the electronic spectra. Similarly a small change in the vibrational energy produces a large change in rotational quantum number as a result of which rotation transition occurs resulting in rotational spectra in the same electronic spectra. Thus we see that each electronic transition includes electronic, vibrational and rotational transitional complications in the spectrum.

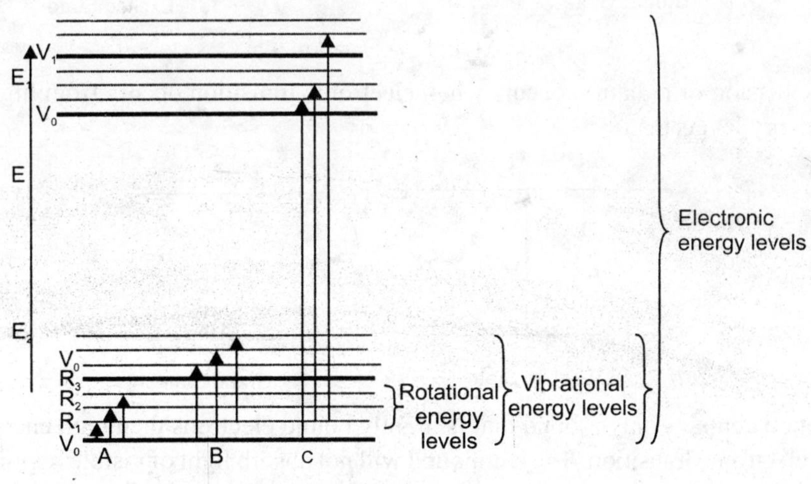

Fig.2.6: Transition of spectrum in various energy levels

U.V-VISIBLE SPECTROSCOPY

UV-visible spectroscopy is reliable and accurate analytical method that measures absorption, transmission and emission of UV and visible light by matter.

Visible region corresponds to 800-400 mm and Ultra violet region corresponds to 400-100mm.

U.V region is further divided into two sub regions :

***(i)* Far U.V/Vacuum UV region :** It ranges from 100-200 nm. The vacumm U.V. is so named becomes molecules of air absorb radiation in this region and thus this region is accessible only with special vacuum equipments.

***(ii)* Near U.V region :** It ranges from 200-400 nm. It deals with transmission of valence electrons from highest occupied molecular obital (HOMO) to lowest unoccupied molecular orbital (LUMO). Consequently, absorption spectroscopy carried out in this region is also called Electronic spectroscopy.

U.V spectrum records the wavelength at which maximum absorption occurs i.e. Kmm and the strength of absorption.

Main basis of UV/VIS SPECTRA : When UV or visible light strikes atoms or molecules, electronic transition between energy levels occur if incident light has energy equal or greater than the energy difference between energy level.

Absorption of UV or visible light results in transition of electron from lower energy level E_1 to higher energy level E_2.

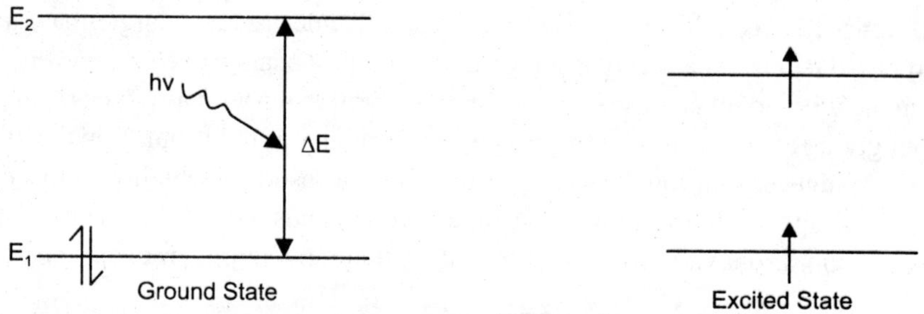

$$\Delta E = E_2 - E_1 = h\nu$$

Similarly, emission of radiation occurs when electronic transition occurs from higher energy level (E_2) to lower energy level (E_1).

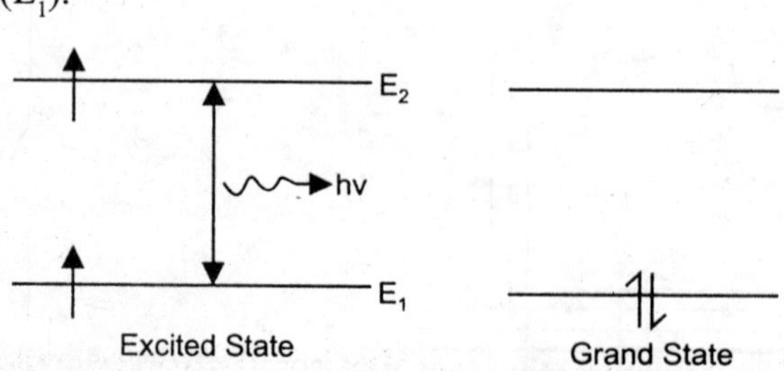

If a compound contains only a-bond that is. tightly bound electrons then high energy radiation will be required for electronic transition. The compound will not absorb light of visible region and will appear colourless as radiation of UV region will be absorbed (high energy and low wavelength).

If a compound is unsaturated that is having n-electrons which are loosely held, then absorption may occur in visible region (low energy and high wavelength) and substance appears coloured.

Types of Electronic transitions :

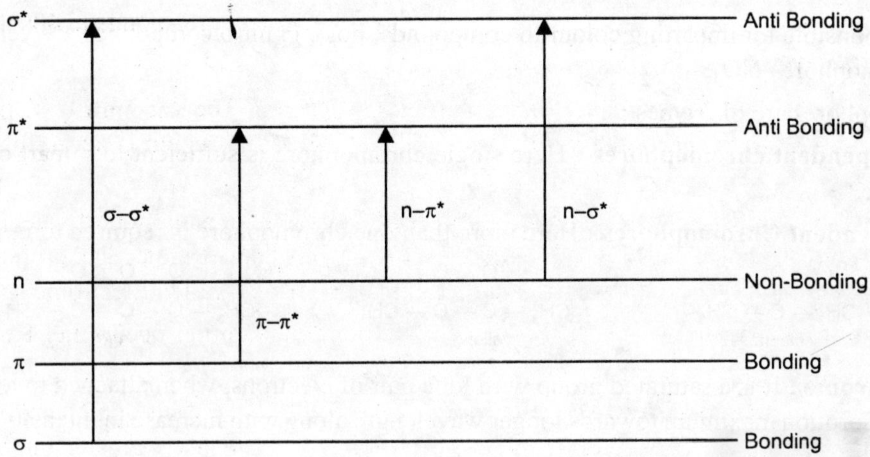

Fig.2.7: Electronic energy levels and Electronic Transitions

The electronic transitions in decreasing order of energy are :

$$\sigma\text{–}\sigma^* > n\text{–}\sigma^* > \pi\text{–}\pi^* > n\text{–}\pi^*$$

***(i)* σ to σ* transition :** A electronic transition from a bonding sigma orbital to antibonding sigma orbital (higher energy) is designated as σ to σ*. These transition occur in compound having a bond only. In alkanes, this is only transition available. Since a bond is very strong bond and a electrons are tightly held, therefore, σ – σ* transition is high energy process. Absorption band occurs in far U.V region ~ 150 nm. For CH_4, λ_{max} = 121.9 nm and C_2H_6, λ_{max} = 135 nm.

Such transition is studied in vacuum U.V region since below 200 nm, Oxygen present in air begins to absorb.

***(ii) n* → σ* transition :** Saturated compounds having lone pair of electrons undergo *n* → σ* transition along with σ – σ* transition *e.g.* saturated halides, alcohols, ethers, aldehydes, ketones, amines, etc.

These transitions require comparatively less energy than σ → σ* transition. Absorption bands appear at longer wavelength ~ 180-200 nm. (near U.V region).

***(iii)* π to π* transition :** These transitions occur in unsaturated compound. Since n electrons are loosely held, hence electronic transition require comparatively less energy. Absorption bands appear at longer wavelength. In simple alkene, although several transitions are available, lowest transition is π → π* and absorption band around 170-190 nm in unconjugated alkenes is due to this transition e.g. dienes, trienes, aromatic compound.

***(iv) n* → π* transition :** Unsaturated compound having hetero atom with lone pair of electrons like S, N, O show this type of transition. This transition involves least amount of energy than all transition. and occur at longest wavelengths (> 200 nm).

e.g. : Ketones, aldehydes. Since *n* → π* transition is symmetry forbidden, thus intensity of band due to this transition is low.

CHROMOPHORE

It is a covalently unsaturated group responsible for electronic absorption *e.g.*

$$C = O, C = C, NO_2.$$

It is responsible for imparting colour to compounds. For *e.g.* nitro compounds are yellow in colour due to chromophore $-NO_2$.

Chromophores are of 2 types :

***(i)* Independent chromophores :** Here single chromophore is sufficient to impart colour for *e.g.* nitro, azo etc.

***(ii)* Dependent Chromophores :** Here more than one chromophore is required to impart color.

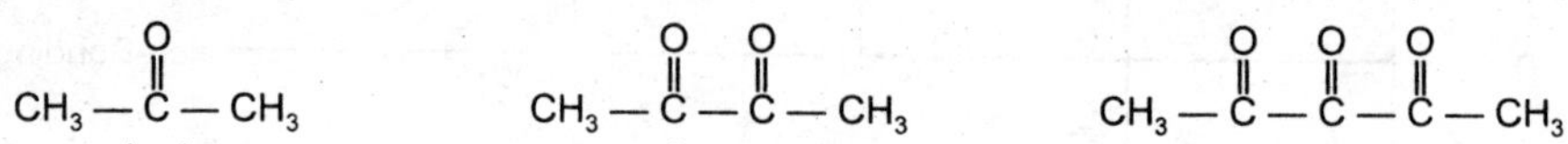

colourless yellow orange

Auxochrome : It is a saturated group with lone pair of electrons, when attached to a chromophore shifts the absorption maximum towards longer wavelength along with increase in intensity of absorption OH, NH_2, Cl.

It does not impart the colour but only deepens the colour. Due to lone pair of electrons they help chromophore in extending the conjugation.

Benzene (Colorless)

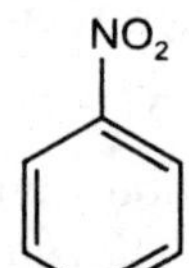

Yellow in Color
$NO_2 \rightarrow$ auxochrome

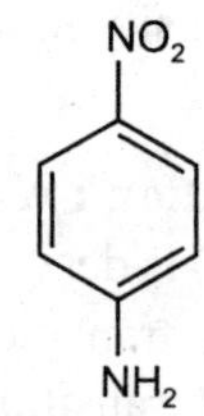

Orange in Color
NH_2, $NO_2 \rightarrow$ auxochrome

Bathochromic shift or Red shift : It is the shifting of absorption maximum towards longer wavelength due to presence of auxochromes' like OH, NH_2. Red shift also occurs due to extended conjugation.

Hypsochromic shift or Blue shift : It is the shifting of absorption maximum towards shorter wavelength. It may be caused due to removal of conjugation or change in solvent. For example acylation of $-NH_2$ and $-OH$ to $-NHCOCH_3$ and $-OCOCH_3$.

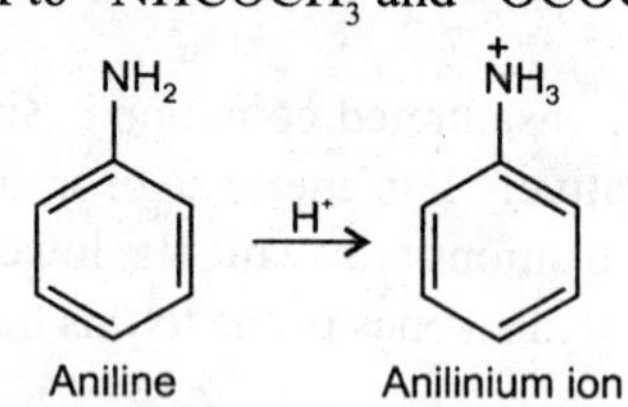

(in anilinium ion due to absence of lone pair of e$^-$, conjugation is no more with auxochrome)

Aniline $\lambda_{max} = 280$ Anilinium ion $\lambda_{max} = 200$ m

Here blue shift is occuring (above)

$$CH_2 = CH_2 \qquad \lambda_{max} = 175 \text{ nm}$$
$$CH_2 = CH—CH = CH_2 \qquad \lambda_{max} = 217 \text{ nm}$$
$$CH_2 = CH—CH = CH—CH = CH_2 \qquad \lambda_{max} = 258 \text{ nm}$$

Here Red shift is occuring.

Hyperchromic Shift : It is the increase in intensity of absorption due to auxochrome or increased conjugation. For *e.g.* in phenol at A = 210 nm, ε_{max} = 6200 but in phenolate on ε_{max} = 9400 at λ_{max} = 210 nm.

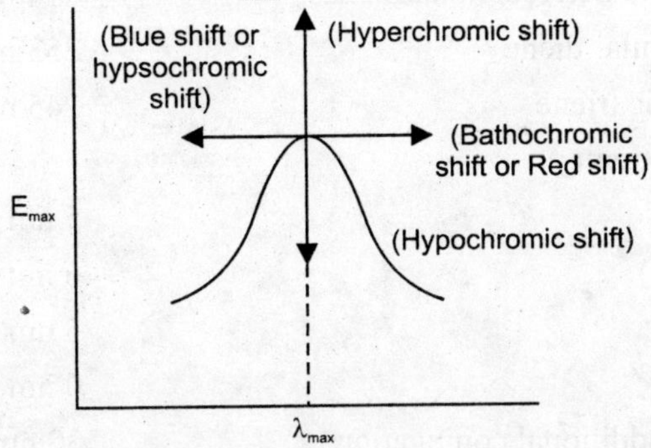

Fig.2.8: λ_{max} and intensity shift

Hypochromic Shift : It is the decrease in intensity of absorption. It is mainly due to the groups which distort the geometry of molecule. For *e.g.* hypochromic shift is observed when methyl group is introduced in position 2 of biphenyl group.

Terms to be applied in Wood ward - Fiser rules :

***(i)* Homo annular dienes :** It is a cyclic diene having conjugated double bonds in same ring.

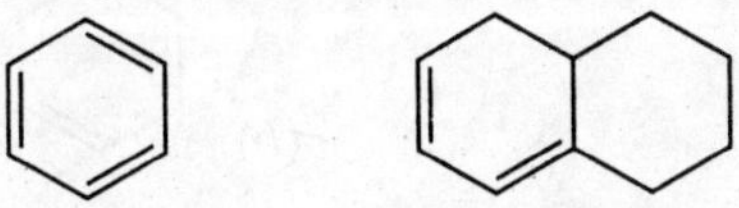

***(ii)* Hetero annular dienes :** It is a cyclic diene in which conjugated double bonds are present in different ring.

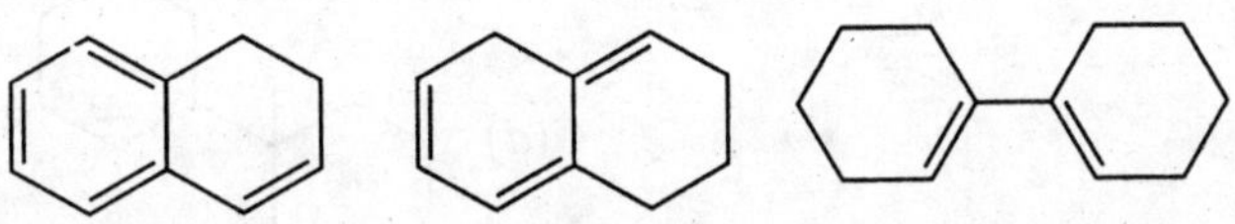

***(iii)* Acyclic diene :** Conjugated double bonds are present in open chain.

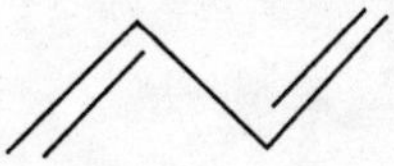

***(iv)* Endocyclic double bond :** It is a double bond present within a ring.

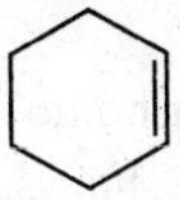

***(v)* Exocyclic double bond :** It is a double bond present outside the ring.

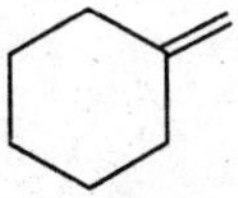

Table 2.3: WOODWARD-FISSER Rules for conjugated dienes, trienes, polyenes

Parent Values of λ_{max} in nm.

(*i*) Acyclic and heteroannular dienes	214 nm
(*ii*) homoannular diene	253 nm
(*iii*) For cyclic triene	245 nm
Addition for each substituent	
–R (alkyl)	5 nm
–OR (alkoxy)	6 nm
–Cl, –Br	5 nm
–OCOR	0 nm
–CH=CH – additional conjugation	30 nm
if one double bond is exocyclic to one ring	5 nm
if exocyclic to two rings simultaneously	10 nm
–SR	30 nm
$-NR_2$	60 nm

Problem : Calculate λ_{max} for following compounds :

(*a*) 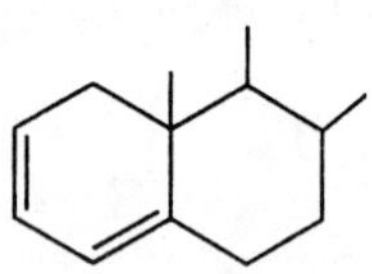

(*b*) 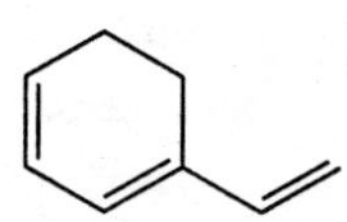

(*c*) 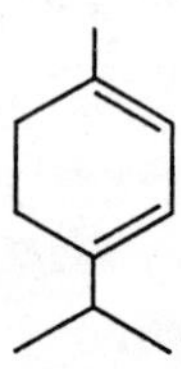

(*d*) 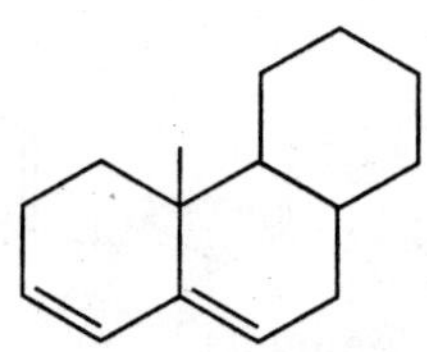

(*e*) 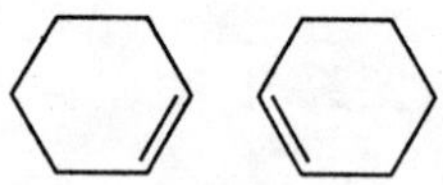

Sol. (*a*) 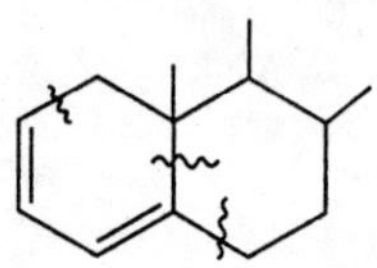

Parent value for homo annular diene	=	253 nm
3 ring residues (3 × 5)	=	15 nm
1 exocyclic double bond	=	5 nm
Calculated λ_{max}	=	273 nm
Observed λ_{max}	=	275 nm

(*b*) 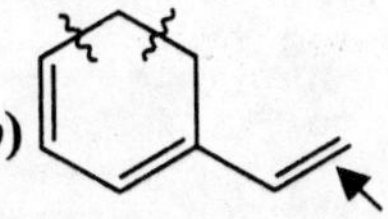

Parent value for homo annular diene = 253 rm
2 ring residues 2 × 5 = 10 nm
additional conjugation = 30 nm

Calculated λ_{max} = 293 nm

(*c*) 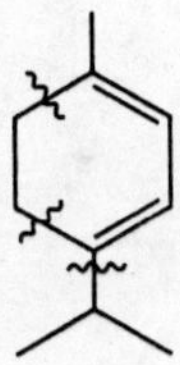

Paent value for homoannular diene = 253 nm
2 ring residues 2 × 5 = 10 nm
2 alkyl residues (2 × 5) = 10 nm

(*d*) 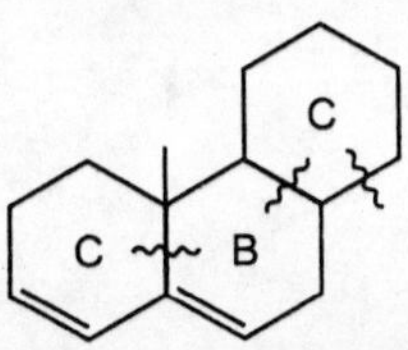

Parent value for homoannular conjugated diene = 253 nm
4 ring residues (4 × 5) = 20 nm
1 extended conjugation (in A) = 30 nm
2 exocyclic double bonds (2 × 5) = 10 nm
Calculated λ_{max} = 313 nm

(*e*) 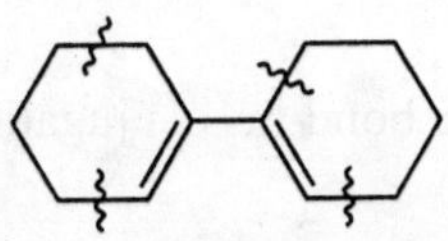

Parent value for conjugated diene = 214 nm
4 ring residues 4 × 5 = 20 nm
Calculated λ_{max} = 234 nm
Observed λ_{max} = 236 nm

Problem : ***A compound having concentration lO⁻³ gll resulted absorbance value 0.20 at AIIUV 510 nm using 1.0 cm cell. Calculate its absorptivity and molar absorptivity values. Molecular weight of compound is 400.***

ε = extinction coefficient or absorptivity
c = concentration = 10^{-3} g/1
l = length of tube = 1.0 cm
A = εcl

Absorbance,

$$A = 0.20$$

$$0.20 = \varepsilon \times 10^{-3} \times 1$$

$$\varepsilon = \frac{0.20}{10^{-3}} = 200 \text{ L/g/cm}$$

To obtain molar absorptivity,

$$\text{Cone} = 10^{-3} \text{ g/1}$$

$$= \frac{10^{-3}}{400} = 0.25 \times 10^{-5} \text{ mol/L}$$

$$0.20 = \varepsilon \times 1 \times 0.25 \times 10^{-5}$$

$$\varepsilon = \frac{0.20}{0.25} \times 10^{5}$$

$$= 8 \times 10^{5} \text{ L/mol/com}$$

Problem : Which of the following compounds will have higher λ_{max} and why?

(*a*) CH_3—CH_2—Cl and CH_3CH_2—F

(*b*) 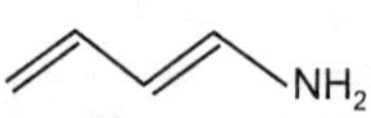

and 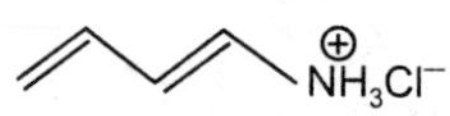

(*c*) 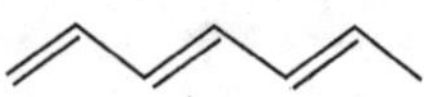 and 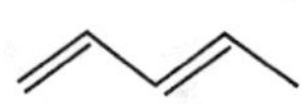

Sol. (*a*) Since F is more electronegative than Cl, hence F will hold its electron more tightly so, it will require higher energy for electronic excitation. Consequently λ_{max} for CH_3CH_2C1 is more than that of CH_3CH_2F.

(*b*) 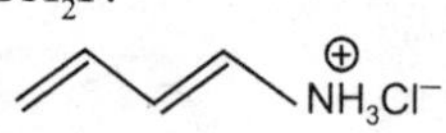

has its lone pair of e~ get involved in protonation, hence lone pair of electron donot contribute to conjugation. Consequently λ_{max} for 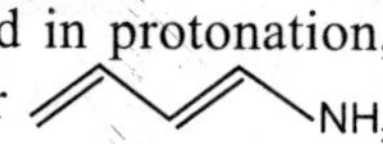

is more in that for 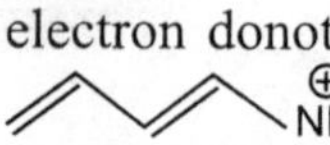

.

(*c*) 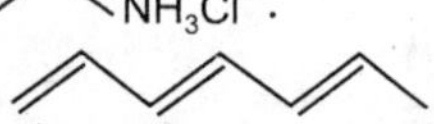 has more λ_{max} value due to an additional double bond in conjugation than 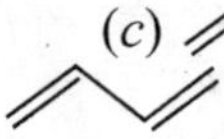

Table 2.4 :Wood Ward-Fieser rules for α, β-unsaturated Carbonyl compounds

Parent Values	**λ_{max} in nm**
(*i*) For α-β unsaturated acyclic or 6 membered ring Ketone	215 nm
(*ii*) For α-β unsaturated five-membered ring ketone	202 nm
(*iii*) For α-β unsaturated aldehydes	210 nm
(*iv*) For α-β unsaturated esters and acid	197 nm
Increments for	
(*i*) extended conjugation	+30 nm
(*ii*) For each exocyclic double bond	+5 nm
(*iii*) For homoannular conjugated iene	+39 nm

(*iv*) For each alkyl substituent or ring residue

at α position	10 nm
at β position	12 nm
at γ and higher position	18 nm

(*v*)

Auxochromes	α	β	γ	δ
–OR	35 nm	30 nm	17 nm	31 nm
–OH	35 nm	30 nm	30 nm	50 nm
–SR	—	80 nm	—	—
–Cl	15 nm	12 nm	12 nm	12 nm
–Br	25 nm	30 nm	25 nm	25 nm
–OCOR	6 nm	6 nm	6 nm	6 nm
$–NF_2$, –NHR, $–NR_2$	—	95 nm	—	—

Problems : *Calculate λ_{max} for following compounds*

(1) 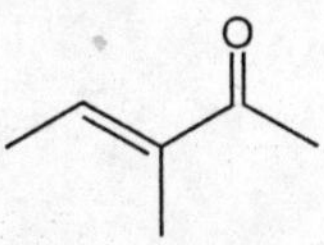 **(2)** 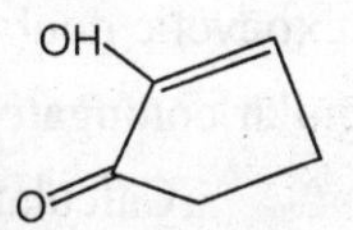 **(3)** 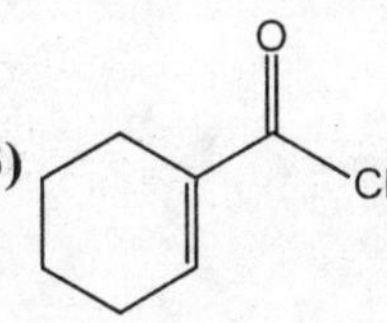

(4) 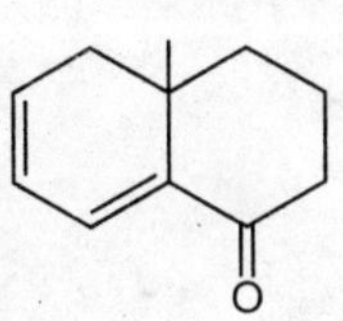 **(5)** 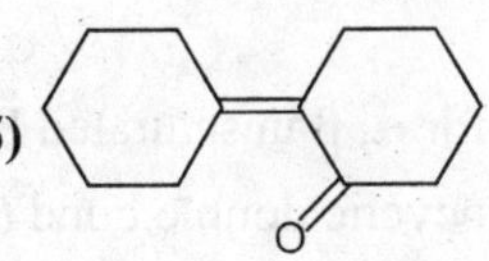 **(6)** 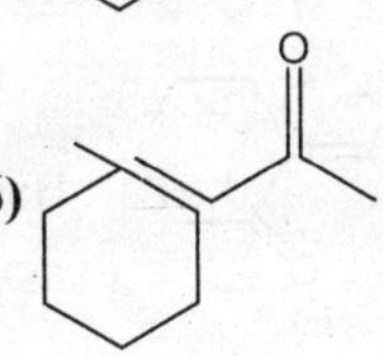

Sol. (1) 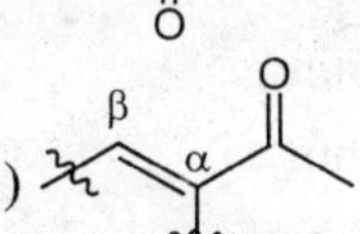

Parent value for α-β unsaturated ketone = 215 nm
2 alkyl residues, 1 at α = 10 nm
other at β = 12 nm
Calc. λ_{max} = 237 nm
Observed λ_{max} = 232 nm

(2) 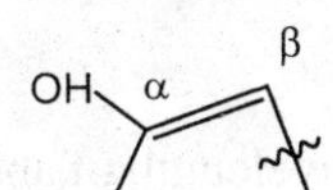

Parent value for α-β unsaturated 5 membered ring ketone 1 at α = 202 nm
OH group at α = 35 nm
1 ring residue at β position = 12 nm
Cale. λ_{max} = 249 nm
Observed λ_{max} = 247 nm

(3) 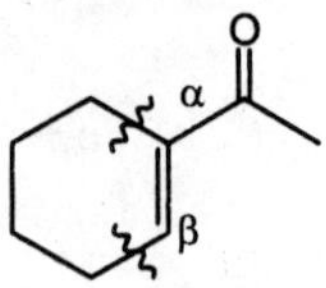

Parent value for α-β unsaturated 6-membered ring Ketone = 215 nm
a ring residue at α = 10 nm
a ring residue at β = 12 nm
Calc. λ_{max} = 237 nm
Observed λ_{max} = 222 nm

(4) 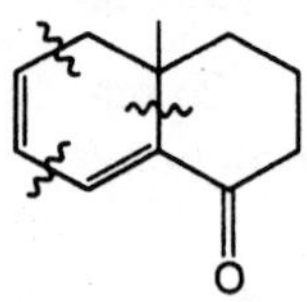

Base value for α-β unsaturated Ketone = 215 nm
a ring residue at α-position = 10 nm
a ring residue at δ-position = 18 nm
1 extended conjugation = 30 nm
1 exocyclic double bond = 5 nm
One homoannular conjugated diene = 39 nm
Calculated λ_{max} = 317 nm
Observed λ_{max} = 319 nm

(5) 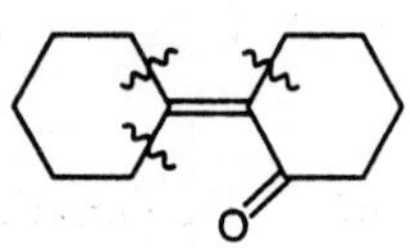

Base value for α, β unsaturated Ketone = 215 nm
Two exocyclic double bond (2 × 5) = 10 nm
One α ring residue = 10 nm
2 β ring residue (2 × 12) = 24 nm
Calc. λ_{max} = 259 nm

(6) 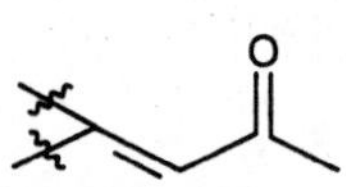

Base value for α, β unsaturated Ketone = 215 nm
2-β alkyl residue (2 × 12) = 24 nm
Calc. λ_{max} = 239 nm

APPLICATIONS OF U-V SPECTROSCOPY

***(a)* Extent of conjugation :** Longer the conjugation, longer the maximum wavelength of absorption spectrum (λ_{max}). Sufficient conjugation shifts λ_{max}. to visible region of spectrum. Compound becomes coloured on sufficient conjugation. Tomato is red coloured due to Kcoptni which has 11 conjugated double bonds.

***(b)* Distinction between conjugated and non conjugated compounds :** U.V spectroscopy helps to distinguish conjugated dienes from non-conjugated one, conjugated dienes from conjugated trienes etc.

(*c*) Detection of functional group : U.V spectroscopy help us to detect certain functional group. For example absence of any absorption above 200 nm shows the absence of conjugation, benzene ring, carbonyl compound.

(*d*) Study of Geometrical isomerism : Trans isomer absorbs at higher wavelength than Cis isomers.

(*e*) Study of Tautomerism : U.V spectroscopy can be used to identify the stable tautomeric species. For *e.g.*

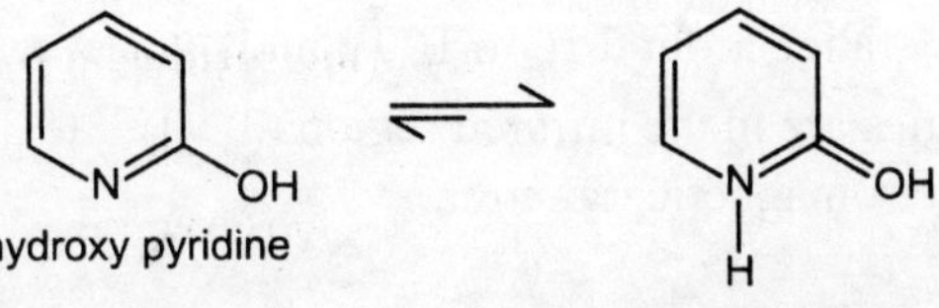

Above equilibrium is more shifted towards right side i.e. U.V spectrum of solution resembles that of a solution of pyrid-2-one and is different from that of 2-hydroxypyridine. Thus U.V spectrum shows ketone isomer predominates.

(*f*) Identification of an unknown compound : An unknown compound can be identified by comparing its spectrum with known spectra. If 2 spectra coincide, the two compounds must be identifical. If 2 spectra dont coincide, the two compounds must be different.

(*g*) Study of structural features in different solvents : λ_{max} for Chloral hydrate in hexane is 290 nm, this band is not seen when chloral hydrate is taken in aqueous solution. It shows that CCl_3. CHO. H_2O in hexane changes to CCl_3 $CH(OH)_2$ in aqueous medium.

SELECTION RULES FOR ELECTRONIC TRANSITION

Light is absorbed by the diatomic molecules, the electronic energy level is changed, the vibrational levels in the same electronic level may change and the rotational levels of the same vibrational level will changed.

1. The Selection Rules for diatomic molecules formed from light elements, the electronic transitions is governed by following selection rules,
2. $\Delta\Lambda = 0, \pm 1$ Where Λ is the component of orbital angular momentum along the Z axis. It can have 0, 1, 2, 3... L electronic states are designated as $\Sigma, \pi, \Delta, \Lambda$, corresponding to $\Lambda = 0, 1, 2, 3...$ The allowed transitions are $\Sigma \rightarrow \Sigma$, $\pi \rightarrow \Sigma$, $\Delta \rightarrow \pi$
3. $\Delta S = 0$ The Spin Rule
4. $\Delta\Sigma = 0$ the quantum number Σ in a molecule is analogues to M in the atom. Dipole allowed transition can take place only between states of the same multiplicity.
5. $\Delta\Omega = 0, \pm 1$
6. $\Sigma + \rightarrow \Sigma +$ and $\Sigma - \rightarrow \Sigma -$ transitions are allowed but $\Sigma + \rightarrow \Sigma -$ and $\Sigma - \rightarrow \Sigma +$ are not. (+)sign refers to symmetric and (–)sign refers to anti symmetric.

$g \leftrightarrow u$	$g \leftrightarrow g$	$u \leftrightarrow u$
allowed	forbidden	forbidden

ROTATIONAL AND VIBRATIONAL SPECTROSCOPY

Free atoms do not rotate or vibrate. For an oscillatory or a rotational motion of a pendulum, one end has to be tied or fixed to some point. In molecules such a fixed point is the centre of mass. The atoms in a molecule are held together by chemical bonds. The rotational and vibrational energies are usually much smaller than the energies required to break chemical bonds.

- The rotational energies correspond to the microwave region of electromagnetic radiation (3×10^{10} to 3×10^{12} Hz; energy range around 10 to 100 J/mol).
- The vibrational energies are in the infrared region (3×10^{12} to 3×10^{14} Hz; energy range around 10 kJ/mol) of the electromagnetic radiation.

Infrared Spectroscopy

Infrared region covers the range of electromagnetic radiation between 0.75 (im and 1000 μm or 3×10^{12} to 3×10^{14} Hertz. In IR spectroscopy wavelength is measured in wavenumbers which have units cm^{-1}

$$\left(\text{wave number} = \frac{1}{\text{wave length}}\right)$$

Table 2.5: Infrared region is divided into three regions

Region	Wavelength range (μm)	Wave number cm^{-1}
Near IR	0.78–2.5	12,800–4000
Mid IR	0.25–25	4000–400
Far IR	25–1000	400–10

Infrared region results in changes in vibrational energy of molecules. Since changes in vibrational energy levels are accompanied by transitions between rational energy levels also, as a result, each line in vibrational structure is accompanied by rotational fine structure. Thus this spectrum is known as vibrational-rotational spectrum which is highly complicated. This complication can be avoided by taking IR spectrum for liquid or in solution or solid sample. Under such conditions rotational fine structure get vanished and we get broad peak for vibrational change only.

Requirement for IR Spectra : All molecules cannot interact with IR radiation. Only those molecules which exhibit change in dipole moment during vibration can give IR spectra. For *e.g.* hetero nuclear diatomic molecule like HCl, HF and another molecules like H_2O, NH_3 etc.

Homonuclear diatomic molecules like H_2, N_2, Cl_2 and O_2 do not exhibit change in dipole moment during vibration hence they don't exhibit IR spectra.

According to quantum theory, permitted vibrational energies are given by expression

$$E_{vib} = \left(\upsilon + \frac{1}{2}\right) h\upsilon_0$$

$\upsilon \rightarrow$ vibrational quantum number, it may have values 0, 1, 2 etc.

$\upsilon_0 \rightarrow$ fundamental frequency.

It may noted that even when $\upsilon = 0$, that is when molecule is at rest, molecules have vibrational energy $= \frac{1}{2}h\upsilon_0$. This is called zero point energy, which signifies that molecules always vibrate and can never be at rest.

Selection rule for vibrational transition : Only those transitions are allowed for which $\Delta V = \pm 1$ that is decrease or increase in vibrational quantum number is by one unit. Thus spacing between 2 successive vibrational energy level will be always $h\upsilon_0$.

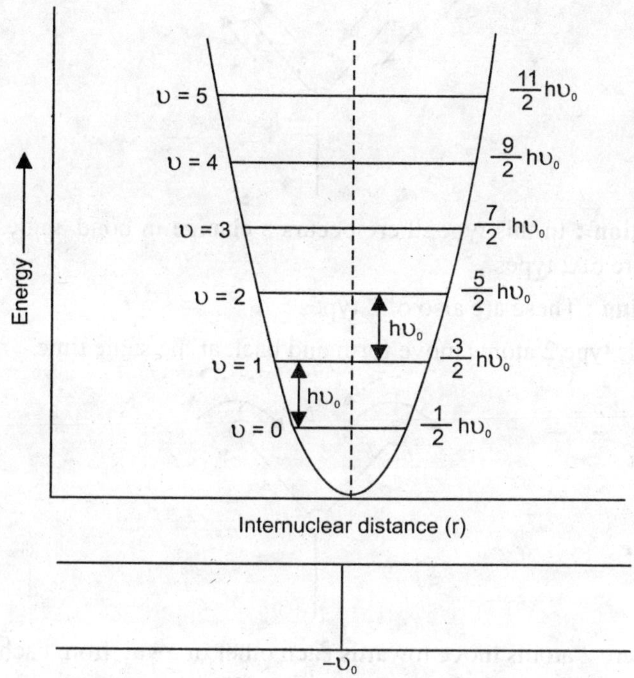

Fig.2.9: The allowed Vibrational energy levels and transitions

Principle of Infra red spectroscopy : For a molecule to absorb IR, their rotation or vibration must be accompanied by a net change in dipole moment. The alternating electric field of radiation interacts with fluctuations in dipole moment of the molecule. If frequency of radiation matches with vibrational frequency of molecule then radiation will be absorbed causing a change in amplitude of molecular vibration.

Molecular vibrations : Molecules are never at rest, they are continuously vibrating even at absolute zero temperature. In general a polyatomic molecule composed of n atoms has 3n degrees of freedom, six of which are translational and rotational. This leaves 3n-6 degrees of vibrational freedom when the molecule is non linear. If molecule is linear, there are *n*–5 vibrational degrees of freedom.

There are mainly two vibrational modes :

(*a*) Stretching (*b*) Bending

(*a*) Stretching Vibration : Atoms move along the bond axis, so that bond length decreases or increases at regular intervals. These are of 2 types *e.g.* symmetric and asymmetric.

(*i*) Symmetric Stretching : Both the bonds'are stretched or contracted at the same time.

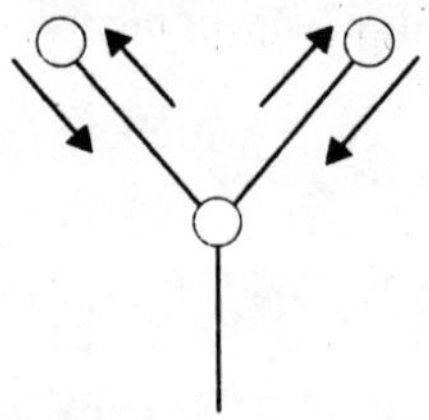

***(ii)* Asymmetric Stretching :** One bond is strecthed while the other is contracted at the same time.

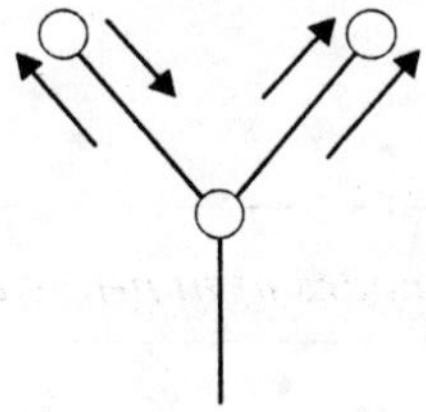

2. Bending Vibration : In this type, there occurs a change in bond angles between bonds with a common atom. These are of 2 types :

***(a)* In-plane bending :** These are also of 2 types :

***(i)* Rocking :** In this type 2 atoms move forth and back at the same time.

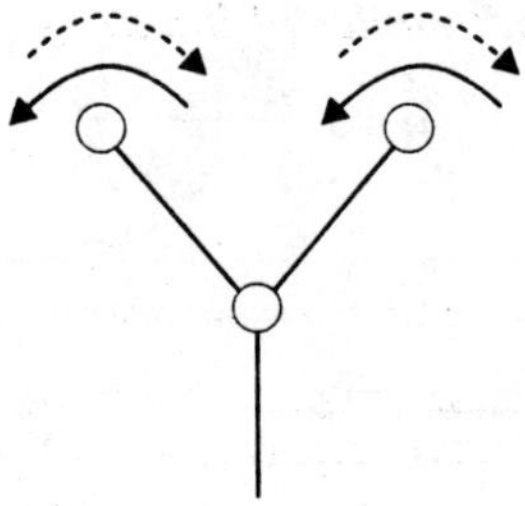

***(ii)* Scissoring :** Here 2 atoms move towards each other or away from each other at the same time.

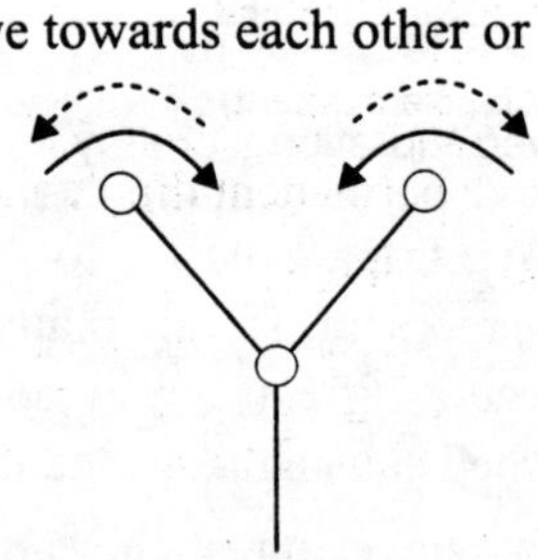

***(b)* Out of plane bending :** These are of 2 types

***(i)* Twisting :** Here one of the atom moves up the plane while the other moves down the plane with respect to the central atom.

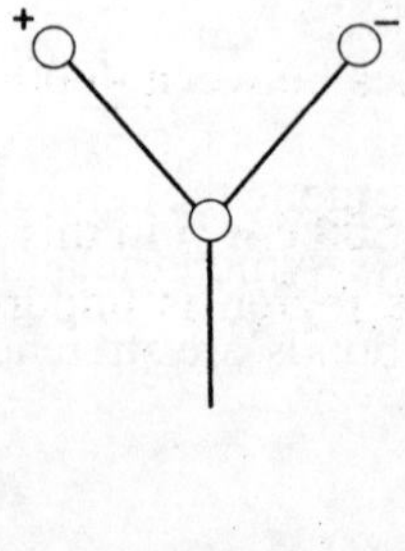

***(ii)* Wagging :** Here the 2 atoms move up and down the plane with respect to central atom at the same time.

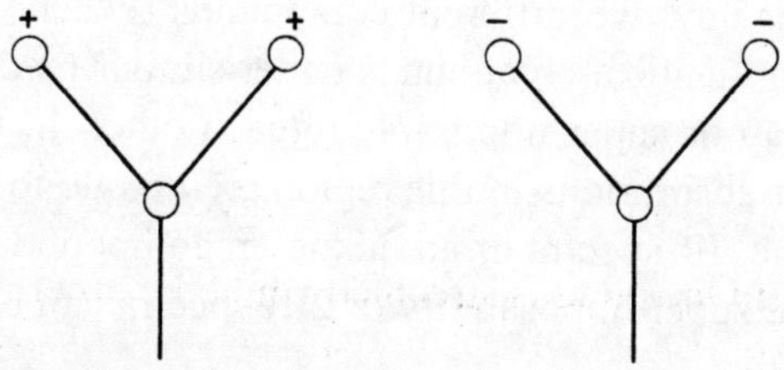

(+) → indicates motion towards reader

(–) → indicates motion away from reader

Example I. ***Fundamental vibrational modes of non-linear triatomic water molecule 3n – 6 = 3 × 3 – 6 = 3 modes.***

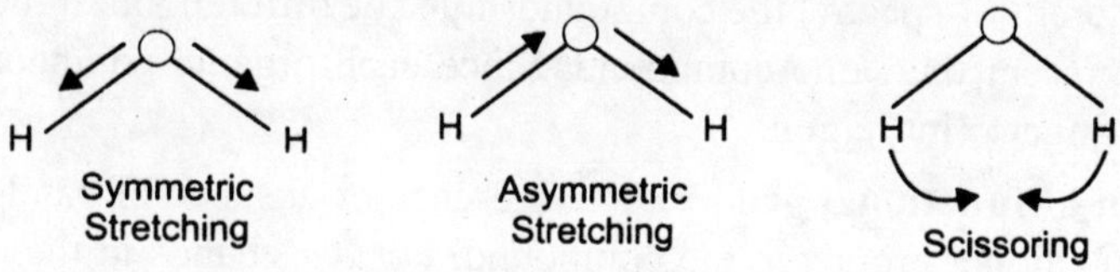

Example II. ***Fundamental vibrational modes of linear molecule, CO_2.***

Vibrational degrees of freedom for CO_2 = 3n – 5

$$3 \times 3 - 5 = 4$$

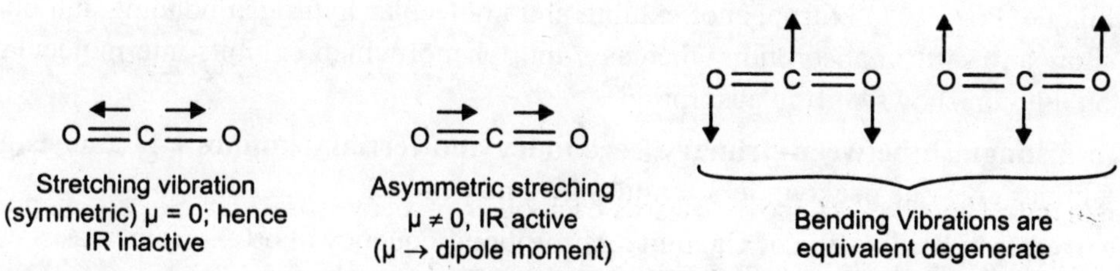

[It is not necessary for molecule to have permanent dipole moment to be IR active]

Vibrational frequency is given by Hooke's low :

$$\upsilon_m = \frac{1}{2\pi}\sqrt{\frac{k}{\mu}} = \frac{1}{2\pi}\sqrt{\frac{k(m_1+m_2)}{m_1 m_2}}$$

$$k = \text{force constant}$$

$$\mu = \text{reduced mass} = \frac{m_1 m_2}{m_1 + m_2}$$

m_1 and m_2 are masses of two atoms (in gms)

Practically the most useful IR region lies between 4000–670 cm^{-1}. Above region may be divided into two regions :

***(i)* Functional group region :** (4000–1500 cm^{-1}) In this region bands are associated with changes in vibrational states of various bonds. This region is important for chracterisation of a molecule or functional group.

***(ii)* Finger print region (1500-600 cm^{-1}) :** In this region bands are associated with changes in complex vibrational and rotational energy of entire molecule. This region is helpful in identification of organic compound. It is impossible for any two different compounds (except enantiomers) to have precisely the same IR spectrum. For two identical compounds under similar medium and condition we get same spectra with every peak overlap or super imposable to each other. In fact this region is named finger print region because pattern of absorptions in this region are unique to any particular compound like a persons finger prints are unique. IR spectra of an unknown compound is compared with set of spectra of possible compounds and the super imposability of 2IR spectra can be taken as a strong evidence for identity of unknown compound.

APPLICATION OF IR SPECTROSCOPY

***(i)* Identification of an organic compound :** An unknown organic compound can be confirmly identified, if its spectra in finger print region superimpose with that of known organic compound. If two pure samples give different IR spectra, the compounds must be different but if they give superimposable spectra, then they represent the same compound. Since each organic compound has its own unique absorption pattern in finger print region.

***(ii)* Identification of functional group :** The presence or absence of bands help in predicting the presence of certain functional groups in the compound. For *e.g.*, bands in the region 1650–1850 cnr1 suspects presence of carbonyl group in the compound.

***(iii)* Distinction between Hydrogen bonding :** In general hydrogen bonding lowers the absorption frequency. Intramolecular hydrogen bonding can be distinguished from intermolecular hydrogen bonding by IR technique. For *e.g.* O Nitrophenol exhibits intramolecular hydrogen bonding and on dilution, it does not show any shift in absorption whereas *p*-introphenol which exhibits intermolecular hydrogen bonding, on dilution shows shift in absorption.

***(iv)* To distinguish between primary, secondary and tertiary amines :** 1° amines give several peaks, 2° amines give one peak while 3° amines give no peaks.

***(v)* To predict hybridization of Carbon :** Absorption frequency of o(C—H) increases with increase in 'S' character.

***(vi)* Detection of impurities :** Spectrum of pure sample always consists of sharp peaks and bands while impure sample has poor bands and some additional bands of impurities.

PROBLEMS

Q. 1. Indicate whether the following molecules are IR active or IR inactive.

$$CH_3—CH_3,\ CHCl_3,\ N_2,\ CO_2,\ CH_2{=}CH_2,\ H_2O.$$

Ans. IR active → $CHCl_3$, CO_2, H_2O.

IR inactive → $CH_3—CH_3$, N_2, $CH_2{=}CH_2$.

Q. 2. Calculate (*a*) the reduced mass and (*b*) the moment of inertia of NaCl using the mean internuclear distance of 2.36 Å. The atomic masses are Cl = 35 × 10^{-3} kg mol^{-1} and Na = 23 × 10^{-3} kg mol^{-1}.

Ans. The reduced mass is given as

$$\frac{1}{\propto} = \frac{1}{m_1} + \frac{1}{m_2} = \frac{m_1 + m_2}{m_1 m_2}$$

$$\mu = \frac{(35\times10^{-3}\ \text{kg mol}^{-1})(23\times10^{-3}\ \text{kg mol}^{-1})}{(35\times10^{-3}\ \text{kg mol}^{-1} + 23\times10^{-3}\ \text{kg mol}^{-1})(6.023\times10^{23}\ \text{mol}^{-1})}$$

$$= 2.304 \times 10^{-26}\ \text{kg}$$

The moment of inertia is given as

$$I = \mu r^2 = (2.304 \times 10^{-26}\ \text{kg})\ (2.36 \times 10^{-10}\ \text{m})^2$$

$$= 12.83 \times 10^{-46}\ \text{kg m}^2$$

ENERGY OF ROTATIONAL TRANSITIONS

When a molecule is irradiated with photons of light it may absorb the radiation and undergo an energy transition. The energy of the transition must be equivalent to the energy of the photon of light absorbed given by: $E = hv$. For a diatomic molecule the energy difference between rotational levels (J to J + 1) is given by:

$$EJ + 1 - EJ = B(J + 1)(J + 2) - BJ(J = 1) = 2B(J + 1) \quad ...(2.2)$$

with J = 0, 1, 2,...

Because the difference of energy between rotational levels is in the microwave region (1 – 10 cm^{-1}) rotational spectroscopy is commonly called microwave spectroscopy. In spectroscopy it is customary to represent energy in wave numbers (cm^{-1}), in this notation B is written as B~. To convert from units of energy to wave numbers simply divide by h and c, where c is the speed of light in cm/s (c=2.998e10 cm/s). In wave numbers B~ = $h8\pi c$I.

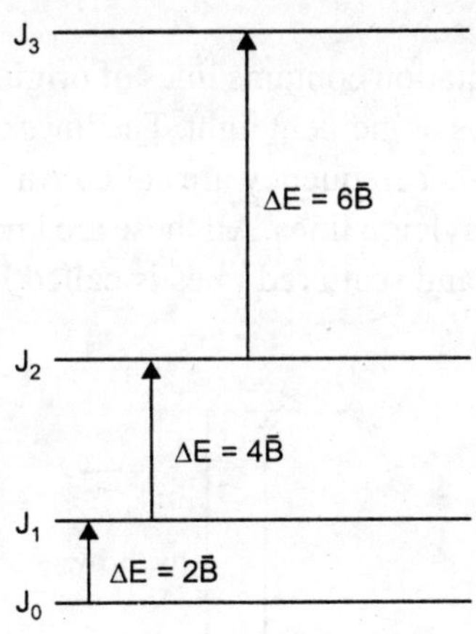

Fig.2.10: Rotational Energy Levels

Figure 2.10 predicts the rotational spectra of a diatomic molecule to have several peaks spaced by 2B~. This contrasts vibrational spectra which have only one fundamental peak for each vibrational mode. From the rotational spectrum of a diatomic molecule the bond length can be determined.

Selection rules only permit transitions between consecutive rotational levels: $\Delta J = J \pm 1$, and require the molecule to contain a permanent dipole moment. Due to the dipole requirement, molecules such as HF and HCl have pure rotational spectra and molecules such as H_2 and N_2 are rotationally inactive.

The selection rule for a rotational transition is,

$$\Delta J = \pm 1$$

In addition to this requirement, the molecule has to possess a dipole moment. As a dipolar molecule rotates, the rotating dipole constitutes the transition dipole operator μ. Molecules such as HCl and CO will show rotational spectra while H_2, Cl_2 and CO_2 will not.

Rotational spectrum values of some rigid diatomic molecules B are in cm^{-1}, 1.92118 (CO), 10.593 (HCl), 20.956 (HF), $^1H_2$ (60.864), $^2H_2$ (30.442), 1.9987 (N_2).

From the value of B obtained from the rotational spectra, moments of inertia of molecules I, can be calculated. From the value of I, bond length can be deduced.

Problem. Calculate the value of I and r of CO. B = 1.92118 cm^{-1}.

Solution: $I = h/(8\pi 2\ BC)$

$= 6.626 \times 10^{-34}/(8 \times 3.14152 \times 1.92118 \times 3 \times 1010)$

Value of $I = 1.45579 \times 10^{-46}$ kg m^2

Since the value of B is in cm^{-1}, the velocity of light c is taken in cm/s.

$$I = \mu r^2.$$

The atomic mass of C ≡ 12.0000 amu, O ≡ 15.9994 amu. 1 amu = 1.6604×10^{-27} kg.

The reduced mass of CO can be calculated to be 1.13836×10^{-27} kg.

Therefore $\quad r^2 = I/\mu = 1.45579 \times 10^{-46}/1.13826 \times -27\ m^2$

Or $\quad r = 1.131$ Å

Raman Scattering

The spectrum of these scattered radiation contains lines of original light frequency alongwith the lines of certain lower and higher frequencies of incident light. The lines of lower frequency than incident light are called stokes lines and those of higher frequency are called Anti-stokes lines and the same frequency as that of incident light is known as Rayleigh lines. All these are known as Raman scattering. The difference $\Delta\nu$ between frequency of incident and scattered lines is called Raman Characteristic frequency.

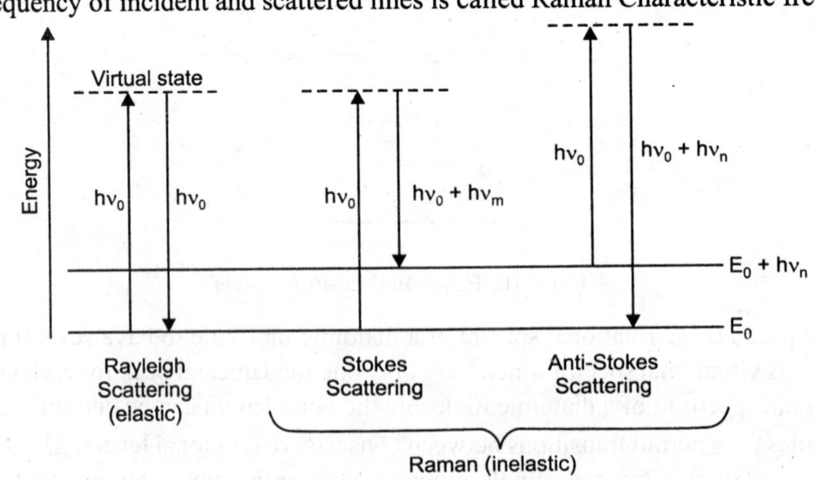

Fig.2.11: Raman Scattering

Suppose v_i and v_s are the frequencies of incident and scattered light, then $v_i - v_s = \Delta v$ is called Raman frequency and the difference $(v_i - v_s)$ is called Raman Shift Δv. Raman shift ranges between 100 – 4,000 cm^{-1}. According the Raman, Raman frequency is a constant for a molecule and its characteristics for that molecule and it does not depend upon the frequency of incident radiation.

Homo nuclear diatomic molecules H_2, N_2, O_2 etc., which do not exhibit IR spectra because these molecules do not have permanent dipole moments, do show Raman spectra. This is because their rotation or vibration is accompanied by a change in polarisability of the molecules, resulting in a change in the induced dipole moment at the vibrational frequency. We shall consider first Rotational and then Vibrational Raman spectra.

FLUORESCENCE SPECTROSCOPY

Fluorescence spectrometry is a fast, simple and inexpensive method to determine the concentration of an analyte in solution based on its fluorescent properties. Fluorescence microscopy has become one of the most widely used techniques, especially for probing specific targets and dynamic processes in chemically fixed or living cells. It has dramatically improved our understanding of the living systems.

Principle

In fluorescence spectroscopy, a beam with a wavelength varying between 180 and ~800 nm passes through a solution in a cuvette. Atom or molecule first absorbs energy – for instance a photon -, this is called excitation. Very shortly (in the order of nanoseconds) after excitation it emits a photon of a longer wavelength named that fluorescence. If the emission takes longer time in the order of milli seconds then will be known as fluorescence. It can be measure from an angle - the light that is emitted by the sample. In fluorescence spectrometry both an excitation spectrum (the light that is absorbed by the sample) and/ or an emission spectrum (the light emitted by the sample) can be measured. The concentration of the analyte is directly proportional with the intensity of the emission.

The figure shows a fluorescence schematic. The light source is in a 90 degree angle with the detector. The sample is located at the intersection of the two beam paths. Only a relatively small number of compounds can fluoresce. Some non-fluorescent compounds can be made fluorescent by adding a fluorescent label. In general molecules that fluoresce have one or more aromatic groups in its structure.

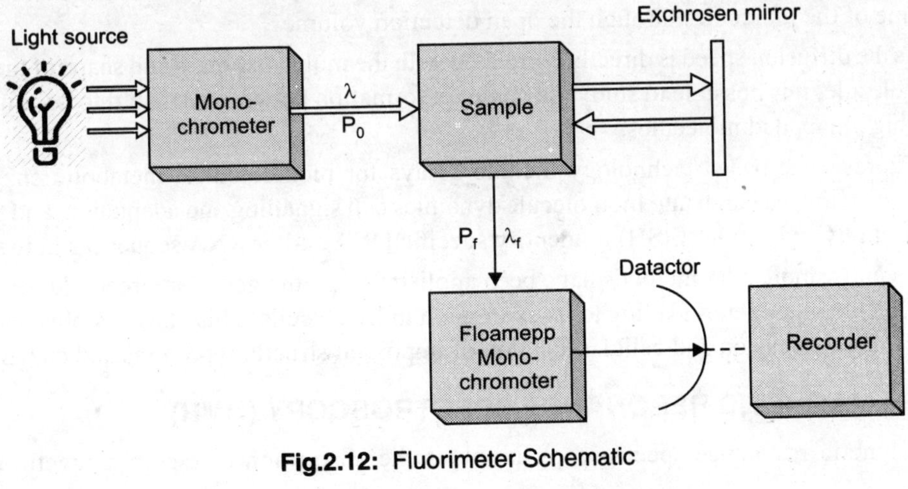

Fig.2.12: Fluorimeter Schematic

Factors affecting intensity of spectra

There are several parameters influencing the intensity and shape of the spectra. When recording an emission spectrum the intensity is dependent on the:

- Excitation wavelength
- Concentration of the analyte solvent
- Path length of the cuvette
- Self-absorption of the sample

Suitable analytes are -

- Fluorescence analysis is suitable for analytes that can be dissolved in solvents like water, ethanol and hexane.
- The analytes need to absorb UV or visible light.
- The analytes need to emit visible or near infra red radiation
- With fluorescence analysis we can do quantitative measurements of a single analyte in solution (Or more than one analytes in solution provided they do not interfere with each other.)

Unsuitable analytes are-

- Analytes that have a photochemical reaction at (or above) the wavelength range of interest.
- Intransparent, not clear or colloidal samples.
- Compounds that do not show fluorescence

Applications

- Fluorescence spectroscopy is being applied in medical microbiology field, for diagnostic technique with high sensitivity and specificity specially for microorganisms associated diseases.
- Fluorescence spectroscopy and Fluorescence correlation spectroscopy (FCS) may be applied to understand various pathophysiological steps of various microorganisms.
- Fluorescence correlation spectroscopy (FCS), a technique basically used for spatial and temporal analysis of molecular interactions of extremely low concentrated biomolecules in solution.
- FCS measures both the average number of molecules in the detection volume and the diffusion time of the molecules through the open detection volume .
- As the diffusion speed is directly correlated with the molecular mass and shape of the fluorescent molecule, it is possible to study the complex formation between a small fluorescent labeled and a big unlabelled molecules.
- Fluorescence-based technology include assays for biomolecules, metabolic enzymes, DNA sequencing, research into biomolecule dynamics cell signalling and adaptation, and fluorescence in situ hybridisation (FISH) to identify specific DNA and/or RNA sequences in tissues.
- Recently, molecular methods have been applied to fuse the gene for Green Fluorescent Protein (GFP) to other genes leading to its expression in living cells. This allows sophisticated analysis of gene expression and cellular location of important structural proteins and enzymes.

NUCLEAR MAGNETIC RESONANCE SPECTROSCOPY (NMR)

Nuclear magnetic resonance spectroscopy involves the absorption of electromagnetic radiation in radiofrequency region. It involves the interaction between oscillating electromagnetic forces and magnetic energy of nuclei usually hydrogen.

Rules for determining net spin (I) of a nucleus

Nuclei of certain atoms behave as if they are spinning charges. Any spinning charge creates a magnetic field and behave as if it were a tiny magnet. |

(*i*) If the number of protons and the number of neutrons are both even, the nucleus has no spin i.e., I = 0, for example, ^{12}C, ^{16}O, ^{32}S etc.

(*ii*) If the sum of number of protons and number of neutrons is odd, the nucleus has half-integer spin i.e. $\frac{1}{2}, \frac{3}{2}, \frac{5}{2}$. For example ^1H, ^{13}C, ^{19}F, ^{31}P etc.

(*iii*) If number of protons and number of neutrons are both odd then the nucleus has an integer spin i.e. I = 1, 2, 3. For example, $_7N^3$.

Nucleus of spin T will have 21 + 1 possible orientations. Thus for ^1H, $I = \frac{1}{2}$, hence number of possible orientation will be $2 \times \frac{1}{2} + 1 = 2$. These 2 orientations are of equal energy in absence of external magnetic field. In presence of external magnetic field, degeneracy get lost and energy level splits and each level has magnetic quantum number '*m*'.

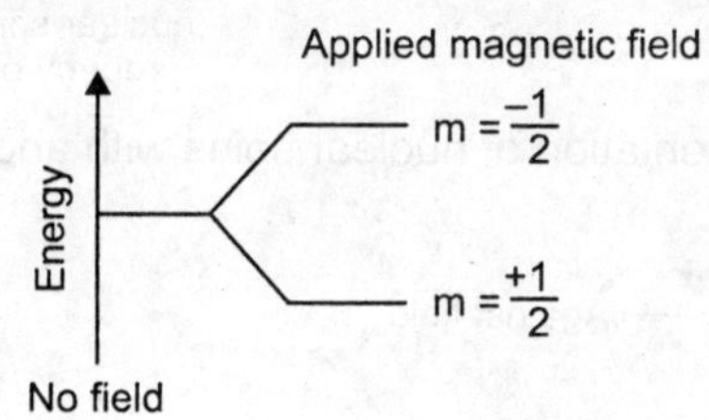

Fig.2.13: Energy levels for a nucleus with spin quantum number $\frac{1}{2}$.

The nuclei which are most commonly study by NMR spectro-scopy are **^1H, ^{13}C, ^{19}F, ^{31}P. Theory of ^1H NMR spectroscopy :**

Spinning nuclei behave just like a bar magnet with small magnetic field whose magnitude and direction can be described by a vector called a magnetic moment.

These magnetic moments are oriented in random fashion in absence of applied field. For ^1H, nuclear spin can have only two values associated with quantum numbers. $+\frac{1}{2}(\alpha) - \frac{1}{2}(\beta)$. There is no difference in energy between two nuclear spin states. But in presence of external magnetic field, two spin states remain no longer degenerate. Their magnetic moments tend either to align with the field (α-spin) or against the field (β-spin). The energy difference between a and β-spin states is very small, although, α spin state is of lower energy. With a NMR spectrometer operating at 90 MHz the energy difference between two spin states is only 9×10^{-6} KCal/mol. Thus energy is needed in order to flip the proton from lower energy state (α) to higher energy state (β). This energy is provided by electromagnetic radiation in radiofrequency region.

Thus a proton of an organic compound absorbs energy when energy difference between two spin states becomes equal the energy defined by operating frequency of NMR spectrophotometer. Under this condition a nucleus is said to be in resonance.

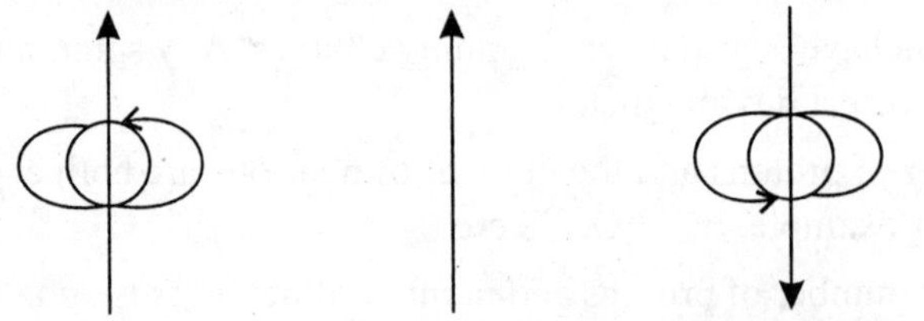

Fig.2.14: Orientation of Nuclear magnetic moment

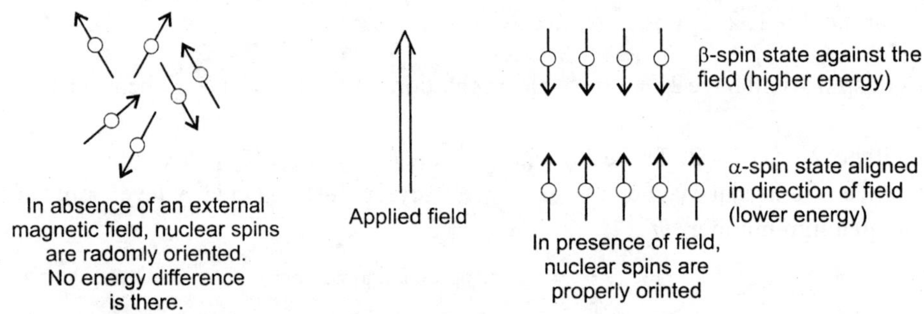

Fig.2.15: [a] Orientation of nuclear spins with and without applied field

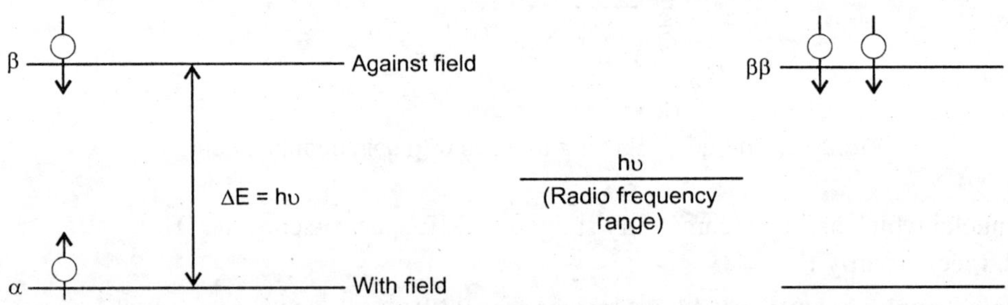

Fig.2.15: [b] Nuclear spins after absorption of RF energy

When the excited nuclei in β orientation start to relax back down to alpha orientation, a fluctuating magnetic field is created which generates a current. This current is electronically converted into peak.

The protons which are having same chemical environment are called equivalent protons. The number of peaks in NMR spectrum give information about types of proton present in a molecule. While the height of peak tells about number of one type of protons, that is larger the height of peak, more be the number of protons of one type.

For example :

→ It nas one type of proton. So it has one peak, but due to six protons, intensity of peak is very high.

→ It has 2 types of proton, so will give 2 peaks.

Chemical Shift : A nucleus does not observe the field equal to applied field as it is surrounded by electrons. The difference between the applied magnetic field and field observed by nucleus is called nuclear shielding.

All protons are not equivalent, they do not precess with same frequency, thus absorptions depend on their environment. Depending on local environment, different protons in a molecule resonate at slightly different frequencies. The shift in peak of NMR spectrum due to chemical environment is called chemical shift.

Example : Ethylenic and aldehydic proton

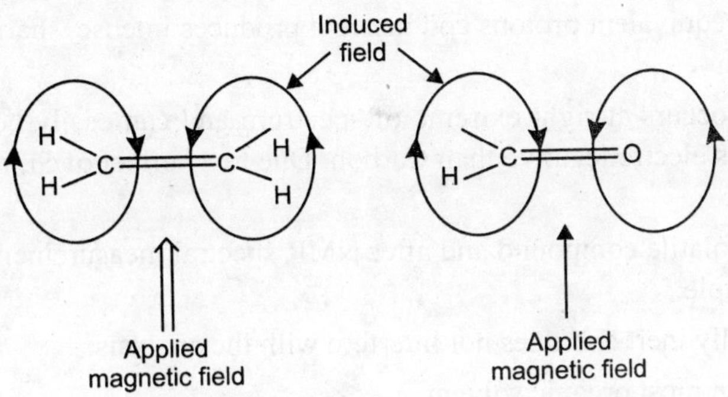

Fig.2.16: Deshielding of protons in alkene and aldehyde

On the other hand if induced magnetic field is in direction opposite to that of applied field then due to opposition, the field observed by proton decreases. This phenomenon is called shielding and proton is called shielded proton. The shielding shift the absorption upfield because now more external field will be required by proton to come in resonance. This is also termed as diamagnetic shift.

Example : Acetylenic protons

$$\text{Chemical Shift} = \frac{\text{Peak position in Hz (relative to TMS)}}{\text{Spectrometer frequency in MHz}}$$

$$\delta = \frac{\upsilon' - \upsilon_{TMS}}{\upsilon_0} \times 10^6$$

υ' = frequency of unknown group of protons in Hz

υ_{TMS} = frequency of reference, TMS in Hz

υ_0 = Operating frequency of spectrometer in MHz

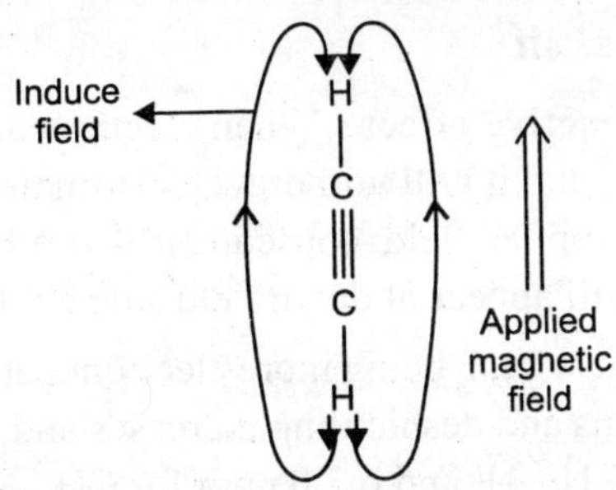

Fig.2.17: Shielding in acetylene

On the δ (delta) scale, the position of TMS signal is taken as 0.0 ppm. On t (tau) scale, the reference is assigned arbit rary position of 10.0.

$$\tau = 10 - \delta$$

Chemical shift is measured relative to a reference compound. For ^{1}H NMR most appropriate reference is tetra methyl silane ($Si(CH_3)_4$, TMS.

Merits of TMS as NMR reference compound

1. TMS has 12 equivalent protons and hence it produces intense, sharp, single signal even at low concentration.
2. TMS signal occurs at right extreme of spectrum and can easily be detected. This is because Silicon is less electronegative than Carbon. Due to +I effect of Si, it produces strong shielding effect.
3. It is highly volatile compound and after NMR spectral measurement it can be easily; removed from the sample.
4. It is chemically inert and does not interfere with the analysis.
5. It is soluble in most organic solvent.

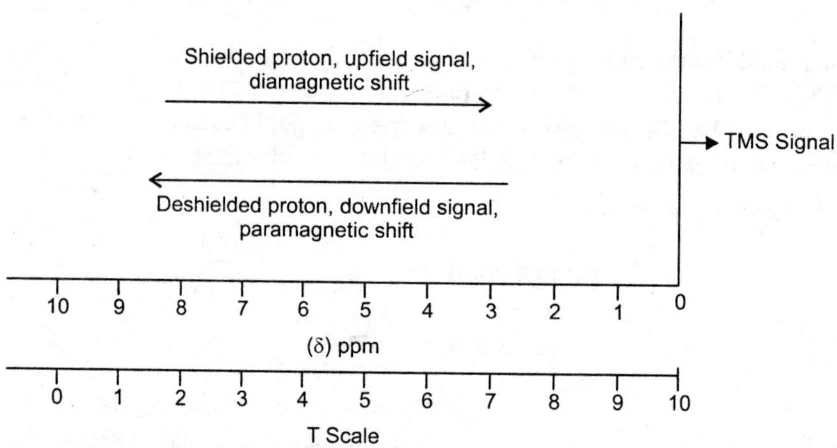

Fig.2.18:

Factors effecting chemical shift

1. Electronegativity and Inductive effect : When electronegative atom or electron with drawing group is attached to organic compound it will withdraw electron density from neighbourhood of proton and thus exposing proton more to applied field, consequently smaller applied field is required by proton to come in resonance and signal will appear at down field side i.e. higher δ value.

For example : CH_3—F and CH_3—Cl, F being more electronegative, it will withdraw electron density from neighbourhood of CH_3 protons and deshielding increases and signal will appear at down field side at higher δ value, δ 4.26 ppm for CH_3—F and δ 3.0 ppm for CH_3—Cl.

(*b*) As distance of electronegative group from proton increases, deshielding decreases.

For example CH_3—Cl δ 3.0 ppm

CH_3—CH_2—Cl δ 1.5 ppm (for CH_3).

(*c*) Electron releasing group increases electron density around the proton, thus shielding increases, as a result signal will appear at up field side, lower δ value.

2. Anisotropic Effects : Induced field, generated by circulation of electron, can either oppose or reinforce applied field at proton, depending on the location of proton and geometry of molecule.

3. Hydrogen bonding : Hydrogen bond shifts the resonance signal to down field side and at higher δ value.

Splitting of Signals : In ^{1}H NMR spectra, the signal due to a particular proton may not be a single but instead appears as a collection of peaks. The signal may split into two peaks, (doublet) 3 peaks (triplet), 4 peaks (quartet) and so on.

Splitting occurs -only between chemically non equivalent protons. Spin-spin coupling (splitting) is due to coupling interaction between proton neighbours and is related to number of possible spin orientation that these neighbours can adopt.

When a molecule is placed in external magnetic field, electrons surrounding nucleus circulate and produce magnetic field of its own called induced field.

If induced magnetic field is in same direction as that of applied field, then it will reinforce the field, as a result field observed by proton increases. This phenomenon is called deshielding and protons are called deshielded protons. This deshielding shifts the absorption downfield because now the value of applied field needed by proton to get in resonance will be less. It is also termed as paramagnetic shift.

The number of peaks into which a proton signal is splitted is equal one more than number of equivalent neighbouring protons. That is for '*n*' number of equivalent neighbouring protons, a peak will split into $n+1$ peaks or multiplet with intensity ratio given by Pascal's triangle. If there are two type of adjacent protons one of '*a*' and other '*b*' type, then peak will split into $(n_a + 1)(n_b + 1)$

n_a = number of protons of '*a*' type.

n_b = number of protons of '*b*' type.

Distance between 2 peaks is given by coupling constant, 'J'.

Intensity ratio of multiplets is given by Pascal's triangle and similar to coefficients of binomial expression.

n = 0						1						singlet
n = 1					1		1					doublet
n = 2				1		2		1				triplet
n = 3			1		3		3		1			quartet
n = 4		1		4		6		4		1		quintet
n = 5	1		5		10		10		5		1	Sexlet

Fig.2.19: Pascal's Triangle

For example

There are two types of proton '*a*' and '*b*'.

Magnetic field that '*a*' proton experiences due to spin orientation of '*b*' protons may be decreased or increased. Spin orientation of CH_3 protons (*b*)

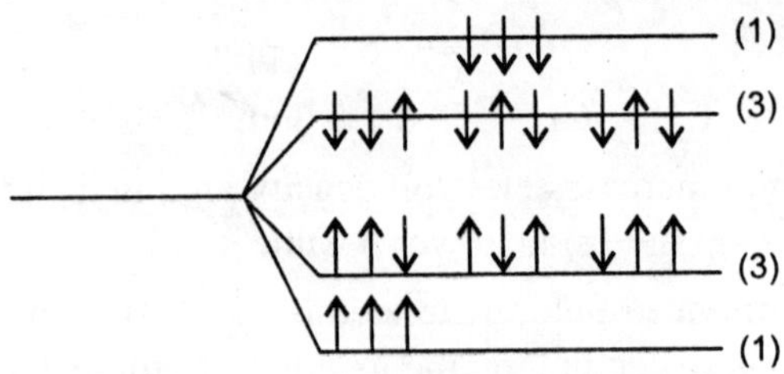

Thus CH_2 protons '*a*' will experience four different field. Consequently, peak for CH_2 protons will split into quartet, with intensity ratio 1 : 3 : 3 : 1. Spin orientation of 'CH_2' protons

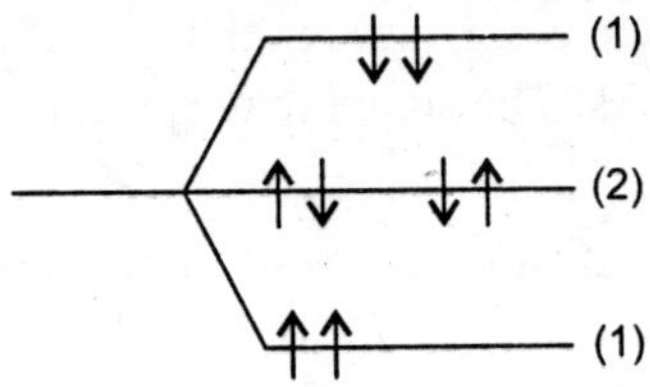

Thus, CH_3 protons '*b*' will experience '3' different field. Consequently peak for CH_3 protons will split into triplet, with intensity ratio 1 : 2 : 1.

Instrumentation :

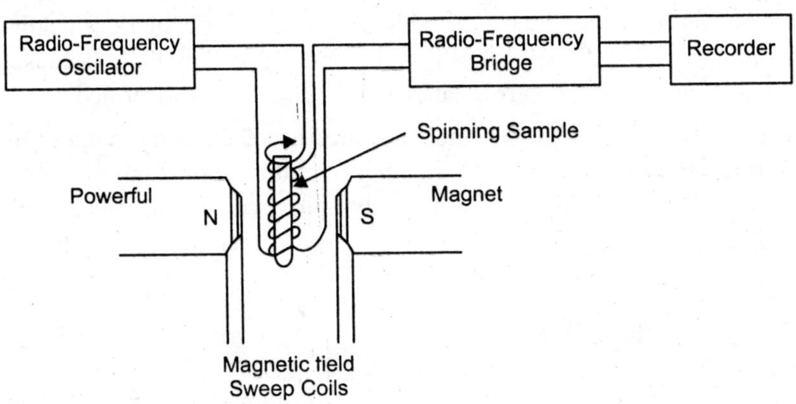

Fig.2.20: Basic arrangement for NMR

(*i*) Magnets : Permanent magnet or electromagnet can be used in NMR spectrometer. Magnet should provide, constant, homogeneous or uniform magnetic field of high field strength.

(*ii*) Sample holder : Sample holder should be durable and transparent to radiofrequency radiation. Glass tubes are commonly used as sample holder.

(*iii*) Magnetic Coils : A pair of coils is used in pole faces of magnet. These coils induce magnetic field that can be varied by varying current flowing through them.

(*iv*) Radio frequency oscillator : It is used to generate radiofrequency. The oscillator is wound around the sample container to maximise the interaction between sample and rf radiation.

Table 2.6: Typical proton chemical shifts

	Type of Protons	Chemical Shift δ (ppm)
Primary	R—CH_3	0.9
Secondary	R_2—CH_2	1.3
Tertiary	R_3—CH	1.5
Vinylic		4.6—5.9
Acetylinic	—C—C—H	2–3
Aromatic	Ar—H	6–8.5
Benzylic		2.2–3
Allylic		1.7
Alkyl flouride		4—4.5
Alkyl chloride		3–4
Alcohol		3.4–4
Phenolic	Ar—O—H	4–12
Aldehydic		9–10
Carboxylic		10.5–12
Amino	R—NH_2	1–5

APPLICATION OF NMR SPECTROSCOPY

1. NMR Spectroscopy tells about types of hydrogens present in the molecule which is always equal to number of signals in NMR spectrum.
2. Intensities'or relative height of peak give information about number of hydrogens of each kind. For *e.g.*, if 3 signals are recorded on NMR spectrum with their relative - heights in ratio 3 : 2 : 1, then there are three protons of one type, 2 protons of other type and 1 proton of another type.
3. Chemical shift give information about c—hemical environment around proton. If electronegative atom is present in neighbourhood of proton, then peak will be obtained at down field side (high δ value). Similarly if electron denating group is present, then peak will appear at upfield side (low δ value).
4. MRI (Magnetic Resonance Imaging) is based on ^1H NMR. MRI is image of section of human body. Brightness of various regions of an image is related to two factors :

   (*a*) The number of protons in the tissue at particular place.

   (*b*) The relaxation times of protons.

   For example, H of H_2O in some cancerous tumours has shorter relaxation time than in healthy cells. Thus MRI can detect brain tumour.
5. From coupling constant, J, we can distinguish between cis and trans isomers $J_{trans} > J_{cis}$.
6. We can study hydrogen bonding. Hydrogen bonding increases deshielding and produces a downfield signal at higher δ-value.

Problem–1 : *Calculate the NMR signals and splitted signals in the following molecules.*

(*i*) CH_3—O—CH_2—CH_3 (*ii*) CH_3—C_6H_5

(*iii*) CH_2=CHCl (*iv*) CH_3— CH_2— CH_2—Cl

(*v*) Fig.

Sol. → There are three types of proton, hence 3 main peaks will be obtained :

'*a*' peak will not split.\

'*b*' peak will split into quartet (3+1)

'*c*' peak will split into triplet (2+1)

(*ii*) 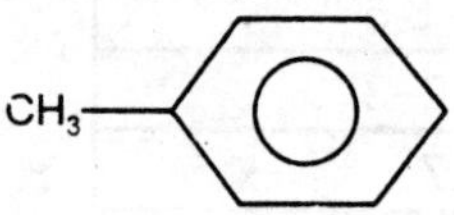

Two peak will be obtained.

Both will not split.

(*iii*) 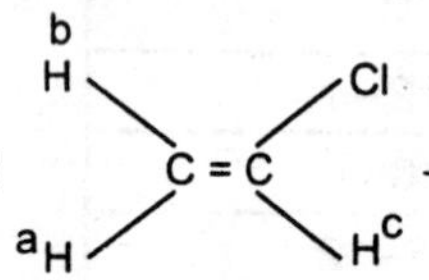

→ Three peaks will be obtained. Each peak will split into doublet.

(*iv*) 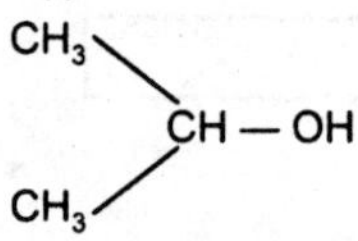

There are 3 signals.

'*a*' peak will split into (2+1) = 3, triplet.

'*b*' peak will split into sextet.

'*c*' peak will split into triplet.

(*v*) → There are three signals.

'*a*' peak will not split

'*b*' peak will split into quartet.

'*c*' peak will split into doublet.

Problem-2. *Predict number ofNMR peaks with relative intensity in low resolution spectrum of following compounds.*

(*i*) CH_3CH_2—CH_2—OH (*ii*) 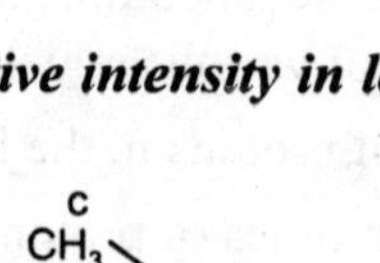

(*iii*) Toluene (*iv*) CH_3—O—CH_3

(*v*) 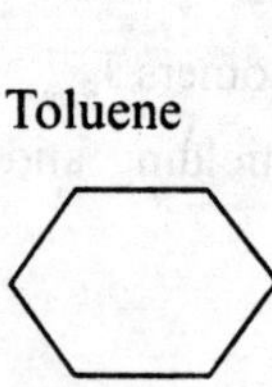

Sol.

(*i*) $CH_3CH_2—CH_2—OH$ → 4 peaks in intensity ratio 3 : 2 : 2 : 1

(*ii*) $\begin{matrix} CH_3 \\ CH_3 \end{matrix}\!\!>CH—OH$ → 3 peaks in intensityh ratio 6 : 1 : 1

(*iii*) 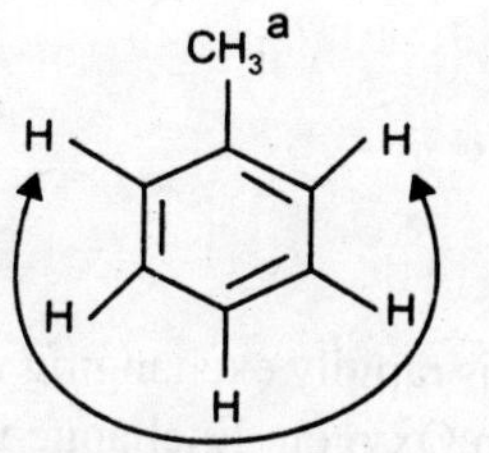

→ 2 peaks in intensity ratio 3 : 5

(*iv*) $CH_3—O—CH_3$ → 1 peak

(*v*)  → 1 peak

Problem 3. ***Predict the appearance of high resolution NMR spectrum of following compounds :***

(*i*) CH_3CHO (*ii*) CH_3COOH (*iii*) Fig. (*iv*) $CH_3—CH_2—OH$

Sol. (*i*)

Number of peaks = 2

Intensity ratio of signal = 3 : 1

Splitting and position of signal :

a = singlet at upfield side (δ 2.4 ppm),

b = singlet at downfield side (δ 9.8 ppm)

(*ii*)

Number of peaks = 2

Intensity ratio of signal = 3 : 1

Splitting and position of signal :

a = singlet at upfield side (δ 2.2 ppm),

b = singlet at downfield side (δ 11.2 ppm)

(*iii*) 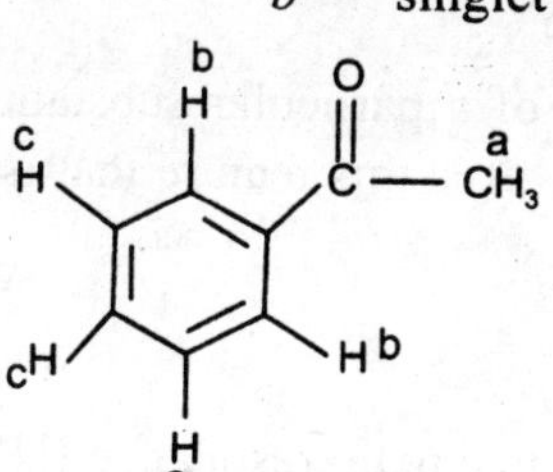

Number of peaks = 3

Intensity ratio of signal = 3 : 2 : 3

Splitting and position of signal :

a = singlet at δ 2.45 ppm

b = multiplet at δ 8.1 ppm

c = multiplet at δ 7.65 ppm

(*iv*)

Number of peaks = 3

Intensity ratio = 1 : 2 : 3

Splitting and position of signal :

a = singlet at 4.8 ppm (proton of hydroxyl is rapidly exchanging with'H' of acid or impurities and remain no longer on Oxygen. Exchange is so rapid that neighbour proton is not affected by spin orientation of hydroxyl proton).

b = quartet at δ 3.6 ppm

c = triplet at δ 1.2 ppm

Uses of NMR spectroscopy

- Nuclear Magnetic Resonance (NMR) spectroscopy is an important analytical technique that is widely used.
- NMR is routinely employed to: determine high-resolution, three-dimensional structures of biological macromolecules in solution, characterize protein-protein, protein-DNA and protein-small molecule interactions.
- Major application of NMR spectroscopy is the study of hydrogen atom in organic molecule. In order to achieve the desired results, a variety of NMR techniques are available.
- Nuclear Magnetic Resonance (NMR) spectroscopy is an analytical chemistry technique used in quality control and research for determining the content and purity of a sample as well as its molecular structure.
- Once the basic structure is known, NMR can be used to determine molecular conformation in solution as well as studying physical properties at the molecular level such as conformational exchange, phase changes, solubility, and diffusion.
- Most common types of NMR are proton and carbon-13 NMR spectroscopy, but it is applicable to any kind of sample that contains nuclei possessing spin.
- A key feature of NMR is that the resonance frequency of a particular substance is directly proportional to the strength of the applied magnetic field. It is this feature that is exploited in imaging techniques.

MAGNETIC RESONANCE IMAGING (MRI)

Magnetic resonance imaging is a medical application of nuclear magnetic resonance (NMR) develop during period of 1970s and 1980s. MRI is a medical imaging technique used in radiology to form pictures of the anatomy and the physiological processes of the body in both health and disease. MRI scanners use strong magnetic fields, electric field gradients, and radio waves to generate images of the organs in the body. MRI does not involve X-rays and the use of ionizing radiation, which distinguishes it from CT or CAT scans.

Theory of MRI

- Certain atomic nuclei are able to absorb and emit radio frequency energy when placed in an external magnetic field. In clinical and research MRI, hydrogen atoms are most often used to generate a detectable radio-frequency signal that is received by antennas in close proximity to the anatomy being examined.
- Hydrogen atoms exist naturally in people and other biological organisms in abundance, particularly in water and fat. For this reason, most MRI scans essentially map the location of water and fat in the body.
- Pulses of radio waves excite the nuclear spin energy transition, and magnetic field gradients localize the signal in space.
- By varying the parameters of the pulse sequence, different contrasts may be generated between tissues based on the relaxation properties of the hydrogen atoms therein.

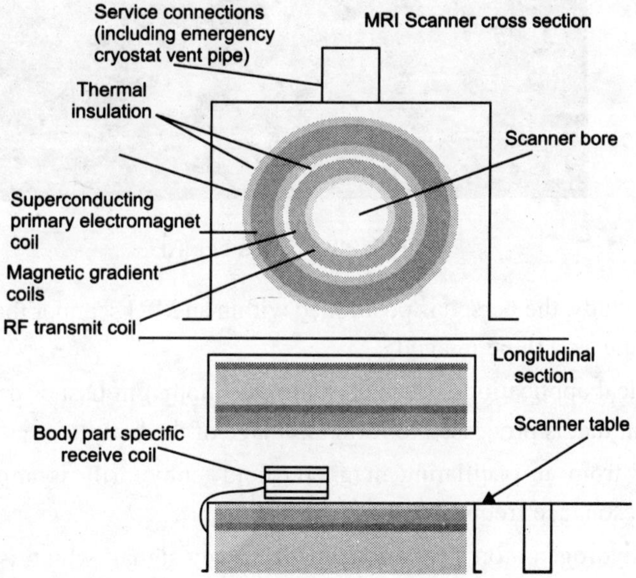

Fig.2.21: Schematic of construction of a cylindrical superconducting MR scanner.

Components of an MRI Scanner

The major components of an MRI scanner are: the main magnet, which polarizes the sample, the shim coils for correcting shifts in the homogeneity of the main magnetic field, the gradient system which is used to localize the MR signal and the RF system, which excites the sample and detects the resulting NMR signal. The whole system is controlled by one or more computers.

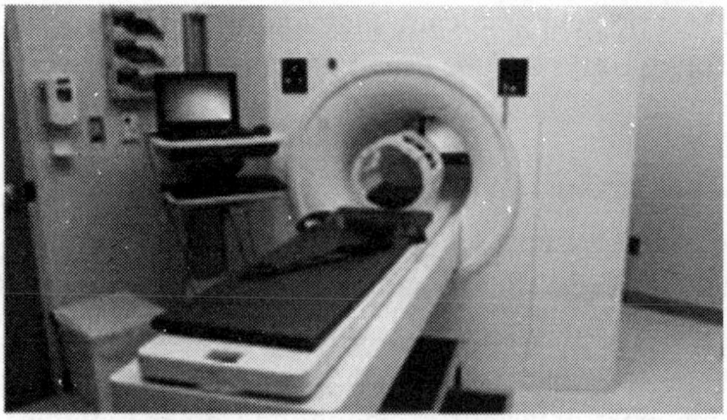

Fig.2.22: Mechanism of MRI

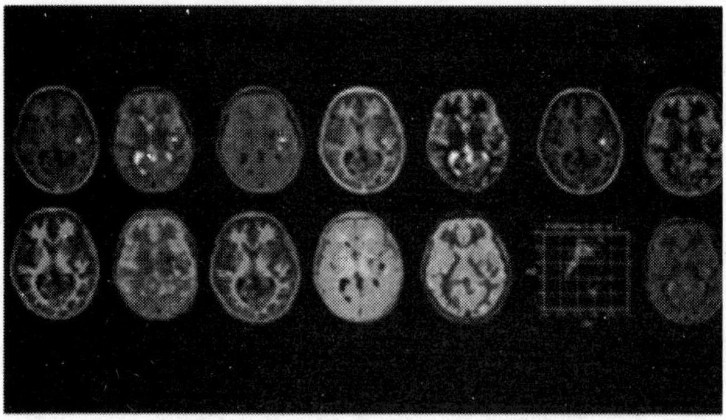

Fig.2.23: Images of MRI

1. To perform a study, the person is positioned within an MRI scanner that forms a strong magnetic field around the area to be imaged.
2. In most medical applications, protons (hydrogen atoms) in tissues containing water molecules create a signal that is processed to form an image of the body.
3. First, energy from an oscillating magnetic field temporarily is applied to the patient at the appropriate resonance frequency.
4. The excited hydrogen atoms emit a radio frequency signal, which is measured by a receiving coil.
5. The radio signal may be made to encode position information by varying the main magnetic field using gradient coils.
6. As these coils are rapidly switched on and off they create the characteristic repetitive noise of an MRI scan.
7. The contrast between different tissues is determined by the rate at which excited atoms return to the equilibrium state.
8. Exogenous contrast agents may be given to the person to make the image clearer.

Applications of MRI

MRI is widely used in hospitals and clinics for medical diagnosis, staging of disease and follow-up without exposing the body to radiation. Compared with CT scans, MRI scans typically take longer and are louder, and they usually need the subject to enter a narrow, confining tube. In addition, people with some medical implants or other non-removable metal inside the body may be unable to undergo an MRI examination safely.

- **Neuroimaging :** MRI is the investigative tool of choice for neurological cancers, as it has better resolution than CT and offers better visualization of the posterior fossa.
- **Cardiovascular :** Cardiac MRI is complementary to other imaging techniques, such as echocardiography, cardiac CT, and nuclear medicine. Its applications include assessment of myocardial ischemia and viability, cardiomyopathies, myocarditis, iron overload, vascular diseases, and congenital heart disease.
- **Musculoskeletal :** Applications in the musculoskeletal system include spinal imaging, assessment of joint disease, and soft tissue tumors.
- **Liver and gastrointestinal :** Hepatobiliary MR is used to detect and characterize lesions of the liver, pancreas, and bile ducts. Extracellular contrast agents are used widely in liver MRI and newer hepatobiliary contrast agents also provide the opportunity to perform functional biliary imaging.
- **Angiography :** Magnetic resonance angiography (MRA) is a group of techniques based on magnetic resonance imaging (MRI) to image blood vessels. MRA is often used to evaluate the arteries of the neck and brain, the thoracic and abdominal aorta, the renal arteries, and the legs.

X-Ray Diffraction (XRD Analysis)

Powder XRD provides detailed information on the crystallographic structure and physical properties of materials and thin films.

The sample is irradiated with a beam of monochromatic x-rays over a variable incident angle range. Interaction with atoms in the sample results in diffracted x-rays when the Bragg equation is satisfied. Resulting spectra are characteristic of chemical composition and phase. The technique uniquely provides phase identification (*e.g.* graphite or diamond), along with phase quantification, % crystallinity, crystallite size and unit cell size. For layered materials, Grazing Incidence XRD (GIXRD) allows compositional depth profiling of phases within the structure.

- Non-destructive analysis with identification of phase e.g. quartz / cristobalite / tridymite for silica
- Phase quantification, % crystallinity and measurement of crystallite and unit cell size
- Definition of bulk crystal structure including defects and solid solution, using Rietveld Structure Refinement
- Depth profiling of compositional and phase changes within layered materials by Grazing Incidence XRD
- Measurement of layer thickness on the nm scale for highly polished materials using X-Ray Reflectivity (XRR)
- Small area analysis and phase mapping by Micro X-Ray Diffraction (μXRD).

Typical Applications

1. Phase composition determination
2. Characterization of doped cell structures for electroceramics
3. Measurement of hard coating composition and structure e.g. carbides, nitrides on machine tools
4. Characterization of hydroxy-apatite coatings on medical implant materials
5. In-depth analysis of silicon wafers, ITO coated glasses and solar cells.

Typical Industries using XRD are Aerospace, Automotive, Materials, Minerals, Glass, Ceramics and Refractories, Healthcare, Medical Devices, Semiconductors, Electronics.

FUNDAMENTAL PRINCIPLES OF X-RAY POWDER DIFFRACTION (XRD)

Max von Laue, in 1912, discovered that crystalline substances act as three-dimensional diffraction gratings for X-ray wavelengths similar to the spacing of planes in a crystal lattice. X-ray diffraction is now a common technique for the study of crystal structures and atomic spacing.

X-ray diffraction is based on constructive interference of monochromatic X-rays and a crystalline sample. These X-rays are generated by a cathode ray tube, filtered to produce monochromatic radiation, collimated to concentrate, and directed toward the sample. The interaction of the incident rays with the sample produces constructive interference (and a diffracted ray) when conditions satisfy Bragg's Law ($n\lambda = 2d \sin \theta$). This law relates the wavelength of electromagnetic radiation to the diffraction angle and the lattice spacing in a crystalline sample. These diffracted X-rays are then detected, processed and counted. By scanning the sample through a range of 2θ angles, all possible diffraction directions of the lattice should be attained due to the random orientation of the powdered material. Conversion of the diffraction peaks to *d*-spacings allows identification of the mineral because each mineral has a set of unique *d*-spacings. Typically, this is achieved by comparison of *d*-spacings with standard reference patterns.

All diffraction methods are based on generation of X-rays in an X-ray tube. These X-rays are directed at the sample, and the diffracted rays are collected. A key component of all diffraction is the angle between the incident and diffracted rays. Powder and single crystal diffraction vary in instrumentation beyond this.

Fig.2.24: Image of XRD machine

Applications

X-ray powder diffraction is most widely used for the identification of unknown crystalline materials (e.g. minerals, inorganic compounds). Determination of unknown solids is critical to studies in geology, environmental science, material science, engineering and biology.

Other applications include

- Characterization of crystalline materials
- Identification of fine-grained minerals such as clays and mixed layer clays that are difficult to determine optically
- Determination of unit cell dimensions
- Measurement of sample purity

QUESTIONS

Q.1. Why are pure rotational spectra studied only in the gaseous states of atoms and molecule?

Ans. The rotational energy levels are not quantized in condensed phases; i.e., in the solid and liquid states. They are quantized in the gaseous states of chemical systems. Thus the microwave spectra are observed in the gaseous state.

Q.2. How many rotational lines will be observed in the microwave spectrum of HCl gas?

Ans. The rotational energy level population is considerable and the ground rotational level is not the most populated, the number of rotational levels is subjected to the selection rule.

Q.3. How many normal modes of vibrations are possible in the linear molecule ethane, C_2H_6, and non-linear molecule benzene, C_6H_6?

Ans. Ethane has – $3n - 5 = 3(2 + 6) - 5 = 19$ possible normal modes of vibrations.

Benzene has – $3n - 6 = 3(6 + 6) - 6 = 30$ possible normal modes of vibrations.

Q.4. Predict the number and gives the names of fundamental modes of vibrations of HCl.

Ans. As there are only two atoms in HCl, it is linear and has $3n - 5$ or $3 \times 2 - 5$ or 1 fundamental modes of vibration which in the symmetric stretch.

Q.5. How many fundamental vibrational frequencies would you predict during IR absorption spectrum of H_2O?

Ans. The bent H_2O molecule will have $3n - 6$ or $3 \times 3 - 6$ or 3 fundamental modes of vibrations. In order to decide whether all three are IR active and whether any degeneracy exists, it is helpful to diagram the vibrational modes as follows:

No other fundamental modes of vibration can be diagrammed. All these three modes of vibration are IR active and are non degenerate.

Q.6. How many fundamental vibrational frequencies would you predict during IR absorption spectrum of CO_2?

Ans. CO_2 is linear, it has $3n - 5 = 4$ fundamental modes of vibrations, which may be represented as –

Vibrational modes for linear CO_2

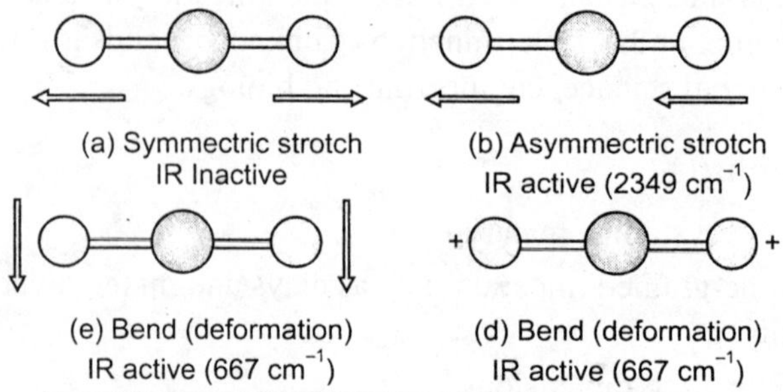

Linear CO_2 – number of modes = 3(3) – 5 = 4

Q.7. The force constant for the band in HF is about 9×10^5 dynes cm^{-1}.
Calculate the vibrational spectrum peak for HF.
Calculate the absorption peak for DF.

Ans. (*a*) 3.99×10^3 cm^{-1} (*b*) 2.99×10^3 cm^{-1}]

Q.8. Calculate the wavenumber of the C-Cl band in CCl_4. The force constant is 3.4×10^5 dynes cm^{-1}.

Ans. 802 cm^{-1}.

Q.9. Calculate the frequency of absorption of the C triple bond Nstretch in HCN.

Ans. 2006 cm^{-1}; observed 2089 cm^{-1}.

Q.10. Calculate the wavelength of absorption of the HCl stretching Frequency.

Ans. – 3.5μ.

Q.11. How will you determine bond length of a linear diatomic molecule?

Q.12. Give the statement of Born-Oppenhier approximation.

Q.13. What type of molecules exhibits rotational spectra? Select which give rotational spectra CO_2, C_6H_6, N_2, H_2, HCl, N_2O.

Q.14. Calculate energy of a proton wavelength 6900A0.

Ans. – 2.88×10^{-12} ergs/photon).

Q.15. Write a short note on –

(*a*) IR spectra
(*b*) Application of IR Spectra
(*c*) Selection rule for Electronic spectroscopy
(*d*) Image of MRI
(*e*) Application of NMR

Q.16. How many fundamental modes of vibration you predict for -

(*a*) Benzene
(*b*) Toluene
(*c*) Methane

Ans. $a = 30$, $b = 39$, $c = 9$

Q.17. How will you differentiate between CH_3CH_2CHO and $CH_{2-} = CH - CH_2OH$ with the help of IR spectrum?

Q.18. Explain the different electronic transitions in UV spectroscopy. with the help of UV spectrum, how will you differentiate between cis and trans isomers?

Q.19. What information can be obtained about following compounds with the help of UV and IR spectroscopy :

(*i*) Acetophenone (*ii*) Cinnamicacide

(*iii*) Phenol (*iv*) Benzamide.

Q.20. Explain the different electronic transitions in UV spectroscopy and arrange them in increasing order of energy. Describe infrared spectrophotometer.

SHORT ANSWER TYPE QUESTIONS

1. Explain the following :

(*i*) Chromophore,

(*ii*) Auxochrome

(*iii*) Effect of Conjugation on UV absorption,

(*iv*) Bathochromic effect,

(*v*) Hypochomic effect,

(*vi*) Electromagnetic spectrum,

(*vii*) Stretching vibration,

(*viii*) Fingerprint area,

(*ix*) Type of Molecular Vibrations,

(*x*) Laws of absorbance of light ,

(*xi*) Hyperchromic effect,

(*xii*) Effect of conjugation on electronic transition,

(*xiii*) Selection rule

2. How is the concentration of an unknown solution determined with the help of UV spectroscopy?

3. What do you mean from emission and absorption spectra?

4. Define and explain molar extinction coefficient.

5. What is Nujol? Why is it used in IR spectroscopy ?

6. Explain the following :

(*i*) λ_{max} 275 of Benzene < λ_{max} 375 of Naphthalence,

(*ii*) Absorption of Mecetyloxide in water is 60 and in ether is 96, Solvation effect

(*iii*) Decrease in the bond frequency due to hydrogen bond. Ethanol vopour No H bond = 37000 cm spectroscopy.

(*iv*) Increase in the bond frequency whilt increasing the conugationinpolyene.

7. Write the factors affecting the –OH absorption in IR –spectroscopy.
8. Describe diagrammatically the various electronic transitions that take place when an organic compound absorbs radiation in ultraviolet or visible region.
9. How will you identify following functional groups with the help of IR spectroscopy?
   (*i*) –OH (*ii*) $-NH_2$ (*iii*) –CN (*iv*) >C = O.
10. Write Short notes on the following :
    (*i*) Applications of IR spectroscopy to detect the presence of – OH and –COOH groups.
    (*ii*) Different types of Bending vibration.
    Represent and differentiate the the $\pi \rightarrow \pi$ and $n \rightarrow \pi$ transitions of $c = o$ group.
    (*iii*) Arrange the following in the increasing order on the basis of their ultraviolet absorption maxima:
    (*a*) Anthracene (*b*) Ethylene,
    (*c*) Naphthalene, (*d*) Buladiene.
11. Explain the molecular vibrations. Give its types.
12. What is the basic difference between ultraviolet spectroscopy and infrared spectroscopy.
13. Describe how many types of molecular vibrations are possible in L.R. Spectroscopy
14. What is vibration frequency ? Write down reasons for deviations from the basic values
15. What does the finger printing region of IR spectrum show ?
16. With the help of Woodward-Fieser rule calculate the value of A_{max} for the following structure:

17. Describe the different types of vibrations in organic compounds due to infrared radiation.
18. How do UV and IR spectrum differ from each other with respect to electron transitions ?
19. How will you differentiate between the following pairs of compounds :
    (*i*) o- and p-HO—C_6H_4—CHO (ii) $C_6H_5COC_2H_5$ and p-$CH_3C_6H_4COCH_3$?
20. Assign and explain the wave numbers (C = C stretching) 1600, 1670 and 1950 cm
    $CH_2 = CH—CH = CH_2$, $CH_2 = CH—CH = CH—CH_3$ and $CH_2 = C = CH—CG_3$

OBJECTIVE TYPE QUESTIONS

1. Which one of the following absorbs ultraviolet light ?
   (*i*) *n*-hexane (*ii*) ethanol
   (*iii*) acetone (*iv*) sucrose.
2. In methyl alcohol which of the following electronic transition absorbs least energy;
   (*i*) $\pi \rightarrow \pi^*$ (*ii*) $n \rightarrow \pi^*$
   (*iii*) $n \rightarrow n^*$ (*iv*) $\sigma \rightarrow \sigma^*$

3. The area of ultraviolet spectrum is :

(*i*) 1000—2000Å (*ii*) 2000—4000A Å

(*iii*) 4000—6000A Å (*iv*) 6000—8000A Å

4. The compound showing abs. orbance at 1690 cm^{-1} is :

(*i*) Acid (*ii*) alcohol

(*iii*) Ether (*iv*) carbonyl.

5. Which of the following electronic transition takes place in saturated hydrocarbons ?

(*i*) $\pi \rightarrow \pi^*$ (*ii*) $n \rightarrow \pi^*$

(*iii*) $n \rightarrow \sigma^*$ (*iv*) $\sigma \rightarrow \sigma^*$

6. In which region of electromagnetic radiation the electronic spectra is found ?

(*i*) Microwave region (*ii*) Visible and BV. region

(*iii*) Infrared region (*iv*) Far infrared region,

7. A compound having molecular formula C_8H_8O exhibits strong IR absorption near1690 cm^{-1}. The compound will be :

(*i*) $C_6H_5COCH_3$ (*ii*) C_6H_s—0—CH = CH_2

(*iii*) C_6H_s—CH—CH_2 (*iv*) HO—C_6H_S—CH = CH_2.

8. The nature of electronic transitions in alkene is :

(*i*) $\pi \rightarrow n^*$ (*ii*) $\pi \rightarrow \pi^*$

(*iii*) $\sigma \rightarrow \pi^*$ (*iv*) $\sigma \rightarrow \sigma^*$

9. Stretching is a vibration in which:

(*i*) Bond length changes

(*ii*) Bond length does not change

(*iii*) The group simultaneously vibrates perpendicular to bond length

(*iv*) None of the above

10. The compound showing absorbance at 1690 cm^{-1} is :

(*i*) Acid (*ii*) Alcohol

(*iii*) Ether (*iv*) Ketone

3 Intermolecular Forces and Potential Energy Surfaces

Ionic, dipolar and van Der Waals interactions. Equations of state of real gases and critical phenomena. Potential energy surfaces of H_3, H_2F and HCN and trajectories on these surfaces.

Introduction

Molecular Interactions are attractive or repulsive forces between molecules and between non-bonded atoms. It is also known as non-covalent interactions or intermolecular interactions. Molecular interactions are important in diverse fields of protein folding, drug design, material science, sensors, nanotechnology, separations, and origin of life.

Molecular Interactions may be two types

1. Intramolecular interaction
2. Intermolecular interaction

Intramolecular interaction: An intramolecular interaction is any force that holds together the atoms making up a molecule or compound. Means it develops in same molecule. They are usually stronger than intermolecular forces, which are present between atoms or molecules that are not bonded.

Intermolecular interaction : Intermolecular interactions occur between all types of molecules or ions in all states of matter. The physical properties of melting point, boiling point, vapour pressure, evaporation, viscosity, surface tension, and solubility are related to the strength of attractive forces between molecules. These attractive forces are called **Intermolecular Forces**.

The interaction between intermolecular forces may be used to describe how molecules interact with each other. The strength or weakness of intermolecular forces determines the state of matter of a substance

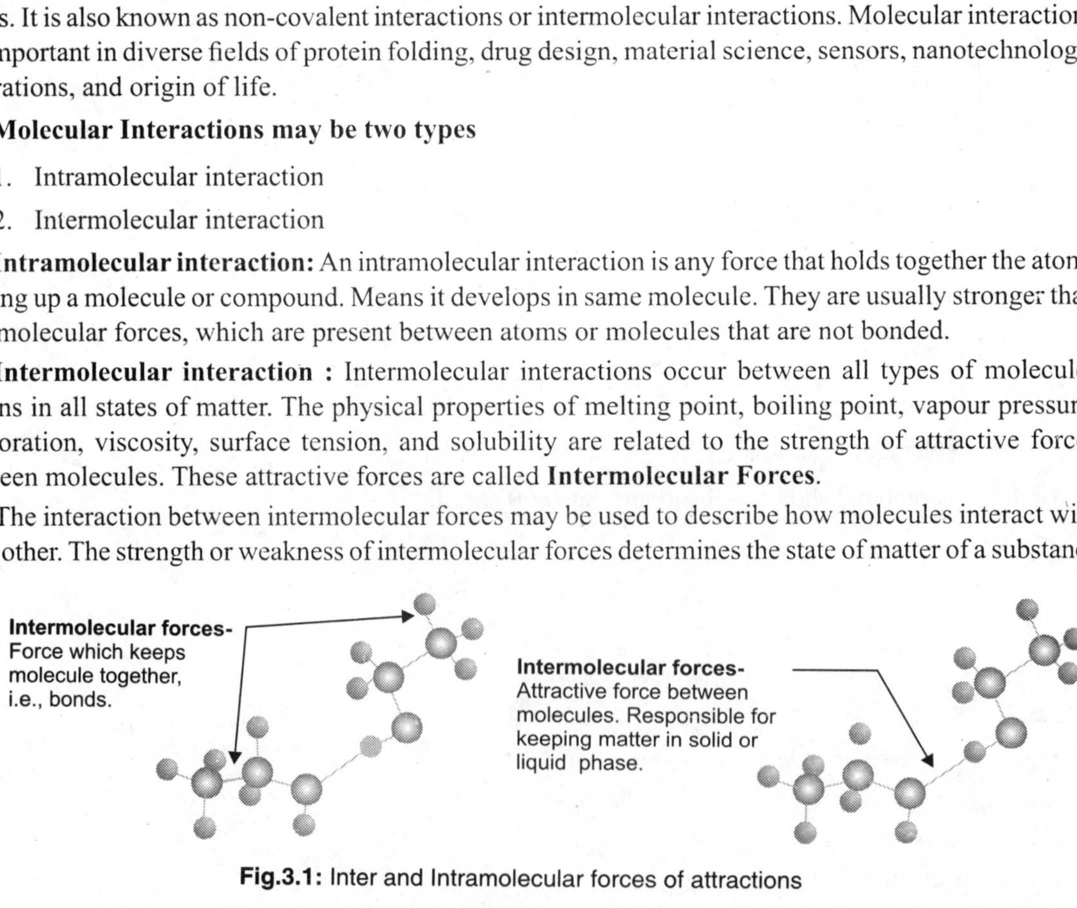

Fig.3.1: Inter and Intramolecular forces of attractions

(*e.g.*, solid, liquid, gas) and some of the chemical properties (*e.g.*, melting point, structure).The amount of "stick togetherness" is important in the interpretation of the various properties listed above.

There are three types of intermolecular forces. Most of the intermolecular forces are identical to bonding between atoms in a single molecule. Intermolecular forces just extend the thinking to forces between molecules and follows the patterns already set by the bonding within molecules.

(a) Ionic interaction

(b) Dipole-dipole interaction

(c) Van der waals interactions

Table 3.1: Weak Intermolecular Interactions

Force	Strength (KJ/mol)	Distance (nm)
Van der Waals	0.4-4.0	0.3-0.6
Hydrogen Bonds	12-30	0.3
Ionic Interactions	20	0.25
Hydrophobic Interactions	<40	varies

Ionic Forces or Ionic interaction are interactions between charged atoms or molecules. This interactions arise from electrostatic attraction between two groups of opposite charge. Ions are of two types according to charges –

- Cations - Positively charged ions, such as Na(+) , Li(+), and Ca(2^+), are termed cations.
- Anions - Negatively charged ions, such as Cl(–), Br(–), HO(–) are called anions .

The attractive forces between oppositely charged ions is described by **Coulomb's Law,** in which the force increases with charge and as decreases as the distance between these ions is increased. The highly polarized (charged) nature of ionic molecules is reflected in their high melting points (NaCl has a melting point of 801°C) as well as in their high water solubility .

Dipole-Dipole Interaction

Dipole-dipole interaction occurs whenever two polar molecules get near each other. The positively charged portion of one molecule is attracted to the negatively charged portion of another molecule.Since many molecules are polar, this is a common intermolecular force.

Dipole-dipole forces. Molecules that have dipole moments tend to align with the positive and negative ends of the dipole toward one another.

- These are highly directional forces, but tend to be disrupted by thermal motion of the molecules especially in the liquid state.
- The magnitude of such interactions is only about 1% that of covalent or ionic bonds.
- These interactions become weak rapidly as the distance between the dipoles increases; E = –2 $\mu_1\mu_2$/or 3 (μ_1 and μ_2 are the individual dipole moments).

Hydrogen bonding may be described as an extreme form of dipole-dipole interaction, although other models can also be applied. Some of the strongest dipole-dipole interactions occur between species that are hydrogen bonded to highly electronegative atoms such as oxygen, fluorine or nitrogen. Because of the small size of the hydrogen atom, these dipoles can typically approach closely and give rise to a larger energy than is often the case.

Example: Hydrogen Bonding

- A dipole-dipole interaction where the H atom is noncovalently attracted to an electronegative atom
- The H must have a large positive delta and the other atom must have a source of electrons to be attracted to H
- Usually negatively charged F, O, N is attracted to the positively charged H
- Use of it is very important in biology.

Strength

- A H bond is a weak bond and is dynamic [3×10^{-12} lifetime]
- In solid phase H-bond is static
- H-bonds between molecules continually form and break in liquid.
- Every water molecule forms 3.4 H bonds with other water molecules.

HYDROGEN BOND

Hydrogen bond is an electrostatic attractive force between covalently bonded hydrogen atom of one molecule and an electronegative atom (such as F, O, N) of another molecule (or same molecule).

For example in H-F, there is a covalent bond between H and F. Since F is more electronegative than H, so it acquires partial negative charge (as it attracts shared pair of electrons more) and Hydrogen acquires partial positive charge.

$$\overset{\delta^+}{H} - \overset{\delta^+}{F}$$

Positive charge on H in a H-F molecule, will be electrostatically attracted by the negative charge on F in another HF molecule.

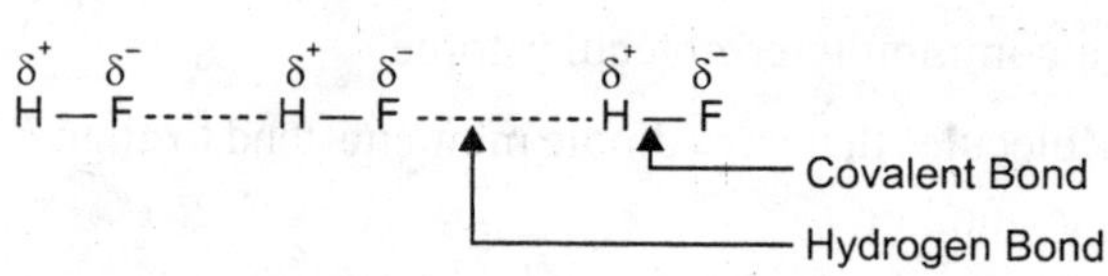

Types of Hydrogen Bonding

1. Intermolecular Hydrogen bonding: This type of Hydrogen bonding occurs between atoms of two same or different molecules.

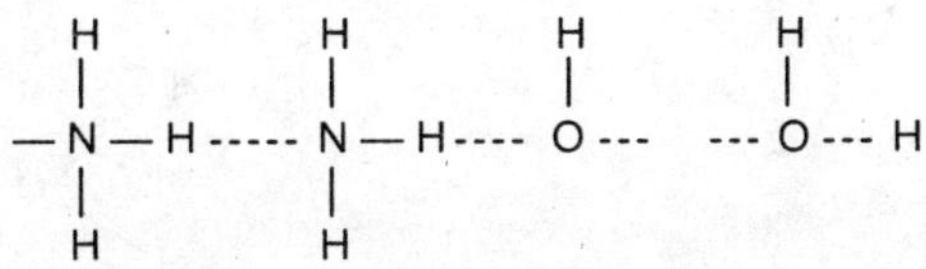

Hydrogen Bonding in NH_3 and H_2O

Intermolecular Hydrogen bonding increases the boiling point of the compound and also its solubility in water.

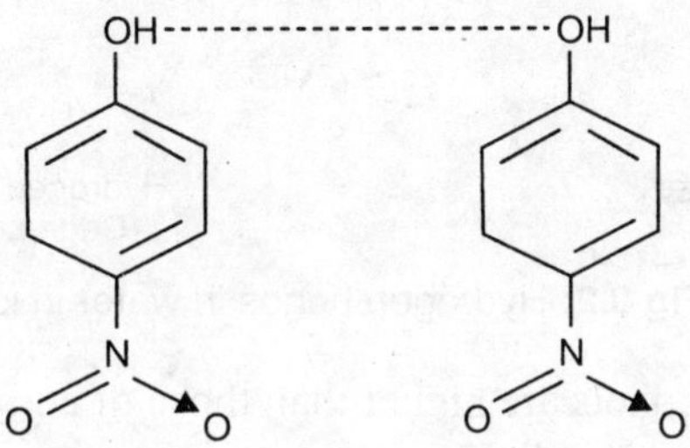

p - nitrophenol

2. Intramolecular Hydrogen bonding: This type of hydrogen bonding occurs between two atoms of the same molecule.

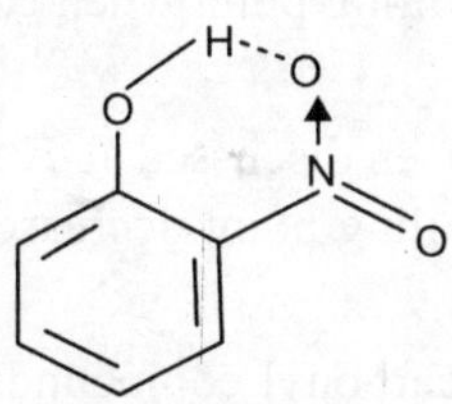

O-Nitrophenol

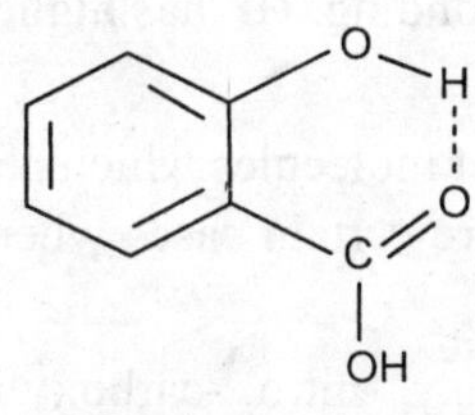

O-hydroxy benzoic acid

Intramolecular hydrogen bonding decreases the boiling point of the compound and also its solubility in water, as it prevents the association between molecules, which would have raised boiling point and it also prevents hydrogen bonding between compound and wate thereby decreasing its solubility in H_2O.

Effects of Hydrogen Bonding

1. Boiling Point: As we know that boiling point increases with increase in intermolecular forces which is greater for heavier molecules.

Although water has the lowest molecular weight among hydrides of 16th group of the periodic table, it has the highest melting and boiling points and is liquid (while hydrides of other members are gases) under ordinary conditions.

This can be explained on the basis that water molecules are associated through hydrogenbonding.

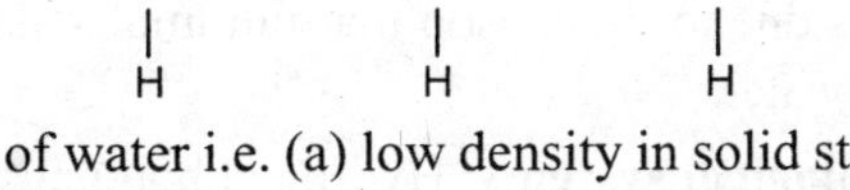

The two unusual properties of water i.e. (a) low density in solid state than that in the liquid state (b) contraction when heated between 0°C and 4°C can be explained on the basis of hydrogen bonding in water.

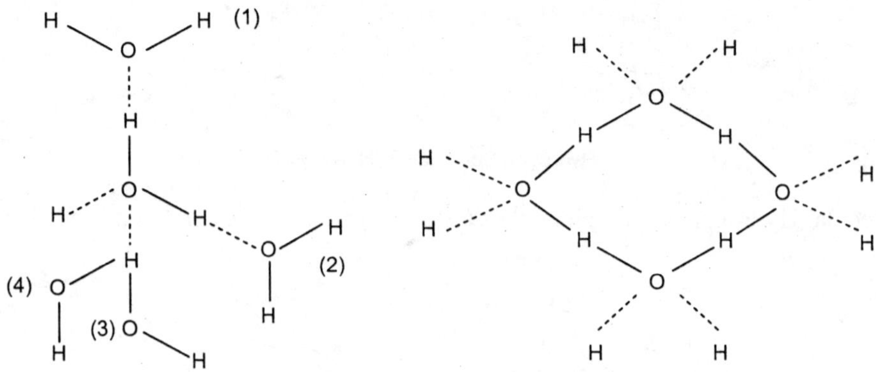

Hydrogen bonds in Water

Hydrogen bonds in in Ice (Cage Like Structure)

Fig.3.2: Hydrogen bonds in water in ice

Similarly boiling points of alcohols are higher than those of corresponding ethers, it is again due to association of alcohols by hydrogen bonding.

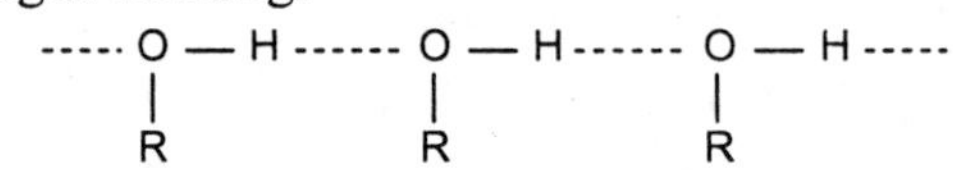

Again due to Hydrogen bonding, HF has highest boiling point when compared among hydrides of halogens.

Increase in boiling point of molecules which have been discussed above are due to (intermolecular) hydrogen bonding but there are certain cases where boiling point decreases due to hydrogen bonding (intramolecular).

For example, ortho-hydroxy, -nitro, -carboxylic, -carbonyl compounds have lower boiling points than that of respective meta or para isomer. The reason is that in ortho substituted isomer there is intramolecular hydrogen bonding, that prevents association of molecules that would have raised the melting and boiling point.

On the other hand, in m and p isomers intramolecular hydrogen bonding is not possible due to geometrical reasons, instead intermolecular H-bonding occurs which cause their association and thereby increasing their boiling points.

2. Water Solubility: A substance, which is capable of forming hydrogen bond with water molecules, are soluble in water. On the other hand substances incapable of forming hydrogen bond with water are insoluble in water. Thus organic molecular like alcohol, acids are soluble in water while alkanes, alkenes, ether which are incapable of forming hydrogen bond are insoluble in water.

But when alkyl group in alcohols has 4 or more carbon atoms, the alkyl group predominates and hydrogen bonds are restricted, with the result their solubility decreases in water.

Intermolecular hydrogen bonding increases the solubility of compound in water while intramolecular hydrogen bonding decreases. It is due to the reason that intramolecular H-bonding prevents hydrogen bonding between compound and water.

3. Hydrogen bonding in biological systems: Hydrogen bonding also plays very important role in biological system. There are hydrogen bonding between base pairs, (purine and pyrimidine) of D.N.A.

There are 2 hydrogen bonds between Ademine and Thymine, while 3 between Guanine and Cytosine. Primary structure of protein is stabilised by hydrogen bonding.

Vander Waals Forces

Van der Waals forces are the interaction between uncharged atoms or molecules. The van der Waals forces include Keesom interaction, the Debye force, and the London dispersion force. So, van der Waals forces include intermolecular forces and also some intramolecular forces.

The forces are used to explain the universal attraction between bodies, the physical adsorption of gases, and the cohesion of condensed phases. Van der Waals forces (London dispersion forces, permanent dipole - permanent dipole and hydrogen bonding) occur between covalent molecules and monatomic gases. These attractions occur due to the movement of electrons within molecules creating dipoles (temporary or permanent) these dipoles are then weakly attracted to the oppositely charged dipole on a neighboring molecule. Due to these attractions being weak, the substances tend to be gases, liquids or low melting point solids as it does not take a large amount of energy to overcome the weak attractions between molecules.

There are four types .It all flows from this general principle: as bonds become more polarized, the charges on the atoms become greater, which leads to greater intermolecular attractions, which leads to higher boiling points.

Ionic Bonds > H Bonding > Van der Waals Dipole-Dipole Interactions > Van Der Waals Dispersion Forces.

Table 3.2 : Force of attraction strongest to weakest

Type of IMFA	Involves	Occurs Between	Strength of Attraction	Effect on Boiling/Melting Points	Effect on Freezing Points
London/ Dispersion	Temporary Dipoles	Nonpolar molecules	Low	Low elevation (↑BP)	(↑↑↑↑FP)
Dipole-Dipole	Permanent dipoles	Polar molecules	Medium	Medium elevation (↑↑BP)	(↑↑↑FP)
Hydrogn Bonds	Permanent dipoels between H & F, O, N	Polar molecules	Medium-High	Medium-High elevation (↑↑↑BP)	(↑↑FP)
Dipole-Dipole	Full ion and dipole	Polar molecules	High	High elevation (↑↑↑↑ BP)	(↑FP)

While the intramolecular forces keep the atoms in a moleucle together and are the basis for the chemical properties, the intermolecular forces are those that keep the molecules themselves together and are virtually responsible for all the physical properties of a material. The intermolecular forces increase in strength according to the following:

London dispersion < dipole-dipole < H-bonding < ion-ion

Now, as these things increase in strength it becomes harder to remove the molecules from each other. Therefore, one would expect the melting and boiling points to be higher for those substances which have strong intermolecular forces. We know that it takes energy to go from a solid to a liquid to a gas. This energy is directly related to the strength of attraction between molecules in the condensed phases. Since energy is directly proportional to the temperature, the above trends ought to hold true.

In addition, there are energies associated with making these phase transitions:

solid → liquid $\Delta H \equiv \Delta H_{fusion}$

liquid → gas $\Delta H \equiv \Delta H_{vaporization}$

solid → gas $\Delta H \equiv \Delta H_{sublimation}$

Each of these processes are endothermic, and scale with the magnitude of the intermolecular forces. Thus, as these intermolecular forces increase, so do the energies requires to melt, vaporize, or sublime (go from solid to a gas) a species.

London Dispersion Force

The London dispersion force is also known as LDF, London forces, dispersion forces, instantaneous dipole forces, induced dipole forces, or the induced dipole-induced dipole force

The London dispersion force is the weakest of the intermolecular forces. This is the force between two nonpolar molecules. The electrons of one molecule are attracted to the nucleus of the other molecule, while repelled by the other molecule's electrons. A dipole is induced when the electron clouds of the molecules are distorted by the attractive and repulsive electrostatic forces.

Example: Interaction between two methyl ($-CH_3$) groups, nitrogen gas (N_2) and oxygen gas (O_2) molecules..

EQUATION OF REAL GAS AND CRITICAL PHENOMENA

The states of matter are liquid, solid, and gas which can be recognized through their key characteristics. The gases that are strong enough to react with other matter, there are even with very strong odour, and some can be dissolved in water. Here we will be able to note some differences between ideal gas and real gas. The behaviour of real gases is very much complex while the behavior of ideal gases is much simpler.

Ideal gases-A gas which obeys Boyle's law, Charles' law etc. under all condition of temperature and pressure is known as an ideal gas or a perfect gas.

Gas equation ($PV = nRT$) derived from Boyle's law and Charles' law so an ideal gas may be defined as –

A gas which obeys the gas equation ($PV = nRT$)under all conditions of temperature and pressure is known as ideal gas.

It is observed that no gas which obeys the gas law or the gas equation under all condition of temperature and pressure. Hence concept of ideal gas is hypothetical or only theoretical so,

Real gas : A real gas is one which obeys the gas law fairly well under low pressure or high temperature.

Real gases are non-hypothetical gases whose molecules occupy space and have interactions; consequently, they adhere to gas laws. To understand the behavior of real gases, the following must be taken into account:

- Compressibility effects
- Variable specific heat capacity
- Van der waals forces

- Non-equilibrium thermodynamic effects
- Issues with molecular dissociation and elementary reactions with variable composition

Difference between ideal and real gases

The main differences ideal and real gases are summed up as under in Table 3.3

Table 3.3: Differences between ideal and real gases

S. No.	Characteristics	Ideal gas	Real gas
1.	Definition	A gas which obeys the gas equation under all conditions of Temperature and pressure is known as ideal gas.	A real gas is one which obeys the gas law fairly well under low pressure or high temperature
2.	Equation	It obeys ideal gas equation (PV = nRT)	It obeys Vander Waal's equation : $\left(P^2 + \frac{an^2}{V^2}\right)$(V-b) = nRT
3.	Definite volume	Ideal gas has no definite volume, volume of gas will be zero at –273°C	Real gas has definite volume as gas will liquefy before reaching +273°C.
4.	Mass	Ideal gas has no mass	Real gas has mass
5.	Collision	Collision of ideal gas particles is elastic	Collision for real gas is non-elastic.
6.	Thermal effect	There will be no thermal Effect on expansion or compression without external pressure	There will be some thermal Effect
7.	Compressibility Factor	Compressibility Factor Z (PV/nRT) is always 1	Value of Z is not always 1 for real gases
8.	Force of attraction	The force of attraction among the molecules of the gas are negligible	The force of attraction among the molecules of the gas are not negligible at all temperature and pressure

Deviation of real gases and Van der waal's equation

The ideal gas laws derived from the kinetic theory of gases which is based on the following important assumptions:

- The volume occupied by the molecules is negligible in comparison to the total volume of the gas.
- The molecules exert no force of attraction upon one another.

These assumptions can not be regarded as applicable to real gases. Therefore real gases show deviation from the ideal behaviour. The molecules of a gas, do occupy a certain volume as the gases can be liquefied and solidified at low temperatures and high pressures. On decreasing the temperature of a gas, and applying high pressure molecules come closer to one another, thereby increasing the forces of attraction between them. Both these factors favour liquefaction and solidification. It is, therefore, apparent that the molecules of a gas must have an appreciable volume.

The molecules in gases also have weak forces of attraction (called van der Waals attractions) between them. This is supported by the fact that gases can be liquefied.

Van der Waals was the first to introduce systematically the correction terms due to the above two invalid assumptions in the ideal gas equation $p_iV_i = nRT$. The corrections are given below.

Correction for Volume

In the ideal gas equation $p_iV_i = nRT$, V_i represents an ideal volume where the molecules can move freely. In real gases, a part of the total volume is, however, occupied by the molecules of the gas. Hence, the free volume V_i is the total volume V minus the volume occupied by the molecules. If b represents the effective volume occupied by the molecules of 1 mole of a gas, then for n moles of the gas V_i is given by

$$V_i = V - nb \qquad \text{...(i)}$$

where b is called the excluded volume. The numerical value of b is four times the actual volume occupied by the gas molecules. This can be shown as follows:

Excluded Volume

If we consider only bimolecular collisions, then the volume occupied by the sphere of radius 2r represents the excluded volume per pair of molecules as shown in Fig.

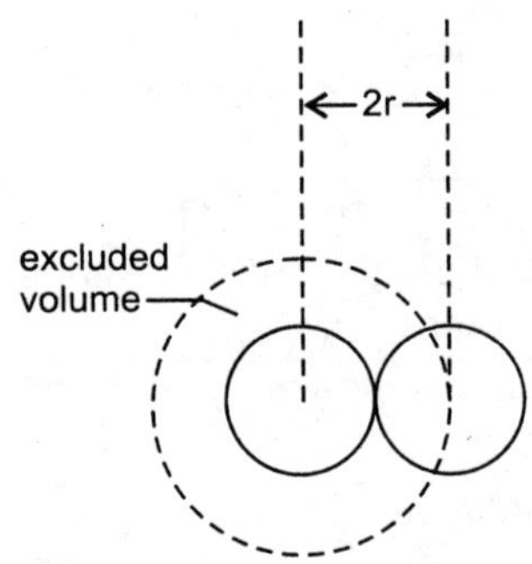

Fig.3.3: Excluded Volume Per Pair of Molecules

Thus, excluded volume per pair of molecules.

$$= \frac{4}{3}\pi(2r)^3 = 8\left(\frac{4}{3}\right)\pi r^3$$

$$\text{Excluded volume per molecule} = \frac{1}{2}\left[8\left(\frac{4}{3}\pi r^3\right)\right] = 4\left(\frac{4}{3}\pi r^3\right)$$

$$= 4(\text{volume occupied by a molecule})$$

Since b represents excluded volume per mole of the gas, it is obvious that

$$b = N_A\left[4\left(\frac{4}{3}\pi r^3\right)\right] \quad (N_A = \text{the number of molecules in a mole})$$

Correction for forces of attraction

Consider a molecule A in the bulk of a vessel as shown in Fig. This molecule is surrounded by other molecules symmetrically in all directions with the result that this molecule on the whole experiences no net force of attraction.

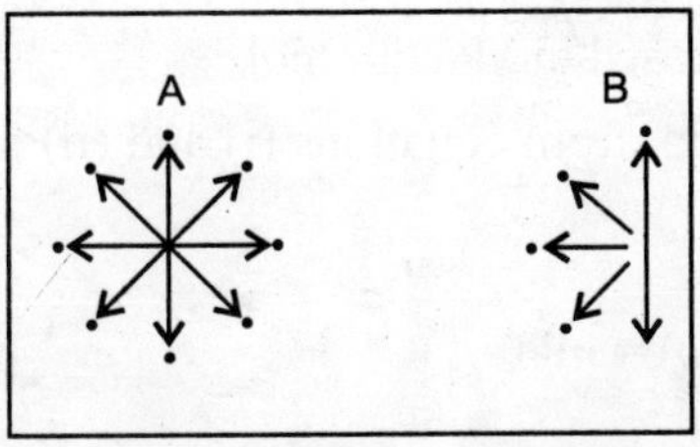

Fig.3.4: Forces of attraction between molecules of a gas

Now, consider a molecule B near the side of the vessel, which is about to strike one of its sides, thus contributing towards the total pressure of the gas. There are molecules only on inner side of the wall. This molecule experiences a net force of attraction towards the centre of the vessel. This results in decreasing the velocity of the molecule. Thus, the molecule does not contribute as much force as it would have, had there been no force of attraction towards the centre. Thus, the pressure of a real gas would be smaller than the corresponding pressure of an ideal gas, *i.e.*

$$p_i = p + \text{correction term}$$

This correction term depends upon two factors:

(i) Th number of molecules per unit volume of the vessel. Larger this number, larger will be the net force with which the molecule B is dragged towards the centre. This decreases the velocity of the molecule B and hence the rate of change of momentum. Consequently, the correction term also has a large value. If n is the number of molecules of the gas present in the volume V of the container.

$$\therefore \quad \text{Correction term} \propto \frac{n}{V}$$

(ii) The number of molecules striking the side of the vessel per unit time. The number of molecules present in unit volume of the container, and hence

$$\text{Correction term} \propto \frac{n}{V}$$

Taking both these factors together, we have

$$\text{Correction term} \propto \left(\frac{n}{V}\right)\left(\frac{n}{V}\right)$$

$$\text{or} \quad \text{Correction term} = \frac{n^2}{V^2}$$

where 'a' is the proportionality constant and is a measure of the forces of attraction between the molecules. Thus,

$$p_i = p + a\frac{n^2}{V^2} \qquad \text{...(ii)}$$

The unit of the term $\frac{an^2}{V^2}$ will be the same as that of the pressure. The SI unit of 'a' will be Pa m^6 mol^{-2}. It may be conventiently expressed in kPa dm^{6}mol^{-2}.

Substituting the value of V. andp. from equations (i) and (ii) in the ideal gas equation $p_i V_i = nRT_i$ we get,

$$\left(p+\frac{an^2}{V^2}\right)(V-nb)=nRT$$

This equation is applicable to real gases and is known as the van der Waals equation. The constants a and b in van der Waals equation are called van der Waals constants which may vary from gas to gas. p. V and T are the observed pressure, volume and temperature for the gas under study.

Significance of Vander Waals Constants

(*i*) The value of 'a' is higher for easily liquefiable gases (as SO_2, NH_3, H_2S, CO_2 etc.) while lower for permanent gases (as H_2, N_2, O_2, He etc.). This is due to the fact that liquefiable gas has greater intermolecular forces of attraction, therefore the value of '*a*' is said to be a measure of the intermolecular forces of attraction. Order of liquification of some gases:

Gas $SO_2 > Cl_2 > NH_3 > CO_2 > N_2 > O_2 > H_2 > Ne > He$

$\alpha \rightarrow$ 6.71 6.5 4.17 3.58 1.39 1.36 0.24 0.03

(*ii*) Since the constant '*b*' is the effective volume of the gas molecules hence it indicates that the gas molecules are incompressible.

Liquefaction of Gases and Critical Temperature

The liquefaction of a gas takes place when the intermolecular forces of attraction become so high that they bind the gas molecules together to form the liquid state. The intermolecular forces of attraction can be increased either by increasing the pressure so that the molecules come close together or by cooling the gas so that the kinetic energy of the molecules decreases and they become slower.

Gases like hydrogen, helium, oxygen, nitrogen could not be liquefied at room temperature by application of pressure alone. For each gas, there is a particular temperature above which it cannot be liquefied, howsoever, high pressure may be applied on the gas. This temperature is called critical temperature.

Critical temperature : Critical temperature of a gas may be defined as that temperature above which it cannot be liquefied however high pressure may be applied on the gas.

Critical Pressure : The pressure required to liquefy the gas at the critical temperature is called critical pressure .

Critical volume : The volume occupied by 1 mole of the gas at critical temperature and critical pressure is called critical volume.

All the three are called critical constants of the gas and are represented by T_c, T_p, T_v.

1 Andrew's experiments on CO_2 gas

From 1861 to 1870 Andrews carried out his famous experiments on the effect of temperature and pressure on the volume of carbon dioxide. Using the experimental arrangement as shown in figure 3.3, he plotted a series of isothermals (curves) to test the validity of Boyle's law over a wide range of pressures. The gases were compressed by tightening the screws and the pressure was estimated using air, assuming that under the conditions of the experiment it still obeyed Boyle's law.

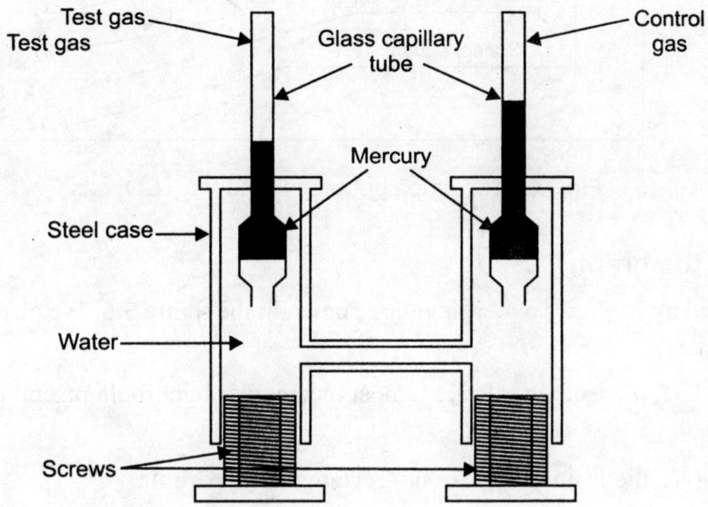

Fig.3.5: The setup for Andrew's experiment

The experimental arrangement for CO_2 gas is shown in the figure 3.5.

Experiment

There are two glass tubes A and B, having thick capillary tubes at the top and bulbs in the middle. Initially pure dry air as the control gas, is passed through the tube A for a long time and then the tube was sealed. Similarly, CO_2 gas is passed through glass tube B again for a long time and then sealed. The lower ends of the tubes are dipped in mercury. Small pellet of mercury is drawn in both the tubes by alternately heating and cooling. Both the tubes are fixed in H-shaped steel case having steel stoppers at the ends. The vessel was filled with water. The tube contained water and hence by screwing in, pressure, as high as 400 atmospheres could be applied.

Since the pressure of air in A and CO_2 in B are the same, from the volume of air in A, pressure of CO_2 can be calculated assuming air to obey Boyle's law. The volume of CO_2 is read from tube B (both the tubes are calibrated to read the volume directly). Air is kept at constant temperature. The temperature of CO_2 could be maintained at any desired temperature between 0°C to 100°C.

Andrews isothermals were drawn at temperatures 13.1°C, 21.5°C, 31.1°C, 32.5°C, 35.5°C and 48.1°C

The experimental result can be depicted in the form of the following graph: Figure (3.6).

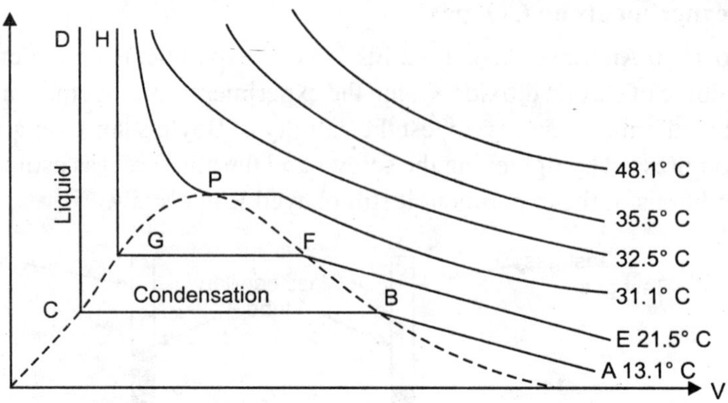

Fig.3.6: The graph between P and V for CO_2 gas.

Explanation of the graph

The curves obtained by Andrews experiment are shown in the figure 3.6. We observe that there are two types of isothermals:

(*a*) Above 31.1°C, the isothermals are almost rectangular hyperbola resembling the behaviour of a perfect gas.

(*b*) Below 31.1°C, the isothermals are not rectangular hyperbola.

At 13.1°C, the portion AB of this curve shows that with the increase of pressure, volume decreases up to the point B showing gaseous behaviour of CO_2 (Boyle's law is obeyed). From B to C, there is change of state of CO_2 from gaseous to liquid (condensation) and the curve becomes parallel to the volume axis. This indicates that volume decreases without appreciable increase in pressure. After the point C, the curve is very steep indicating that after C, the substance becomes highly incompressible. At C, the gas gets liquefied completely. The portion CD represents the liquid phase of CO_2.

Conclusions

1. Thus the portion AB represents the gaseous phase of the substance and BC represents the conversion from vapor to liquid state. At B conversion starts and at C, it gets completed.
2. At 21.5°C, the curve is similar but only the horizontal portion BC has decreased. This shows that compression produces liquefaction.
3. At 31.1°C, the horizontal portion just disappeared which suggests that gas can be liquefied under compression.
4. Above this temperature i.e. 32.5°C, 35.5°C and 48.1°C, the horizontal portion vanished and the behaviour of the gas resembles that of a perfect gas.
5. So, CO_2 cannot be liquefied by exerting pressure alone but also by cooling it below a certain temperature. This temperature is called as the **'critical temperature'**. This behavior is not specific of CO_2 only but it is common to all gases. Thus from experimental curves, it becomes clear that

for every gas there exists a temperature above which it can never be liquefied by exerting pressure alone, however great that pressure may be.

Importance of Critical Temperature

- The critical temperature of a gas is a measure of the strength of the intermolecular forces of attraction.
- Weaker are the intermolecular forces, more difficult it is to liquefy that gas and hence lower would be the critical temperature of that gas.
- Helium and hydrogen have weak intermolecular forces, thus they are difficult to liquefy and hence have low critical temperature.
- Carbon dioxide and ammonia have strong intramolecular forces of attraction, they can be easily liquefied and their critical temperature are high which are above room temperature.
- Van der waals constant a is also a measure of intermolecular forces of attraction. Hence it is found that the values of constant a increase in the same order as the critical temperature.
- As a gas can be liquefied below the critical temperature by applying pressure. Therefore above the critical temperature, it is a gas but below the critical temperature it is a vapour.

For example: Carbon dioxide below its critical temperature is called **carbon dioxide vapor.**

Table 3.4: Van der Waals constants for some gases

Substance	Molar mass (g)	a (L^2- atm $mole^{-2}$)	b (L mol^{-1})
Hydrogen H_2	2	0.244	0.0266
Helium He	4	0.034	0.0237
Methane CH_4	16	2.25	0.0428
Water H_2O	18	5.46	0.0305
Nitrogen N_2	28	1.39	0.0391
Carbon dioxide CO_2	44	3.59	0.0427

The van der Waals equation is only one of many equations of state for real gases. More elaborate equations are required to describe the behavior of gases over wider pressure ranges. These generally take account of higher-order nonlinear attractive forces, and require the use of more empirical constants. Although we will make no use of them in this course, they are widely employed in chemical engineering work in which the behavior of gases at high pressures must be accurately predicted.

The Critical Point

Liquid and gas can coexist only within the regions indicated by the green-shaded area in the diagram above. As the temperature and pressure rise, this region becomes more narrow, finally reaching zero width at the critical point. The values of P, T, and V at this juncture are known as the critical constants P_c, T_c, and V_c. The isotherm that passes through the critical point is called the critical isotherm. Beyond this isotherm, the gas and liquids become indistinguishable; there is only a single fluid phase, sometimes referred to as a supercritical liquid figure 3.7.

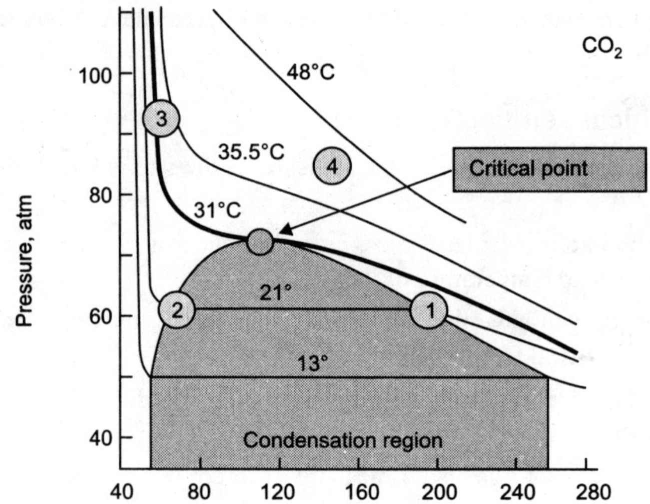

Fig.3.7: Critical Points

At temperatures below 31°C (the critical temperature), CO_2 acts somewhat like an ideal gas even at a rather high pressure (4). Below 31°, an attempt to compress the gas to a smaller volume eventually causes condensation to begin. Thus at 21°C, at a pressure of about 62 atm (1), the volume can be reduced from 200 cm^3 to about 55 cm^3 without any further rise in the pressure. Instead of the gas being compressed, it is replaced with the far more compact liquid as the gas is essentially being "squeezed" into its liquid phase. After all of the gas has disappeared (2), the pressure rises very rapidly because now all that remains is an almost incompressible liquid. Above this isotherm (3) CO_2 exists only as a supercritical fluid.

What happens if you have some liquid carbon dioxide in a transparent cylinder at just under its P_c of 62 atm, and you then compress it slightly? Nothing very dramatic until you notice that the meniscus has disappeared. By successively reducing and increasing the pressure, you can "turn the meniscus on and off".

Critical Phenomena

The essential condition for the liquefaction of the gas is described by the study of critical temperature, critical pressure and critical volume and their inter relationships.

When a gaseous system is transformed to its liquid state, there is a tremendous decrease in the volume. This decrease in volume can be effectively brought about by lowering of temperature, or by increasing pressure (or) by both. In both these effects the gaseous molecules come closer to each other and experience an increase in force of attraction which results in liquefaction of gases. At any constant temperature when pressure is increased volume is decreased and vice versa. Such P-V curves at constant temperature are known as isotherms. There is a definite deviation exhibited by the real gas from ideal gas behaviour at high pressure and low volumes.

Critical Temperature (T_c)

It is defined as the characteristic temperature of a gas at which increase in pressure brings in liquefaction of gas above which no liquefaction occurs although the pressure may be increased many fold. For instance Tc of CO_2 is 31.1°C. This means that it is not possible to liquefy CO_2 by applying pressure when its temperature is above 31.1°C.

Critical Pressure (P_c)

It is defined as the minimum pressure required to liquefy 1 mole of a gas present at its critical temperature.

Critical Volume (V_c)

The volume occupied by 1 mole of a gas at its critical pressure and at critical temperature is the critical volume (V_c) of the gas.

A gas is said to be at its critical state when its pressure, volume and temperature are P_c, V_c and T_c.

1. Continuity of state

Andrews isotherms of carbondioxide: The importance of critical temperature of a gas was first discovered by Andrews in his experiments on pressure - volume isotherms of carbon dioxide gas at a series of temperature. The isotherm of carbon dioxide determined by him at different temperatures are shown in Fig. 3.8.

Consider first the isotherm at the temperature 13.1°C. The point A represents carbon dioxide in the gaseous state occupying a certain volume under a certain pressure. On increasing the pressure its volume diminishes as is indicated by the curve AB. At B, liquefaction of gas commences and thereafter a rapid decrease in the volume takes place at the same pressure, since more and more of the gas is converted into the liquid state. At C, the gas becomes completely liquefied. After 'C' the increase of pressure produces only a very small decrease in volume. This is shown by a steep line CD which is almost vertical. Thus, along the curve AB, carbon dioxide exists as gas. Along BC, it exists in equilibrium between gaseous and liquid state. Along CD it exists entirely as a liquid. The isotherm at 21.5°C shows that the decrease in volume becomes smaller because higher the temperature greater is the volume. Therefore more pressure is applied to decrease the volume. This effect makes liquefaction to commence at higher pressure compared to the previous isotherm at 13.1°C.

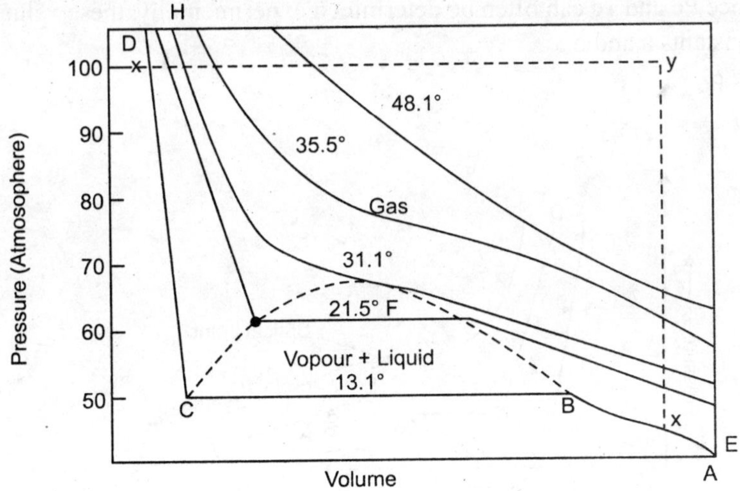

Fig.3.8: Andrews isotherms of carbon dioxide

At still higher temperature, the horizontal portion of the curve becomes shorter and shorter until at 31.1°C it reduces to a point. The temperature 31.1°C is regarded as the critical temperature of CO_2. At this temperature, the gas passes into liquid imperceptibly. Above 31.1°C the isotherm is continuous. CO_2 cannot be liquefied above 31.1°C no matter how high the pressure may be. The portion of area covered by curve H with *zyx* portion always represents the gaseous state of CO_2.

Thomson's experiment

Thomson (1871) studied the isotherm of CO_2 drawn by Andrews. He suggested that there should be no sharp points in the isotherms below the critical temperature. These isotherms should really exhibit a complete continuity of state from gas to liquid. This, he showed as a theoretical wavy curve. The curve MLB in represents a gas compressed in a way that it would remain stable. The curve MNC represents a superheated liquid because compression above Tc, leads to heating effects. This type of continuity of state is predicted by Vanderwaal's equation of state which is algebraically a cubic equation. The Vanderwaal's equation may be written as

$$PV^3 - (RT + Pb)\,V^2 + aV - ab = 0$$

Thus, for any given values of P and T there should be three values of V. These values are indicated by points B, M and C of the wavy curve. The three values of V become closer as the horizontal part of the isotherm rises. At the critical point the three roots of Vanderwaal 'V' become identical and there is no longer any distinction between the gas and liquid states. Here, the gas is said to be in critical state. This effect enables the calculation of T_c, P_c and V_c in terms of Vanderwaal's constants.

2. Derivation of critical constants from Vanderwaal's constants

Let us derive the values of critical constants T_c (critical temperature), V_c (critical volume) and P_c (critical pressure) in terms of the Vanderwaal's constants '*a*' and '*b*'. The Vanderwaal's equation is given by

$$T_c = 8a / 27Rb$$

Hence the critical constants can be calculated using the values of Vanderwaal's constants of a gas and vice versa. Since Pc and Tc can often be determined experimentally, these values may be employed to calculate the constants a and b.

$$a = 3Vc^2\ Pc$$

$$b = V_c/3$$

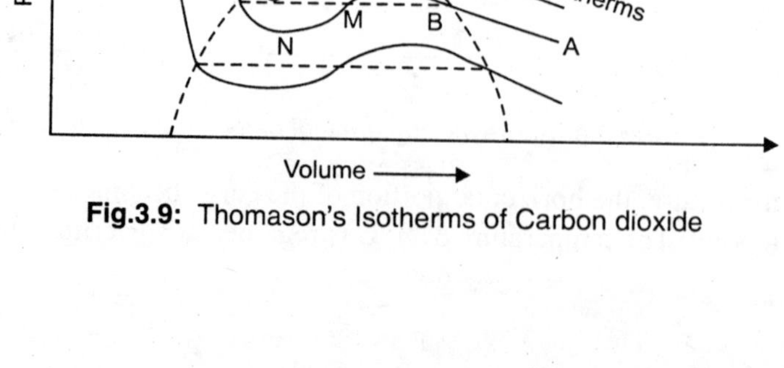

Fig.3.9: Thomason's Isotherms of Carbon dioxide

Based on the critical temperature values, gases are categorised as "permanent" and "temporary" gases. H_2, N_2, He, ... gases having very low critical temperature belong to the permanent type. Gases like NH_3, CO_2, SO_2, HCl etc. having critical temperature in the ordinary range of temperatures belong to the temporary gases type.

JOULE-THOMSON EFFECT AND INVERSION TEMPERATURE (T_i)

Joule-Thomson Effect

Joule-Thomson showed that when a compressed gas is forced through a porous plug into a region of low pressure, there is appreciable cooling. The phenomenon of producing lowering of temperature when a gas is made to expand adiabatically from a region of high pressure into a region of low pressure, is known as Joule - Thomson effect.

When the gas is allowed to escape into a region of low pressure the molecules move apart rapidly against the intermolecular attractive forces. In this case work is done by the gas molecules at the expense of internal energy of the gas. Therefore cooling occurs as the gas expands. This reduction in the temperature is generally referred as Joule-Thomson effect and is used in the liquefaction of gases.

Inversion temperature (T_i)

The Joule-Thomson effect is obeyed by a gaseous system only when its temperature is below a characteristic value. The characteristic temperature below which a gas expands adiabatically into a region of low pressure through a porous plug with a fall in temperature is called as inversion temperature (T_i).

T_i is characteristic of a gas and it is related to the Vander waal's constant 'a' and 'b',

$$2a\ T_i = 2a/Rb$$

At the inversion temperature there is no Joule Thomson effect (i.e.) there is neither fall nor rise in temperature. Only when the temperature of the gas is below the inversion temperature there is a fall in temperature during adiabatic expansion. If the temperature of the gas is above T_i there is a small rise in temperature. For gases like H_2 and He whose T_i values are very low –80°C and –240°C respectively, these gases get warmed up instead of getting cooled during the Joule-Thomson experiment. These gases will obey Joule-Thomson effect only when they are cooled to a temperature below these T_i values.

Conditions of liquefaction of gases: Linde's Method, Claude's process: Many industrial processes require large quantities of liquid air, liquid ammonia, liquid carbondioxide etc. The production of liquids from various gases is therefore an important commercial operation.

Conditions of liquefaction of gases

- Many industrial processes require large quantities of liquid air, liquid ammonia, liquid carbondioxide etc. The production of liquids from various gases is therefore an important commercial operation.
- There are different methods of liquefaction of gases, such as (*i*) based on the concept of critical temperature followed by the compression (*ii*) based on Joule-Thomson effect (*iii*) Adiabatic demagnetisation.

- In the case of gases like NH_3, Cl_2, SO_2 and CO_2 whose T_c values are near and below the ordinary temperatures, they can be liquefied easily by increasing the pressure alone at their respective T_c values.
- Gases like H_2, O_2, N_2 and He have very low T_c values and hence Joule Thomson effect may be applied to bring in effective cooling.
- Helium is cooled by Joule-Thomson effect to a lower temperature and further cooling for its liquefaction, is carried out by the method of adiabatic demagnetisation.

Linde's Method

This method makes use of Joule Thomson effect and is used to liquify air or any other gas. Pure air or any gas is first compressed to about 200 atmospheres and is allowed to enter the innertube of the concentric pipes as shown in Fig. 3.10. The valve *v* of jet J is then opened and the gas is allowed to epand suddenly into the wider chamber C.

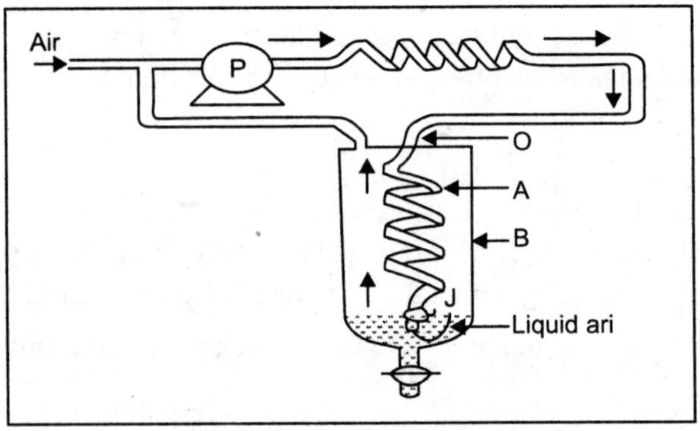

Fig.3.10: Linde's Experiment

The gas gets cooled due to expansion and its pressure is reduced to about 50 atm. The gas is now allowed to pass through the outer tube 'O'. At this stage the incoming gas is initially cooled by the outgoing gas. Further cooling of the incoming gas occurs during expansion in the chamber C. The cooled gas is again compressed and is circulated in. By repeating the process of cooling and compression followed by expansion, the gas is liquefied and finally the liquid air drops out from the jet into the bottom of chamber C.

Claude's Process

In this method compressed air is allowed to do mechanical work of expansion. This work is done at the expense of the kinetic energy of the gas and hence a fall of temperature is noted. This principle is combined with Joule-Thomson effect and utilised in Claude's process of liquefaction of air. Air is compressed to about 200 atmospheres and is passed through the pipe ABC Fig. 3.11. At C, a part of the air goes down the spiral towards the jet nozzle J and a part of the air is led into the cylinder D provided with an air tight piston. Here the air moves the piston outwards and expands in volume as a result of which considerable cooling is produced. The cooled air passes up the liquefying chamber during which process it cools the portion of the incoming compressed air. The precooled incoming compressed air then experiences Joule-Thomson expansion when passed through Jet nozzle J and gets cooled further. The above process takes place repeatedly till the air is liquefied.

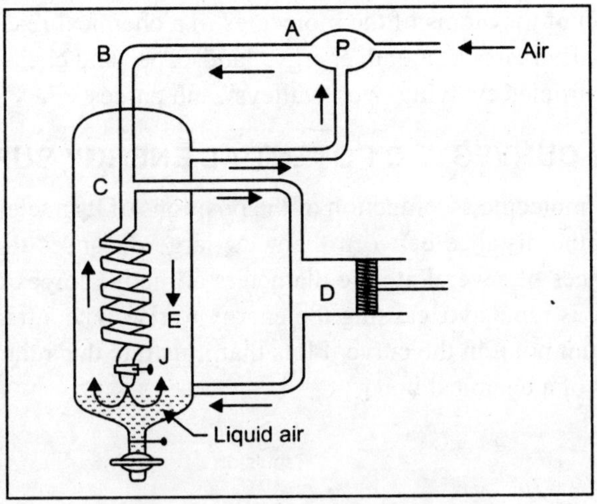

Fig.3.11: Claude's Experiment

Critical Constant and Van der Waal's Constant

The relationship between critical constants of the gases and their Van der Waal constants is as follows:

Through equation it is possible to calculate the values of crticicle constants P_c, T_c and V_c of a gas in terms of the Van der Waals constants a and b.

(*i*) $V_c = 3b$

(*ii*) $P_c = a/27b^2$

(*iii*) $T_c = 8a/27Rb$

(*iv*) The critical compressibility factor Z_c is given by,

$$Z_c = \frac{P_cV_c}{RT_c} = \frac{(a/27b^2)(3b)}{R[8a/27Rb]} = \frac{a \times 3b \times 27Rb}{R \times 8a \times 27b^2} = \frac{3}{8} = 0.375$$

These relations have been derived from the calculations based on Van der Waal's equation.

Potential energy surfaces of H_3, H_2F, HCN

A **potential energy surface** is a mathematical function that gives the **energy** of a molecule as a function of its geometry. Molecular Mechanics provides this **energy** as a function of stretches, bends, torsions, etc

A potential energy surface (PES) describes the potential energy of a system, especially a collection of atoms, in terms of certain parameters, normally the positions of the atoms.

The surface might define the energy as a function of one or more coordinates; if there is only one coordinate, the surface is called a potential energy curve or energy profile. It is helpful to use the analogy of a landscape: for a system with two degrees of freedom (*e.g.* two bond lengths), the value of the energy (analogy: the height of the land) is a function of two bond lengths (analogy: the coordinates of the position on the ground). The Potential Energy Surface represents the concepts that each geometry (both external and internal) of the atoms of the molecules in a chemical reaction is associated with it a unique potential energy. This creates a smooth energy "landscape" and chemistry can be viewed from a topology perspective (of particles evolving over "valleys" and passes").

POTENTIAL ENERGY CURVES (1-D POTENTIAL ENERGY SURFACES)

The PES is the energy of a molecule as a function of the positions of its nuclei r. This energy of a system of two atoms depends on the distance between them. At large distances the energy is zero, meaning "no interaction". At distances of several atomic diameters attractive forces dominate, whereas at very close approaches the force is repulsive, causing the energy to rise. The attractive and repulsive effects are balanced at the minimum point in the curve. Plots that illustrate this relationship are quite useful in defining certain properties of a chemical bond.

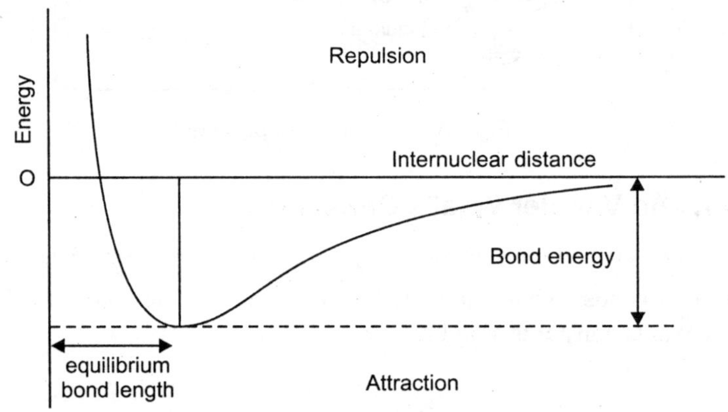

Fig.3.12: A potential Energy Curve for a covalent bond.

The internuclear distance at which the potential energy minimum occurs defines the bond length. This is more correctly known as **the equilibrium bond length**, because thermal motion causes the two atoms to vibrate about this distance. In general, the stronger the bond, the smaller will be the bond length.

Attractive forces operate between all atoms, but unless the potential energy minimum is at least of the order of RT, the two atoms will not be able to withstand the disruptive influence of thermal energy long enough to result in an identifiable molecule. Thus we can say that a chemical bond exists between the two atoms in H_2. The weak attraction between argon atoms does not allow Ar_2 to exist as a molecule, but it does give rise to the van Der Waals force that holds argon atoms together in its liquid and solid forms.

Potential, Kinetic, and Total Energy for a System

Potential energy and kinetic energy Quantum theory tells us that an electron in an atom possesses kinetic energy K as well as potential energy V, so the total energy E is always the sum of the two:

$$E = V + K.$$

The relation between them is surprisingly simple: $K = -0.5\ V$.

This means that when a chemical bond forms (an exothermic process with $\Delta E < 0$), the decrease in potential energy is accompanied by an increase in the kinetic energy (embodied in the momentum of the bonding electrons), but the magnitude of the latter change is only half as much, so the change in potential energy always dominates. The bond energy $-\Delta E$ has half the magnitude of the fall in potential energy.

THE DIMENSIONALITY OF A POTENTIAL ENERGY SURFACE

To define an atom's location in 3-dimensional space requires three coordinates (*e.g.*, x, y, and z or r, θ and phi in Cartesian and Spherical coordinates) or degrees of freedom. However, a reaction and hence the corresponding PESs do not depend of the absolute position of the reaction, only the relative positions (internal degrees). Hence both translation and rotation of the entire system can be removed (each with 3 degree of freedom, assuming non-linear geometries). So the dimensionality of a PES is

$$3\,N-6$$

where N is the number of atoms involves in the reaction, i.e., the number of atoms in each reactants). The PES is a hypersurface with many degrees of freedom and typically only a few are plotted at any one time for understanding. See Calculate Number of Vibrational Modes to get a more details picture of how this applies to calculating the number of vibrations in a molecule

For very simple chemical systems or when simplifying approximations are made about inter-atomic interactions, it is sometimes possible to use an analytically derived expression for the energy as a function of the atomic positions. An example is

$$H + H_2 \rightarrow H_2 + H$$

system as a function of the three H-H distances. For more complicated systems, calculation of the energy of a particular arrangement of atoms is often too computationally expensive for large scale representations of the surface to be feasible.

APPLICATION OF POTENTIAL ENERGY SURFACES

A PES is a conceptual tool for aiding the analysis of molecular geometry and chemical reaction dynamics. Once the necessary points are evaluated on a PES, the points can be classified according to the first and second derivatives of the energy with respect to position, which respectively are the gradient and the curvature. Stationary points (or points with a zero gradient) have physical meaning: energy minima correspond to physically stable chemical species and saddle points correspond to transition states, the highest energy point on the reaction coordinate (which is the lowest energy pathway connecting a chemical reactant to a chemical product). Three

PES do not show kinetic energy, only potential energy.

At $T = {}^0K$ (no KE), species will want to be at the lowest possible potential energy, (i.e., at a minimum on the PES). Between any two minima (valley bottoms) the lowest energy path will pass through a maximum at a saddle point, which we call that **saddle point** a transition-state structure. The PES concept finds application in fields such as chemistry and physics, especially in the theoretical sub-branches of these subjects. It can be used to theoretically explore properties of structures composed of atoms, for example, finding the minimum energy shape of a molecule or computing the rates of a chemical reaction.

QUESTIONS

1. Which of the following molecules have a permanent dipole moment?

(*a*) H_2O (*b*) CO_2

(*c*) CH_4 (*d*) N_2

(*e*) CO (*f*) NH_3

Ans. CO_2 CH_4 and N_2 are symmetric, and hence they have no permanent dipole moments. A molecule with polar bonds unsymmetrically arranged will possess a permanent dipole.

2. Which has the higher boiling point, Br_2 or ICl?

Ans. They have similar molecular weights: Br_2, 160; ICl, 162. Their boiling points are 332 K and 370 K respectively.

3. An atom or molecule can be temporarily polarized by a nearby species. Polarization separates centers of charge giving

(*a*) permanentdipole
(*b*) temporarycharges
(*c*) hydrogenbonding
(*d*) induceddipole
(*e*) inducedions
(*f*) radicals

Ans. Induced dipoles are responsible for the London dispersion forces. The heavier the molecule, the larger the induced dipole will be. Thus, London dispersion forces are strong for heavy molecules.

4. Which has a higher boiling point, I_2 or Br_2?

Ans. Atomic weights for Br and I are 80 and 127 respectively. The higher the molecular weight, the stronger the London dispersion forces.

5. If only London dispersion forces are present, which should have a lower boiling point, H_2O or H_2S?

Ans. The b.p. for H_2O is 100 deg C, and that of H_2S is –70 deg C. Very strong hydrogen bonding is present in liquid H_2O, but no hydrogen bonding is present in liquid H_2S.

6. Contrary to most other substances, the density of water decreases as temperature decreases between 4 and 0 deg C. This is due to

(*a*) dipole-dipole interaction
(*b*) London dispersion
(*c*) decreasing number of hydrogen bonds
(*d*) increasing number of hydrogen bonds formed

Ans. As more hydrogen bonds form when the temperature decreases, the volume expands, causing a decrease in density. Above 4 deg C, the thermal expansion is more prominent than the effect of hydrogen bonds.

7. Ethanol (C_2H_5OH, molar mass 46) boils at 351 K, but water (H_2O, molar mass 18) boils at higher temperature, 373 K. This is because:

(*a*) water is denser
(*b*) water has stronger London dispersion forces
(*c*) water has stronger hydrogen bonds
(*d*) water molecules contain no carbon

Ans. A hydrogen atom between two small, electronegative atoms (such as F, O, N) causes a strong intermolecular interaction known as the hydrogen bond. The strength of a hydrogen bond depends upon the electronegativities and sizes of the two atoms.

8. Ethanol C_2H_5OH and methyl ether CH_3OCH_3 have the same molar mass. Which has a higher boiling point?

Ans. R-OH group is both proton donor and acceptor for hydrogen bonding. Methyl groups have very weak hydrogen bonding, if any.

SHORT ANSWER QUESTIONS

1. Explain the liquefaction of gases based on Joule- Thomson effect.
2. What are the limitations of Van der waal's equations.
3. Write 4 differences between ideal and real gases.
4. Write the units of Van der waal's constants a and b.
5. Explain Andrew's isotherms.
6. Write short notes on
   (*a*) Hydrogen Bonding
   (*b*) Dipole – Dipole Interactions
   (*c*) Strength of molecular interactions
   (*d*) Real and Ideal Gases
   (*e*) Gas equation and correction in Vander waal's.

LONG ANSWER QUESTIONS

1. In what manner are ideal gases are different from real ? Derive Van der waal's equation of state and explain the behavior of real gases.
2. What are critical phenomenon of gases. Deduce the relationship between critical constants and Vander waal's equation for gas.
3. What are critical constants? Calculate the values of Vc, Pc and Tc in terms of a and b by applying Van der waal's equation on the critical phenomenon.
4. What are the reason of deviation from ideal behavior of gas equation PV = *n*RT? Write Van der waal's equations and gives its application.
5. What you mean by molecular interaction? How many types of it? Explain any two in detail.
6. Discuss Vander waal's interactions in details.

4 Use of Free Energy in Chemical Equilibria

Thermodynamic functions: energy, entropy and free energy. Estimations of entropy and free energies. Free energy and emf. Cell potentials, the Nernst equation and applications. Acid base, oxidation reduction and solubility equilibria. Water chemistry. Corrosion. Use of free energy considerations in metallurgy through Ellingham diagrams.

The branch of chemistry which concern with the study of inter-relations of the various forms of energy in a system is called thermodynamics.

Why Thermodynamics?

Let us think about some everyday experiences which related with chemical reactions.

- Frying an egg is an example of an irreversible reaction.
- Why does the NaCl dissolve? Does it give off heat? Does it require energy?
- How is it that diamond, a high-pressure form of C, can coexist with the low pressure form, graphite, at Earth's surface?
- Melt and refreeze of ice shows that H_2O has two phases due to temperature.

Chemical thermodynamics provides us with a means of answering these questions and more.

FREE ENERGY (ΔG)

Free energy has the dimensions of energy, and its value is determined by the state of the system and not by its history. Free energy is used to determine how systems change and how much work they can produce. It is expressed in two forms:

1. Helmholtz free energy or work function F
2. Gibbs free energy G.

If,U is the internal energy of a system, PV the pressure-volume product, and TS the temperature-entropy product (T being the temperature above absolute zero), Then

$$F = U - TS$$

and, $$G = U + PV - TS$$

$$H = U + PV \text{ is the enthalpy}$$

So, equation can also be written as

$$G = H - TS \quad ...(4.1)$$

The Gibbs free energy (ΔG) of a reaction is a measure of the thermodynamic driving force that makes a reaction occur. A negative value for ΔG indicates that a reaction can proceed spontaneously without external inputs, while a positive value indicates that it will not. Free energy is an extensive property, meaning that its magnitude depends on the amount of a substance in a given thermodynamic state.

APPLICATION OF FREE ENERGY

The changes in free energy, ΔF or ΔG, are useful in determining the direction of spontaneous change and evaluating the maximum work.

1. **To calculate Work Done**

   a. **Work done at constant temperature and constant volume:** In a reversible process the maximum useful work that can be obtained from a system under constant temperature and constant volume is equal to the (negative) change in the Helmholtz free energy,

   $$-\Delta F = -\Delta U + T\Delta S \quad ...(4.2)$$

   b. **Work done at constant temperature and constant pressure:** The maximum useful work done under constant temperature and constant pressure (other than work done against the atmosphere) is equal to the (negative) change in the Gibbs free energy,

   $$-\Delta G = -\Delta H + T\Delta S \quad ...(4.3)$$

2. **Changes in free energy can be used to judge whether changes of state can occur spontaneously.**

It $\Delta G = \Delta H - T\Delta S$ is zero, indicating that the two phases are in equilibrium and that no useful work can be extracted from the phase transition (other than work against the atmosphere due to changes in pressure and volume).

ΔG is negative for $T > 273.16$ K, indicating that the direction of spontaneous change is from ice to water.

ΔG is positive for $T < 273.16$ K, where the reverse reaction of freezing takes place.

ENTHALPY (ΔH)

The enthalpy (ΔH) is a measure of the actual energy that is liberated when the reaction occurs (the "heat of reaction"). If it is negative, then the reaction gives off energy, while if it is positive the reaction requires energy.

It is another measure of energy called enthalpy 'H' a kind of measure of the thermal energy of a crystal. As we will see below,

$$dH = dQ + VdP \quad ...(4.4)$$

Recall that we interpreted

$$dU = dQ - PdV \quad ...(4.5)$$

to mean that the internal energy change is the heat change minus the energy lost to relaxation of the crystal. Thus,

$$dH = dQ + VdP \quad ...(4.6)$$

means that the enthalpy change is the heat change plus the energy the crystal gains by virtue of not being allowed to expand.

Sign of dH – When heat absorb during chemical reaction dH will have positive sign and when heat is evolved , then will have negative sign.

ENTROPY (ΔS)

A very important thermodynamic function known as entropy was introduced by Clausius in 1950. Moreover, there are other processes that proceed in the apparent absence of any heat change: e.g., mixing of gases or the spreading of dye in water. What is it that causes these reactions to proceed spontaneously even if the heat change is zero or endothermic?

The answer is entropy S, which is a measure of the order or disorder.

The entropy (ΔS) is a measure of the change in the possibilities for disorder in the products compared to the reactants. For example, if a solid (an ordered state) reacts with a liquid (a somewhat less ordered state) to form a gas (a highly disordered state), there is normally a large positive change in the entropy for the reaction.

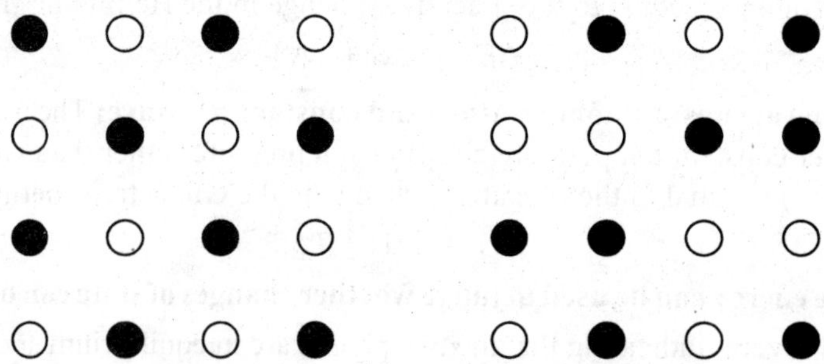

Ordered Disordered

Fig.4.1: Entropy, degree of randomness

The entropy of a system always increases during irreversible processes; i.e., for a reversible process, $dS = 0$, whereas for irreversible processes $dS > 0$. This is the Second Law of Thermodynamics.

The change in entropy ΔS is defined as the ratio between the total heat absorbed in a change carried out isothermally and reversibly and the absolute temperature at which the heat was absorbed.

$$\Delta S = \frac{Q_{Rev}}{T} \qquad \text{...(4.7)}$$

Unit of entropy ΔS is an extensive property. Therefore its value depends upon the quantity of substances.Its unit is calorie per degree per mol (Cal deg^{-1} mol^{-1}).

Entropy of a cyclic process– entropy is a state function in cyclic process , there will be no change in entropy of the system so, change in entropy will be

$$\Delta S = 0 \qquad \text{...(4.8)}$$

DEFINITION OF SECOND LAW OF THERMODYNAMICS IN TERMS OF ENTROPY

Since, spontaneous processes or naturally occurring processes are all reversible, therefore they all are associated with a net increase of entropy. This is one of the most important consequences of Second Law of Thermodynamics and therefore this law may be stated as follows–

All processes, physical or chemical, during in nature takes place in a way that the sum of entropies of the system and surroundings increases. In the limiting case of a reversible process the sum of entropies remains constant.

This is the most general statement of second law of thermodynamics. Clausius summed up the above statement by saying that the entropy of the world tends to a maximum.

Entropy change in adiabatic process in an adiabatic change heat is neither given to the system nor taken out of the system. Therefore, q = 0, and the change in entropy during the process, will be

$$\Delta S = \frac{q_{rev}}{T} = \frac{q}{T} = 0 \qquad \text{...(4.9)}$$

Such process are known as iso-entropic.

ISO-ENTROPIC PROCESS

The process during which the entropy of the system remains constant is called as iso-entropic process. During iso-entropic process the value of entropy of the system at initial and final state remains constant means,the value of $\Delta S = 0$.

The iso-entropic process can be reversible or irreversible. For the isentropic process to be reversible, it is essential that it is adiabatic process also, if it is not adiabatic, then the iso-entropic process cannot be reversible. This also implies that the process which is reversible and adiabatic is always an isentropic process.

In actual practice the reversible iso-entropic process never really occurs, it is only an ideal process. In actual practice whenever there is change in the state of the system the entropy of the system increases. Here are the various causes of the increase in entropy of the closed system are:

(1) Due to external interaction: In closed system the mass of the system remains constant but it can exchange the heat with surroundings. Any change in the heat content of the system leads to disturbance in the system, which tends to increase the entropy of the system.

(2) Internal changes in the system: Due to internal changes in the movements of the molecules of the system there is further disturbance inside the system. This causes irreversibility inside the system and an increase in its entropy.

In a perfectly ordered, pure crystalline material the entropy is zero. This is a simple statement of the Third Law of Thermodynamics, which follows from the fact that heat capacities approach zero at zero K:

$$\lim_{T \to 0} C = 0$$

However, because the rate of atomic diffusion also goes to zero at 0 K, all compounds have some zero-point entropy S°0.

Entropy is thus the only thermodynamic potential for which we can calculate an absolute value. What we typically do is determine the heat capacity from near absolute zero to ambient conditions and then integrate it to get the (absolute) entropy (in fact this gives us only the vibrational entropy and ignores configurational and electronic contributions to entropy).

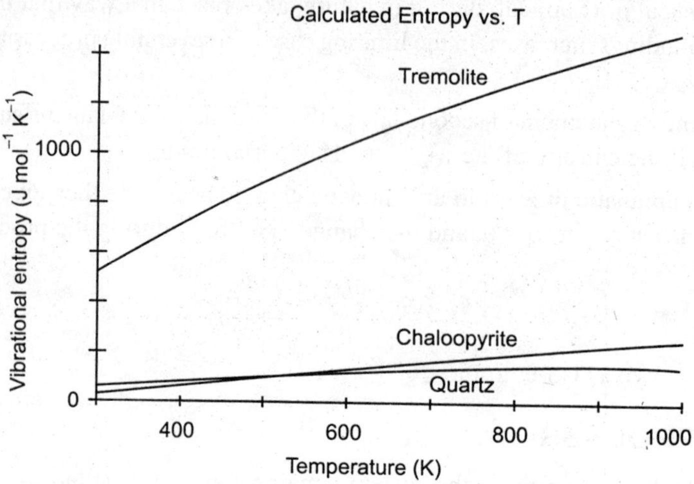

Fig.4.2: Correlation of entropy with temperature

ENTROPY CHANGE OF AN IDEAL GAS WITH CHANGE IN TEMPERATURE, PRESSURE AND VOLUME

Entropy is a thermodynamic property which depends upon the state of the system and therefore its value for a pure substance will depend upon any two out of the three independent state variables, viz., pressure, volume and temperature. Entropy changes will be considered under two heads:

(*i*) When T and V are considered variables.

(*ii*) When T and P are considered variables.

(*i*) **When T and V are Variables**–We have defined increase in entropy of a gas for an infinitesimal change by the expression,

$$dS = dq_{rev}/T$$

where, dq_{rev} is the small amount of heat energy absorbed by the given substance or system reversible with respect to the surroundings at constant temperature T. From first law of thermodynamics,

$$dq_{rev} = dU + PdV \qquad ...(4.10)$$

we also know that for an ideal gas the change of internal energy with temperature at constant volume is heat capacity, C_v

Thus,
$$C_v = \left(\frac{\delta U}{\delta T}\right)_v$$

$$C_v dT = dU$$

And for n moles of ideal gases $nC_v dT = dU$

Finally we get expression for change in entropy is

$$\Delta S = nC_v \ln\frac{T_2}{T_1} + nR\ln\frac{V_2}{V_1} \qquad ...(4.11)$$

Where,

R = Gas constant,

T_1 = Temperature in the initial state

V_1 = Volume in the initial state

T_2 = Temperature in the final state

V_2 = volume in the final state,

In the above equation the first term on the right hand side of the equation gives the changes in entropy of n moles of an ideal gas on change of temperature at constant volume and the second term gives the ΔS due to change of volume at constant temperature. Thus , it is evident that the entropy change for change of state of an ideal gas is independent on the initial and final temperature as well as on initial and final volumes of the system. By changing the logarithm to the base 10, the equation will be

$$\Delta S = 2.303n\ C_v \log\frac{T_2}{T_1} + nR\,2.303\log\frac{V_2}{V_1} \qquad ...(4.12)$$

Problem 1: Calculate the change of entropy suffered by 2 moles of an ideal gas on being heated from 500°C and a volume of 100 litres to 150°C and a volume of 150 litres. C_v for gas is 7.88 cal per mol per degree.

Solution : Using

Formula–
$$\Delta S = nC_v \ln\frac{T_2}{T_1} + nR\ln\frac{V_2}{V_1}$$

Given
$$n = 2, Cv = 7.88, T_2 = 273 + 150 = 423K$$
$$V_2 = 150 \text{ litres}, V_1 = 100 \text{ litres, and } T_1 = 273 + 50 = 323K$$

Therefore,
$$\Delta S = 2 \times 7.88 \times 2.303 \log\frac{423}{323} + 2 \times 2 \times 2.303 \times \log\frac{150}{100}$$
$$= 2 \times 7.88 \times 2.303 \times 0.1173 + 2 \times 2 \times 2.303 \times 0.1761$$
$$= 4.25 + 1.63$$
$$= \mathbf{5.88\ e.u.}$$

***(ii)* When T and P are considered variables**

Let P_1 be the pressure and V_1 be the volume in the initial state and P_2 be the pressure and V_2 be the volume in the final state, then for n moles of ideal gase,

$$P_1V_1 = nRT_1$$
$$P_2V_2 = nRT_2$$

and
$$\frac{V_2}{V_1} = \frac{P_1T_2}{P_2T_1}$$

Finally we get expression for change in entropy is

$$\Delta S = nC_p \ln\frac{T_2}{T_1} - nR\ln\frac{P_2}{P_1} \qquad ...(4.13)$$

Where,

R = Gas constant,

T_1 = Temperature in the initial state

P_1 = Pressure in the initial state

T_2 = Temperature in the final state

P_2 = Pressure in the final state,

Or

$$\Delta S = nC_p 2.303 \log \frac{T_2}{T_1} - nR 2.303 \ln \frac{P_2}{P_1} \quad ...(4.14)$$

The first term in equation gives the change in entropy suffered by n moles of an ideal gas due to variation in temperature at constant pressure while the second term gives the entropy change due to variation in pressure when temperature kept constant.

Thus entropy change for the change of states of an ideal gas is dependent upon the initial and final temperature as well as on the initial and final pressure of the system.

Problem 2: Calculate the entropy change suffered by 2 moles of an ideal gas on being heated from 50°C and pressure 5 atm. To 100°C and pressure 10 atm. The C_p for the gas is 9.88 cal. Mol^{-1} $degree^{-1}$.

Solution : Using

Formula–
$$\Delta S = nC_p \ln \frac{T_2}{T_1} - nR \ln \frac{P_2}{P_1}$$

Given :
$$n = 2, C_p = 9.88, T_2 = 273 + 100 = 373°K, P_1 = 5 \text{ atm.}$$
$$T_1 = 273 + 50 = 323°K, P_2 = 10 \text{ atm.}$$

Therefore,
$$\Delta S = 2 \times 9.88 \times 2.303 \log \frac{373}{323} - 2 \times 2 \times 2.303 \times \log \frac{10}{5}$$
$$= 2 \times 9.88 \times 2.303 \times 0.0625 - 2 \times 2 \times 2.303 \times 0.3010$$
$$= 2.8442 - 2.7728$$
$$= \mathbf{0.0714 \text{ e.u.}}$$

ENTROPY CHANGES FOR OTHER PROCESS

Isochoric Process – In this process volume does not change *i.e.* it remains constant–

$$\Delta S = nC_v \ln \frac{T_2}{T_1} + nR \ln \frac{V_2}{V_1}$$

Putting
$$V_1 = V_2$$

We get,

$$\Delta S_v = n C_v \ln \frac{T_2}{T_1} \quad ...(4.15)$$

Isobaric Process – In this process Pressure does not change i.e. it remains constant. By equation

$$\Delta S = nC_p \ln \frac{T_2}{T_1} - nR \ln \frac{P_2}{P_1}$$

Putting $P_1 = P_2$

We get,

$$\Delta S_p = nC_p \ln \frac{T_1}{T_1} \quad ...(4.16)$$

Isothermal Process – In this process Temperature remains constant, means does not changes *i.e.* hence by putting $T_1 = T_2$ inequation (11) and (13) both

$$\Delta S = nC_v \ln \frac{T_2}{T_1} + nR \ln \frac{V_2}{V_1}$$

We get,

$$\Delta S_T = nR \ln \frac{V_2}{V_1} \quad ...(4.17)$$

and

$$\Delta S = nC_p \ln \frac{T_2}{T_1} - nR \ln \frac{P_2}{P_1}$$

$$\Delta S_T = -nR \ln \frac{P_2}{P_1}$$

From the equation it is evident that in isotheral expansion of an ideal gas there is always an increase in entropy while in isothermal compression the entropy of gas decreases.

Adiabatic Process – In this process neither heat is given nor heat is taken in, therefore, q = 0. Hence $\Delta S = q/T = 0$

Putting this value in above equation

$$\Delta S = nC_v \ln \frac{T_2}{T_1} + nR \ln \frac{V_2}{V_1}$$

$$nC_v \ln \frac{T_2}{T_1} = nR \ln \frac{V_2}{V_1} \quad ...(4.18)$$

$$\Delta S = nC_p \ln \frac{T_2}{T_1} - nR \ln \frac{P_2}{P_1}$$

$$nC_p \ln \frac{T_2}{T_1} = nR \ln \frac{P_2}{P_1} \quad ...(4.19)$$

Problem 3: Calculate the entropy change when 2.5 moles of an ideal gas at 270C expands from 2 litres to 20 litres.

Solution : Using equation: $\Delta S_T = nR \ln \frac{V_2}{V_1}$

$$= 2.303 \text{ n R} \log V_2/V_1$$

$$= 2.303 \times 2.5 \times 1.99 \times \log 20/2$$

$$= \mathbf{11.457 \text{ cal deg}^{-1}}.$$

When, R is taken in Joule (R = 8.314 J) then

$$\Delta S = 2.5 \times 8.314 \times 2.303 \log 20/2$$
$$= \mathbf{47.867\ J\ deg^{-1}}.$$

Problem 4: Calculate the entropy change of one mole of Helium gas at constant pressure heated from 200K to 400 K . Given $C_p = 5$ cal deg^{-1} mol^{-1}.

Solution : Using equation : $\Delta S_P = nC_p 2.303 \log \frac{T_2}{T_1}$

$$= 1 \times 5.0 \times 2.303 \times \log 400/200$$
$$= 1 \times 5.0 \times 2.303 \times \log 2$$
$$= 1 \times 5.0 \times 2.303 \times 0.3010$$
$$= \mathbf{3.466\ cal\ deg^{-1}\ mol^{-1}}$$

Problem 5: Calculate the entropy change of an ideal gas if 3 moles of it at constant pressure suffered a temperature change from 27°C to 327°C. Given C_p of the gas = 5/2 R.

Solution : $\Delta S = nC_p 2.303 \log \frac{T_2}{T_1}$

$$= 3 \times 5/2\ R \times 2.303 \times \log 600/300$$
$$= 3 \times 5/2 \times 1.99 \times 2.303 \times \log 2$$
$$= 3 \times 5/2 \times 1.99 \times 2.303 \times 0.3010$$
$$= \mathbf{10.346\ cal\ deg^{-1}}$$

ENTROPY CHANGE IN PHYSICAL CHANGE

Change in entropy also occurs when some physical transformation such as fusion, vaporisation or transition from one crystalline state to another takes place.

If S_1 is the entropy of initial state or form S_2 that of final state or form then change in entropy in physical change is given by

$$\Delta S = S_2 - S_1 \qquad ...(4.20)$$

For all physical changes entropy change ΔS is given by

$$\Delta S = \frac{q}{T} = \frac{\Delta H}{T} \qquad ...(4.21)$$

Thus change in entropy ΔS is the ratio of the heat necessary to accomplish transition and the temperature at which transition takes place. The equation (4.21) holds good only when equilibrium exists between the two form or phases.

The use of equation (4.21) will be made clear by certain illustrative examples.

1. **Change of solid phase to liquid phase or entropy of fusion** – Consider the melting of one mole of solid substance reversibly at the fusion temperature T_f under a constant pressure. In this process heat is absorbed. The amount of heat absorbed by 1 mole of a substance for melting at constant temperature and pressure is called molar heat of fusion ΔH_f. Thus when one mole of

a substance reversibly melts at its fusion point T_f and the molar heat of fusion is ΔH_f, then the entropy change in transition form solid state to liquid at temperature T_f will be given by

$$\Delta S_f = \frac{\Delta H_f}{T_f} \qquad ...(4.22)$$

Where, the suffix 'f' represents fusion. In this expression (4.22) ΔH_f is positive, therefore the process is fusion involve an increase in entropy.

If the reverse change *i.e.*, the change of state from liquid to solid is considered, then ΔH_f become negative, hence in the process of freezing the decrease in entropy occurs.

2. **Change from liquid phase to vapour phase or entropy of evaporation** – Consider the evaporation of one mole of liquid to its vapour reversibly at its boiling point T_b under a constant pressure. In this process heat is absorbed. The amount of heat absorbed by 1 mole of a liquid at its boiling point T_b at constant pressure is called molar heat of vaporisation ΔH_v. Thus when one mole of a substance reversibly vaporises at its boiling point T_b and the molar heat of vaporisation is ΔH_v, then the entropy change in transition form liquid state to vapour at temperature T_b will be given by

$$\Delta S_v = \frac{\Delta H_v}{T_b} \qquad ...(4.23)$$

Where, the suffix 'v' represents vaporisation. In this expression (4.23) v is positive, therefore the process is vaporisation involve an increase in entropy.

if the reverse change *i.e.*, the change of state from vapour to liquid is considered, then ΔH_v become negative, hence in the process of condensation of vapours to liquidwill be accompanied by the decrease in entropy.

3. **Change from one crystalline form to other crystalline form** – Consider the transition of one mole of crystalline substance from its one crystalline form to other crystalline form, at transition temperature T_t. In this transition there would be either evolution or absorption of heat.the amount of heat evolved or absorbed by one mole of a substance in its transition from one crystalline form to another at a constant transition temperature T_t is called molar heat transition. It is represented by ΔH_t.

Thus, when one mole of a solid substance changes its state from one crystalline state to another at transition temperature T_t the entropy change ΔS_t is given by

$$\Delta S_t = \frac{\Delta H_t}{T_t} \qquad ...(4.24)$$

GIBBS FREE ENERGY

The Gibbs free energy G is the thermodynamic potential. The two measures of energy (enthalpy H and entropic energy TS) are brought together in the Gibbs free energy equation: (the chemical potential is the equivalent for a component) it can write this in a more understandable way

$$G = H - TS \qquad ...(4.25)$$

which says that G is the difference between the heat energy and the entropic energy.

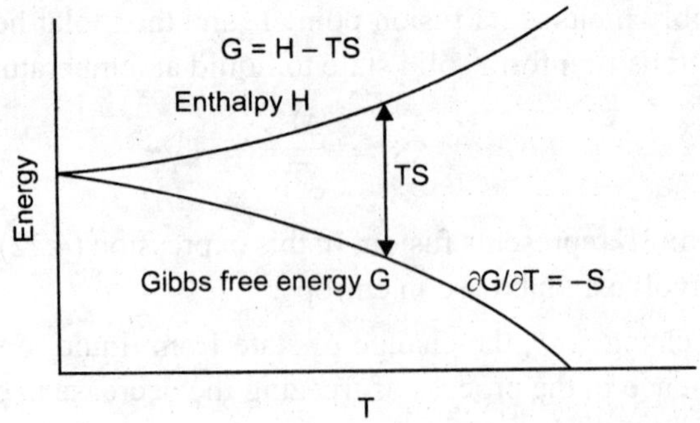

Fig.4.3: Relationship among G, S, and H

Relationship Among G, S, and V

Relation among all can be write as

$$dG = VdP - SdT \quad ...(4.26)$$

It means that changes in Gibbs free energy are produced by changes in pressure and temperature acting on the volume and entropy of a phase.

May be written as

$$dG = \frac{\partial G}{\partial P} dP - \frac{\partial G}{\partial T} dT \quad ...(4.27)$$

which means that

$$\left(\frac{\partial G}{\partial P}\right)_T = V \quad ...(4.28)$$

and

$$\left(\frac{\partial G}{\partial P}\right)_P = -S \quad ...(4.29)$$

These relations indicate that the change in Gibbs free energy with respect to pressure is the molar volume V and the change in Gibbs free energy with respect to temperature is minus the entropy S.

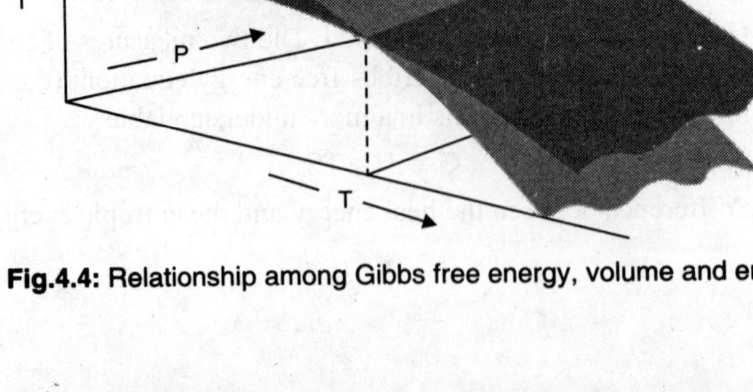

Fig.4.4: Relationship among Gibbs free energy, volume and entropy

Gibbs Free Energy of Formation

Gibbs free energy of formation $\Delta_f G^o$ is

$$\Delta_f G^o = \Delta_f H^o - T\Delta_f S^o \quad ...(4.30)$$

Example– Calculate the Gibbs free energy of formation of anhydrite, can use

$$\Delta_f H^o_{CaSO4} = -1434.11 \text{ kJ/mole}, \; S^o_{CaSO4} = 106.7 \text{ J mol}^{-1} \text{ K}^{-1}$$
$$S^o_{Ca} = 41.42 \text{ J mol}^{-1} \text{ K}^{-1}, \; S^o_S = 31.80 \text{ J mol}^{-1} \text{ K}^{-1}$$
$$S^o_{O2} = 205.138 \text{ J mol}^{-1} \text{ K}^{-1},$$

and we may calculate the entropy of formation of anhydrite

$$\Delta_f S^o = S^o_{CaSO4} - S^o_{Ca} - S^o_S - 2 * S^o_{O2} = -376.796 \text{ J mol}^{-1} \text{ K}^{-1}$$

and then use

$$\Delta_f G^o = \Delta_f H^o - T\Delta_f S^o = -1434,110 - 298.15 * -376.796$$
$$= -1321.77 \text{ kJ/mol}$$

Cell potentials

The cell potential, E_{cell}, is the measure of the potential difference between two half cells in an electrochemical cell. The potential difference is caused by the ability of electrons to flow from one half cell to the other. Electrons are able to move between electrodes because the chemical reaction is a redox reaction. A redox reaction occurs when a certain substance is oxidized, while another is reduced. During oxidation, the substance loses one or more electrons, and thus becomes positively charged. Conversely, during reduction, the substance gains electrons and becomes negatively charged. This relates to the measurement of the cell potential because the difference between the potential for the reducing agent to become oxidized and the oxidizing agent to become reduced will determine the cell potential. The cell potential (E_{cell}) is measured in voltage (V), which allows us to give a certain value to the cell potential.

ELECTROCHEMICAL CELL

An electrochemical cell is comprised of two half cells. In one half cell, the oxidation of a metal electrode occurs, and in the other half cell, the reduction of metal ions in solution occurs. The half cell essentially consists of a metal electrode of a certain metal submerged in an aqueous solution of the same metal ions. The electrode is connected to the other half cell, which contains an electrode of some metal submerged in an aqueous solution of subsequent metal ions. The first half cell, in this case, will be marked as the anode. In this half cell, the metal in atoms in the electrode become oxidized and join the other metal ions in the aqueous solution. An example of this would be a copper electrode, in which the Cu atoms in the electrode loses two electrons and becomes Cu^{2+}.

Components of electrochemical cell: A typical electrochemical cell contains following Components- Two half cells

1. Two metal electrodes
2. One voltmeter
3. One salt bridge
4. Two aqueous solutions for each half cell

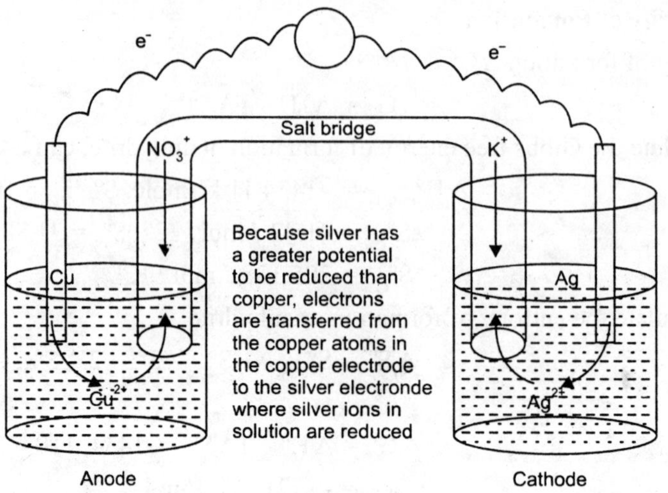

Fig.4.5: Electro - chemical cell

Cell Diagram may be written as

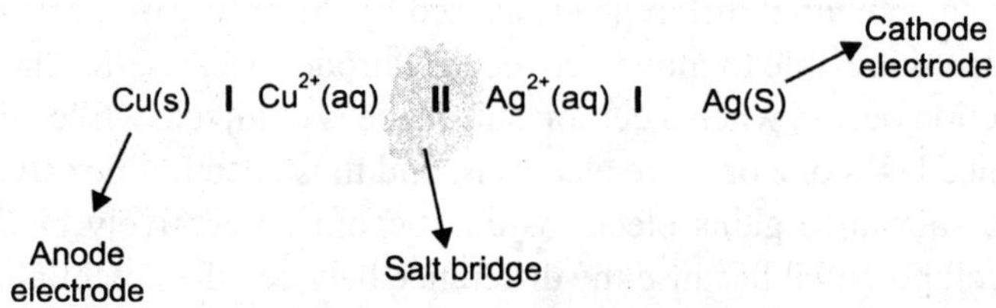

Fig.4.6: Equation on cell

- The image above is called the cell diagram.
- The cell diagram is a representation of the overall reaction in the electrochemical cell.
- In the cell diagram, the anode half cell is always written on the left side of the diagram, and in the cathode half cell is always written on the right side of the diagram.
- Both the anode and cathode are seperated by two vertical lines (ll) as seen in the blue cloud above.
- The electrodes (yellow circles) of both the anode and cathode solutions are seperated by a single vertical line (l).
- When there are more chemicals involved in the aqueous solution, they are added to the diagram by adding a comma and then the chemical.
- For example, in the image above, if copper wasn't being oxidized alone, and another chemical like K was involved, you would denote it as (Cu, K) in the diagram. The cell diagram makes it easier to see what is being oxidized and what is being reduced. These are the reactions that create the cell potential.

Standard Cell Potential

The standard cell potential (E^o_{cell}) is the difference of the two electrodes, which forms the voltage of that cell. To find the difference of the two half cells, the following equation is used:

$$E^o_{Cell} = E^o_{Red, Cathode} - E^o_{Oxd, Anode} \quad ...(4.31)$$

With,

- E^o_{Cell} is the standard cell potential (under 1M, 1 Barr and 298°K).
- E^o_{Red}, Cathode is the standard reduction potential for the reduction half reaction occurring at the cathode
- E^o_{Oxd}, Anode is the standard reduction potential for the oxidation half reaction occurring at the anode

The units of the potentials are measured in volts (V). Note that this equation can also be written as a sum rather than a difference

$$E^o_{Cell} = E^o_{Red, Cathode} + E^o_{Oxd, Anode} \quad ...(4.32)$$

STANDARD CELL POTENTIAL EXAMPLE

Copper and Silver cell diagram.

The oxidation half cell of the redox equation is:

$$Cu(s) \rightarrow Cu^{2+}(aq) + 2e^- \; E^o_{Ox} = -0.340 \text{ V}$$

The reduction half cell is:

$$(Ag^+ + e^- \rightarrow Ag(s)) \text{ x2 } E^o_{Red} = 0.800 \text{ V} \quad ...(4.33)$$

where we have multiplied the reduction chemical equation by two in order to balance the electron count but we have not doubled E^o_{Red} since E° values are given in units of voltage. Voltage is energy per charge, not energy per reaction., The chemical equations can be summed to find:

$$Cu(s) + 2Ag^+ \rightarrow Cu^{2+}(aq) + 2Ag(s)$$

where the potentials of the half-cell reactions can be summed

$$E^o_{Cell} = E^o_{Red, Cathode} + E^o_{Ox, Anode}$$

$$E^o_{Cell} = 0.800 \text{ V} + (-0.340 \text{ V})$$

$$E^o_{Cell} = 0.460\text{V}$$

to find that the standard cell potential of this cell **is 0.460 V.**

Calculating the Cell Potential:

The net reaction of a voltaic cell constructed from a standard zinc electrode and a standard copper electrode is obtained by adding the two half-reactions together:

oxidation	$Zn(s) \longrightarrow Zn^{2+}(aq) + 2e^-$	$E^o = 0.763$ V
reduction	$2\,[1e^- + Ag^+(aq) \longrightarrow Ag\,(s)]$	$E^o = 0.337$ V
		sum
net	$Zn(s) + 2\,Ag^+(aq) \longrightarrow 2\,Ag(s) + Zn^{2+}(aq)$	$E^o = 1.100$ V

NERNST EQUATION

In electrochemistry, the Nernst equation is an equation that relates the reduction potential of an electrochemical reaction (half-cell or full cell reaction) to the standard electrode potential, temperature, and activities (often approximated by concentrations) of the chemical species undergoing reduction and oxidation. It is the most important equation in the field of electrochemistry. It is named after the German physical chemist who first formulated it, Walther Nernst.

A reaction may start at standard-state conditions, but as the reaction proceeds, the concentrations of the solutions change, the driving force behind the reaction becomes weaker, and the cell potential eventually reaches zero.

When the cell potential equals zero, the reaction is at equilibrium.

Nernst Equation – Can be used to find the cell potential at any moment in during a reaction or at conditions other than standard-state.

$$E = E^{o} - \frac{RT}{nF} \ln Q_C \qquad ...(4.34)$$

E^{o} = cell potential (V) under specific conditions

E = cell potential at standard-state conditions

R = ideal gas constant = 8.314 J/mol–K

T = temperature (kelvin), which is generally 25°C (298 K)

n = number of moles of electrons transferred in the balanced equation

F = Faraday's constant, the charge on a mole of electrons = 95,484.56 C/mol

$\ln Q_c$ = the natural log of the reaction quotient at the moment in time

Reaction quotient (Q_c) – The mathematical product of the concentrations of the products of the reaction divided by the mathematical product of the concentrations of the reactants.

Since the temperature is generally 25°C (298 K), three of the terms in the above Nernst equation can be considered constants: R, T, and F. Substituting the values of these constants, results in the following equation:

$$E = E^{o} - \frac{0.02568}{n} \ln Q_c \qquad ...(4.35)$$

For the reaction:

$$Zn + Cu^{+2} \rightarrow Zn^{2+} + Cu$$

There is a transfer of 2 electrons, so n = 2.

$$Q_c = \frac{[Zn^{2+}]}{[Cu^{2+}]} \quad E = E^{o} - \frac{0.02568}{2} \ln\left(\frac{[Zn^{2+}]}{[Cu^{2+}]}\right)$$

At equilibrium E = 0 and $Q_c = K_c$:

$$0 = E^{o} - \frac{RT}{nF} \ln K_c$$

The Nernst equation can be rearranged as follows:

$$nFE^{o} = RT \ln K_c$$

$$\ln K_c = \frac{nFE^{o}}{RT}$$

This equation can be used to calculate the equilibrium constant for any oxidation-reduction reaction from its standard-state cell potential.

Acid Base Theories

Acid – An acid is defined as substance whose water solution has sour in taste, turns blue litmus to red, can neutralise base and evolves hydrogen gas when treated with metals like Zn, Mg, Na. **Example** – HCl, H_2SO_4, HNO_3.

Base – A base is defined as substance whose water solution has bitter in taste, turns blue litmus to red, has soapy touch and can neutralise acids.

Example – NaOH, KOH, NH_4OH.

Limitations

- Above definition based on some observed properties of acids and base.
- Above definition are unable to explain structure, responsible for this properties.so, there is need of some conceptual definitions.

Three principal theories about acid-base are

1. Arrhenius theory
2. Bronsted-Lowry theory
3. Lewis theory

THE ARRHENIUS THEORY OF ACIDS AND BASES

Acid – Acids are substances which produce hydrogen ions (H^+) in solution.

Base – Bases are substances which produce hydroxide ions (OH^-) in solution.

Neutralisation happens because hydrogen ions and hydroxide ions react to produce water.

$$H^+_{(aq)} + OH^-_{(aq)} \longrightarrow H_2O_{(l)}$$

Limitations of the theory

Hydrochloric acid is neutralised by both sodium hydroxide solution and ammonia solution. In both cases, get a colourless solution which can crystallise to get a white salt - either sodium chloride or ammonium chloride.

These are clearly very similar reactions. The full equations are:

$$NaOH_{(aq)} + HCl_{(aq)} \longrightarrow NaCl_{(aq)} + H_2O_{(l)}$$

$$NH_{3(aq)} + HCl_{(aq)} \longrightarrow NH_4Cl_{(aq)}$$

In the sodium hydroxide case, hydrogen ions from the acid are reacting with hydroxide ions from the sodium hydroxide - in line with the Arrhenius theory.

However, in the ammonia case, there don't appear to be any hydroxide ions.

So you can just say about ammonia is not a base as per definition of Arrhenius, it does produces hydroxide ions in solution. But most of the reaction is going to be a direct reaction between ammonia molecules and hydrogen ions - which doesn't fit the Arrhenius definition.

THE BRONSTED-LOWRY THEORY OF ACIDS AND BASES

In 1923, Bronsted and Lowry proposed another theory of acid and base which is applicable for wider range as compared to Arrhenius concept. It is applicable for aqueous and non-aqueous solutions both.

Acid – An acid is a proton (hydrogen ion) donor.

Base – A base is a proton (hydrogen ion) acceptor.

The relationship between the Bronsted-Lowry theory and the Arrhenius theory

The Bronsted-Lowry theory doesn't go against the Arrhenius theory in any way - it just adds to it. Hydroxide ions are still bases because they accept hydrogen ions from acids and form water. An acid produces hydrogen ions in solution because it reacts with the water molecules by giving a proton to them.

When hydrogen chloride gas dissolves in water to produce hydrochloric acid, the hydrogen chloride molecule gives a proton (a hydrogen ion) to a water molecule. A co-ordinate (dative covalent) bond is formed between one of the lone pairs on the oxygen and the hydrogen from the HCl. Hydroxonium ions, H_3O^+, are produced.

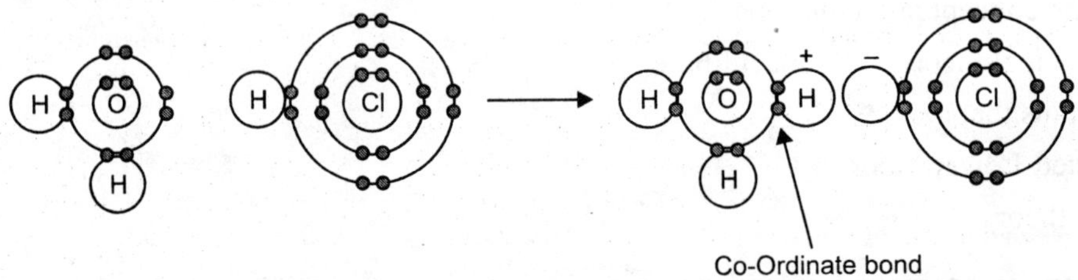

Fig.4.7: Formation of Hydroxonium ions, H_3O^+, through co-ordinate bond

$$H_2O + HCl \longrightarrow H_3O^+ + Cl^-$$

When an acid in solution reacts with a base, what is actually functioning as the acid is the hydroxonium ion. For example, a proton is transferred from a hydroxonium ion to a hydroxide ion to make water.

$$H_3O^+_{(aq)} + OH^-_{(aq)} \longrightarrow 2H_2O_{(l)}$$

Showing the electrons, but leaving out the inner ones:

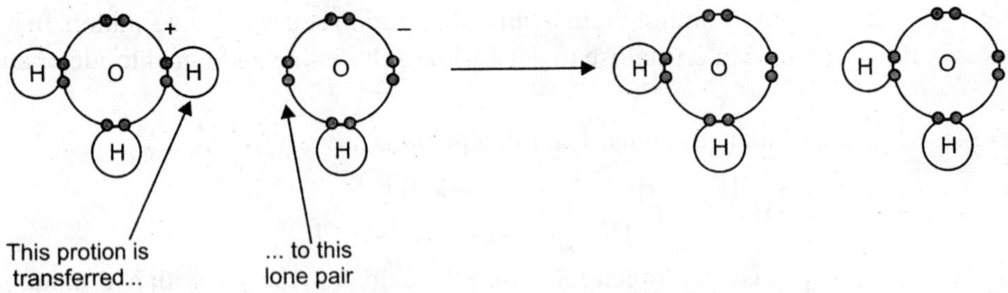

Fig.4.8: Electron transfer

It is important to realise that whenever you talk about hydrogen ions in solution, $H^+_{(aq)}$, what you are actually talking about are hydroxonium ions.

Conjugate pairs

When hydrogen chloride dissolves in water, almost 100% of it reacts with the water to produce hydroxonium ions and chloride ions. Hydrogen chloride is a strong acid, and we tend to write this as a one-way reaction:

$$H_2O + HCl \longrightarrow H_3O^+ + Cl^-$$

In fact, the reaction between HCl and water is reversible, but only to a very minor extent. In order to generalise, consider an acid HA, and think of the reaction as being reversible.

$$HA + H_2O \rightleftharpoons H_3O^+ + A^-$$

Thinking about the forward reaction:

- The HA is an acid because it is donating a proton (hydrogen ion) to the water.
- The water is a base because it is accepting a proton from the HA.

But there is also a back reaction between the hydroxonium ion and the A^- ion:

- The H_3O^+ is an acid because it is donating a proton (hydrogen ion) to the A^- ion.
- The A^- ion is a base because it is accepting a proton from the H_3O^+.

The reversible reaction contains two acids and two bases. We think of them in pairs, called conjugate pairs.

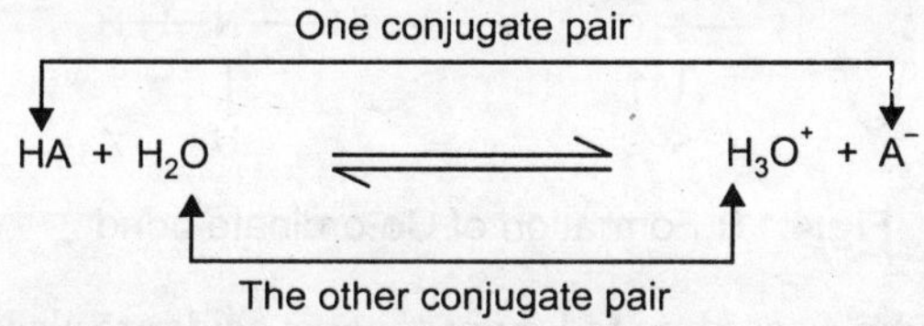

Fig.4.9: Conjugate acid and Conjugate base pairs in water molecule

THE LEWIS THEORY OF ACIDS AND BASES

In 1923 Lewis suggested a theory based on transaction of electrons.

The theory

- An acid is an electron pair acceptor.
- A base is an electron pair donor.

The relationship between the Lewis theory and the Bronsted-Lowry theory

Lewis bases

Ammonia reacts with BF_3 by using its lone pair to form a co-ordinate bond with the empty orbital on the boron.

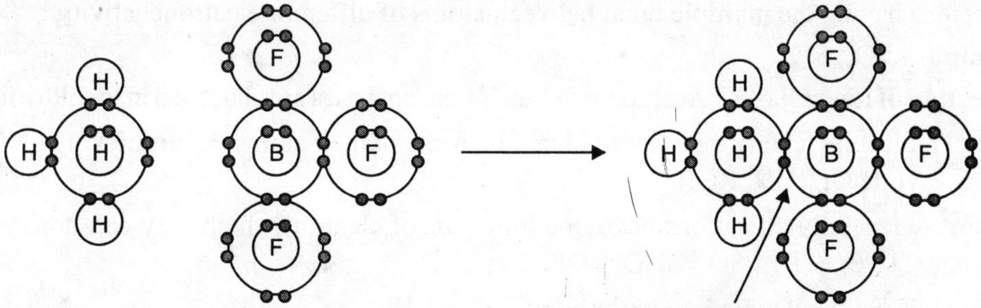

Fig.4.10: Bond formation between NH_3 (Base) and BF_3 (acid) formed salt $NH_3 - BF_3$

As far as the ammonia is concerned, it is behaving exactly the same as when it reacts with a hydrogen ion - it is using its lone pair to form a co-ordinate bond. If you are going to describe it as a base in one case, it makes sense to describe it as one in the other case as well.

Lewis acids

Lewis acids are electron pair acceptors. In the above example, the BF_3 is acting as the Lewis acid by accepting the nitrogen's lone pair.

The lone pair on the nitrogen of an ammonia molecule is attracted to the slightly positive hydrogen atom in the HCl. As it approaches it, the electrons in the hydrogen-chlorine bond are repelled still further towards the chlorine.

Eventually, a co-ordinate bond is formed between the nitrogen and the hydrogen, and the chlorine breaks away as a chloride ion.This is best shown using the "curly arrow" notation commonly used in organic reaction mechanisms.

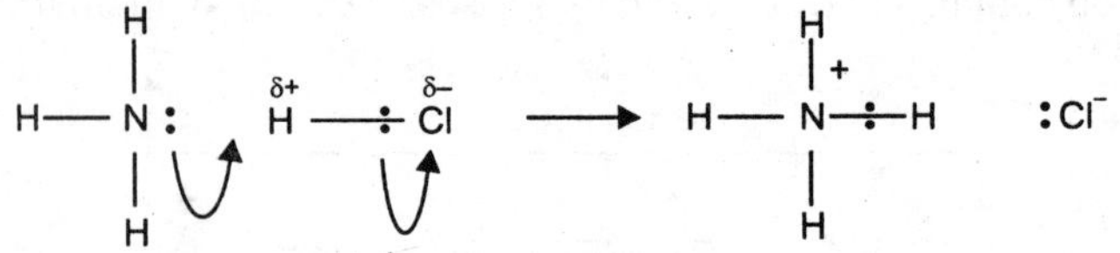

Fig.4.11: Formation of Co-ordinate bond

Classification of Lewis acids: According to Lewis concept acids are classified in to following manner

1. **Simple cations** – Strength of cations increases with–
   - Increase in charge of ion.
   - Decrease in ionic radius
   - Decrease in shielding electrons

   The order of Lewis acid strength of some cations μ

   $$Na^+ < Mg^{++} < Al^{+++}$$

   $$Cs^+ < K^+ < Na^+ < Li^+$$

2. Molecule having incomplete octet in their central atom- Halides of Be, B, and aluminium.
   Example – BF_3, $AlCl_3$, $BeCl_2$ etc.
3. Molecules having vacant 'd' orbital in their central atom
   Example – SiX_4, $SnCl_3$, GeX_4, $TiCl_4$, PX_3, PF_5, SF_4 etc.
4. Molecules having the multiple bond between atoms of different electronegativity
   Example – SO_2, CO_2.

Classification of Lewis bases : According to Lewis concept bases are classified in to following manner

1. Neutral molecules having lone pair of electron on any of their atoms–**Examples** are –NH_3, RNH_2, R_2NH, R_3N, H_2O, ROH, ROR.
2. Anions – Generally all cations have one lone pair of electron which they can donate to Lewis acid. e.g. Cl^-, Br^-, NH_2^-, SO_3^{2-}, NO_2^-, OR^- etc.
3. A molecule having double bond or π bond $CH_2 = CH_2$.
4. Halides – Which can give halogen ions, acts as Lewis base, e.g. XeF_2, CsF, XeF_4, XeF_6, NoF etc.

OXIDATION AND REDUCTION

During oxidation process oxidation number increases than participating species and during reduction oxidation number decreases than original number of participating species.

Oxidation half-reaction: This reaction include the oxidation of Zn to Zn^{++} ion by loss of two electrons. Oxidation number of Zn is 0 (zero) after oxidation it becomes +2.

$$Zn \rightarrow Zn^{++} + 2e^-$$

Reduction half-reaction: This reaction include the reduction of Cu^{++} ion to Cu by gain of two electrons. Oxidation number of Cu^{++} is +2 after reduction it becomes 0 (zero).

$$Cu^{++} + 2e^- \rightarrow Cu$$

After cancelling two electrons from each side complete reaction may be written as

$$Zn + Cu^{++} \rightarrow Zn^{++} + Cu$$

Oxidation and reduction process may be defined in several ways. Some of them are mentioned below-

1. Oxidation and reduction in terms of oxygen transfer

Definitions:

- Oxidation is gain of oxygen.
- Reduction is loss of oxygen.

For example, in the extraction of iron from its ore:

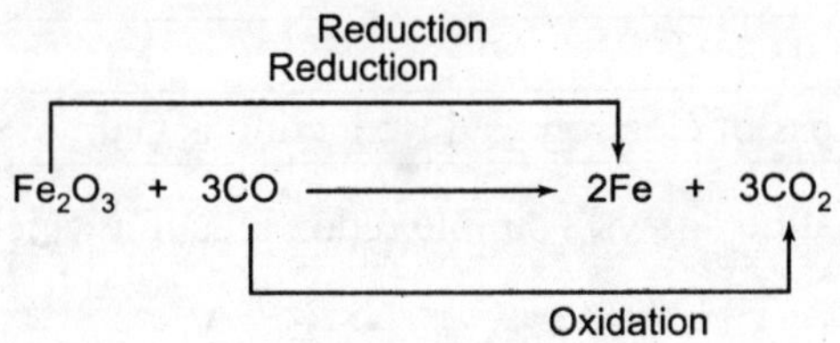

Fig.4.12: Redox reaction between Fe_2O_3 and carbonmono-oxide

Because both reduction and oxidation are going on side-by-side, this is known as a redox reaction.

OXIDISING AND REDUCING AGENTS

An oxidising agent is substance which oxidises something else. In the above example, the iron (III) oxide is the oxidising agent. **A reducing agent** reduces something else. In the equation, the carbon monoxide is the reducing agent. Oxidising agents give oxygen to another substance.agents remove oxygen from another substance.

Oxidation and reduction in terms of hydrogen transfer

These are old definitions which aren't used very much nowadays. The most likely place you will come across them is in organic chemistry.

Definitions

- Oxidation is loss of hydrogen.
- Reduction is gain of hydrogen.

Notice that these are exactly the opposite of the oxygen definitions.

For example, ethanol can be oxidised to ethanal:

$$CH_3CH_2OH \longrightarrow CH_3CHO$$

Oxidation by loss of hydrogen

You would need to use an oxidising agent to remove the hydrogen from the ethanol. A commonly used oxidising agent is potassium dichromate (VI) solution acidified with dilute sulphuric acid.Ethanal can also be reduced back to ethanol again by adding hydrogen to it. A possible reducing agent is sodium tetrahydridoborate, $NaBH_4$.

Another update on oxidising and reducing agents – Oxidising agents give oxygen to another substance or remove hydrogen from it.Reducing agents remove oxygen from another substance or give hydrogen to it.

Oxidation and reduction in terms of electron transfer

This is easily the most important use of the terms oxidation and reduction at A' level.

Definitions

- Oxidation is loss of electrons.
- Reduction is gain of electrons.

It is essential that you remember these definitions. There is a very easy way to do this. As long as you remember that you are talking about electron transfer

OIL	RIG
Oxidation is Loss of electron	Reduction is Gain of electron

A simple example - The equation shows a simple redox reaction which can obviously be described in terms of oxygen transfer.

$$CuO + Mg \rightarrow Cu + MgO$$

Copper (II) oxide and magnesium oxide are both ionic. The metals obviously aren't. If you rewrite this as an ionic equation, it turns out that the oxide ions are spectator ions and you are left with:

$$Cu^{2+} + Mg \rightarrow Mg^{2+} + Cu$$

If you look at the equation above, the magnesium is reducing the copper (II) ions by giving them electrons to neutralise the charge. Magnesium is a reducing agent. The copper (II) ions are removing electrons from the magnesium to create the magnesium ions. The copper (II) ions are acting as an oxidising agent.

We can summarise –

- An oxidising agent oxidises something else.
- Oxidation is loss of electrons (OIL).
- That means that an oxidising agent takes electrons from that other substance.
- So an oxidising agent must gain electrons.

WATER CHEMISTRY

Chemical formula of water is H_2O. A water molecule consists of one atom of oxygen bound to two atoms of hydrogen. The hydrogen atoms are "attached" to one side of the oxygen atom, resulting in a water molecule having a positive charge on the side where the hydrogen atoms are and a negative charge on the other side, where the oxygen atom is. Since opposite electrical charges attract, water molecules tend to attract each other, making water kind of "sticky." The side with the hydrogen atoms (positive charge) attracts the oxygen side (negative charge) of a different water molecule.

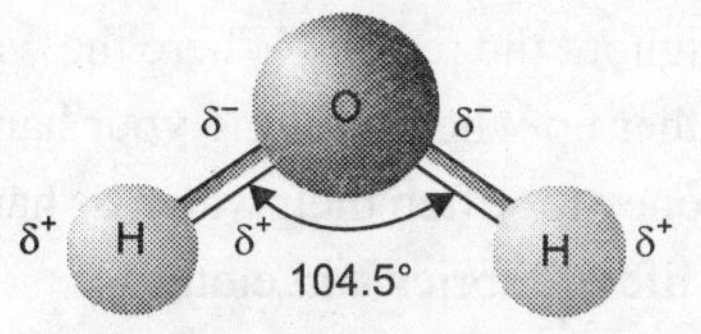

Fig.4.13: Structure of Water

Properties of water

1. Water is called the "universal solvent" because it dissolves more substances than any other liquid. This means that wherever water goes, either through the ground or through our bodies, it takes along valuable chemicals, minerals, and nutrients.
2. Pure water has a neutral pH. Pure water has a pH, of about 7, which is neither acidic nor basic.
3. Water dissociates into H^+ and OH^-; the dissociation constant of water is:

   $K_{water} = [H^+][OH^-] = 10^{-14}$.
4. It can is found in all three states i.e. liquid, solid (ice), and gas (steam). Water freezes at 32° Fahrenheit (F), 0°C on Celsius and boils at 212° (F), 100°C. In fact, water's freezing and boiling points are the baseline with which temperature is measured.
5. Water has a high specific heat index. This means that water can absorb a lot of heat before it begins to get hot. This is why water is valuable to industries and in your car's radiator as a coolant.

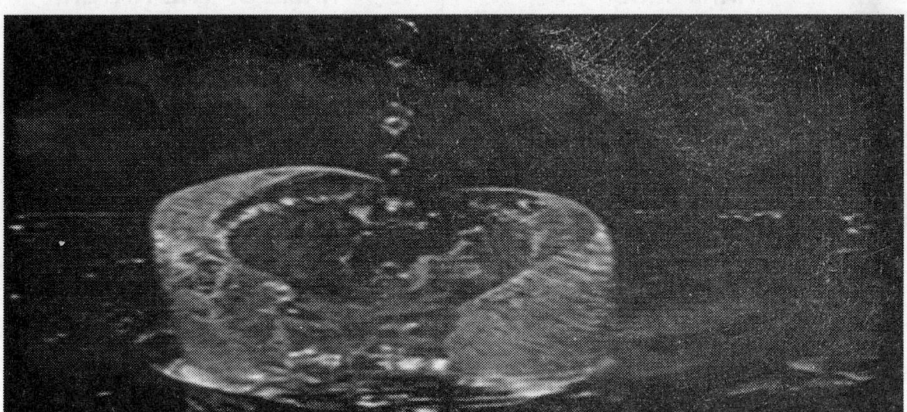

6. Water has a very high surface tension. Means, it is sticky and elastic, and tends to clump together in drops rather than spread out in a thin film. Surface tension is responsible for capillary action, which allows water (and its dissolved substances) to move through the roots of plants and through the tiny blood vessels in our bodies.

7. **Dissolved Oxygen:** Although water molecules contain an oxygen atom, this oxygen is not what is needed by aquatic organisms living in our natural waters. A small amount of oxygen, up to about ten molecules of oxygen per million of water, is actually dissolved in water. This dissolved oxygen is breathed by fish and zooplankton and is needed by them to survive.
8. **Hardness:** The amount of dissolved calcium and magnesium in water determines its "hardness." Water hardness varies throughout the region. Where the water is relatively hard, you may notice that it is difficult to get a lather up when washing your hands or clothes. And, industries in your area might have to spend money to soften their water, as hard water can damage equipment. Hard water can even shorten the life of fabrics and clothes.

Hardness of water – Water which does not produces lather with soap are termed as hard water. On the other hand soft water readily produce a lot of lather when mixed even with little amount of soap. The hardness is mostly due to the presence of Ca and Mg metallic ions which react with soap and form precipitates. Some major cations are Mg^{++}, Ca^{++} Al^{+++} and Fe^{+++}. Some important anions are HCO^{-}_{3}, NO^{-}_{3}, SO_4^{-} etc.

The hard water causes boiler troubles like scale formation etc. Example : A reaction of soap (Sodium stearate) with calcium chloride and Magnesium sulphate is as follows–

$$\underset{\text{Sodium stearate Hardness}}{2\,C_{17}H_{35}COONa + CaCl_2} \rightarrow \underset{\text{Calcium stearate (Scum)}}{(C_{17}H_{35}COO)_2Ca + 2\,NaCl}$$

$$\underset{\text{Sodium stearate Hardness}}{2\,C_{17}H_{35}COONa + MgSO_4} \rightarrow \underset{\text{Magnesium stearate (Scum)}}{(C_{17}H_{35}COO)_2Mg + Na_2SO_4}$$

The water are commercially classified on the basis of degree of hardness as follows–

HARDNESS	NAME OF THE WATER
0-70 mg/liter	Soft water
70-150 mg/liter	Moderate hard water
150-300 mg/liter	Hard water
300 mg/liter and above	Very hard water

Types of Hardness

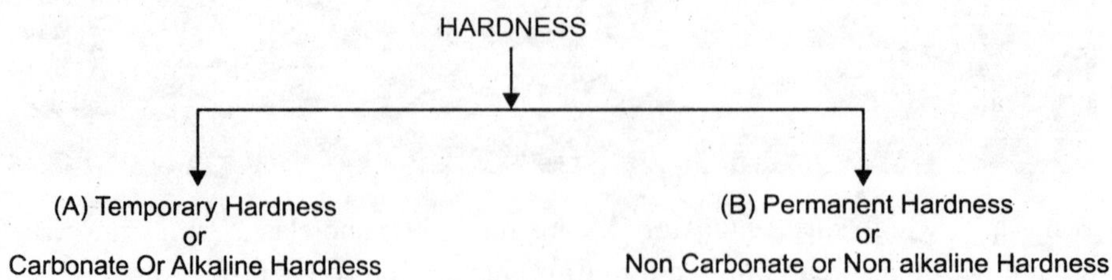

(A) Temporary Hardness	(B) Permanent Hardness
(*i*) It is caused by the presence of dissolved bicarbonates of calcium, magnesium and other heavy metals.	(*i*) It is caused by the presence of soluble chlorides and sulphates of calcium, Magnesium, iron and other heavy metals.
(*ii*) Salts which are responsible for temporary hardness are – $Ca(HCO_3)_2$ and $Mg(HCO_3)_2$.	(*ii*) Salts which are responsible for permanent hardness are – $CaCl_2$, $MgCl_2$, $CaSO_4$, $FeSO_4$, $Al_2(SO_4)_3$.
(*iii*) It can be removing by simple boiling of water.	(*iii*) It can't be removed by simply boiling of water.
(*iv*) Bicarbonates are decomposed and are converted into insoluble carbonates or hydroxides, which precipitate and settle down at the bottom of the vessel.	(*iv*) It is also known as non - carbonate or non - alkaline hardness.

Hardness of water is expressed as equivalent of calcium carbonate ($CaCO_3$).

The reason for choosing $CaCO_3$ as standard is – molecular weight of $CaCO_3$ is exactly 100, so, ease in the calculations occurs. Secondly it forms most insoluble salt that can be precipitated in water treatment.

Degree of hardness

$$= \frac{\text{Mass of hardness producing substance}}{\text{Chemical equivalent of hardness producing substance}} \times 50$$

Temporary hardness can be removed by boiling but Permanent hardness cannot be removed by boiling. It requires softening methods. Theses method are–

1. Lime –Soda method
2. Zeolite Method
3. Ion-Exchange method

CORROSION

Any process of deterioration (or destruction)and consequent loss of a solid metal through an unwanted chemical or electrochemical attack of its environment , starting at its surface is called corrosion.

"Corrosion is a reverse process of extraction"

Most of the metals except Nobel metals (Pt, Au) are exist in combined forms as their oxide, carbonates, sulphide,chlorides etc. In this state, in the form of ore metal have very less energy so this state is a thermodynamically stable state. During metallurgical operation ore is converted into metal. In each step of metallurgy ore become superior and thermodynamically unstable by gaining energy. Formed metal gained more energy than ore and become unstable. To become stable again metal undergo corrosion.

ORE		Pure Metal
Have less energy,	Metallurgical Process →	Have more energy
Thermodynamically stable	← Corrosion Process	Thermodynamically unstable
Not useful in this state		Useful in this state

Examples of corrosion

- Formation of ferric hydroxide on the surface of Iron metal Formula Fe_3O_4
- Formation of green layer of basic copper carbonate on the surface of Copper metal $cuco_3 . H_2O$
- Formation of MoO_3 on surface of Molybdenum

Fig.4.14: (a) Rusting of Iron (Black and yellow rust) (b) Rusting on Copper (green Layer)

Corrosion can be prevented by using below mentioned techniques–

- The selection of appropriate materials.
- The design of the product.
- Protective methods (Coatings, anodes, etc.)
- Correct installation and maintenance.
- Research, development and testing.

Types of corrosion

Corrosion may be divided into two types

1. **Chemical or Dry Corrosion :** Chemical corrosion can be seen as oxidation and occurs by the action of dry gases, often at high temperatures. Example – Oxidation Corrosion.
2. **Electrochemical or Wet corrosion :** Electrochemical corrosion on the other hand takes place by electrode reactions, often in humid environments, i.e. wet corrosion. Wet corrosion covers large area of corrosion.Example Galvanic corrosion, Strees corrosion, Waterline corrosion, Concentration cell corrosion.

PARAMETERS AFFECTING THE CORROSION RATE

Some of the most important parameters affecting the corrosion rate of metals are outlined below.

1. **Oxidizing agents:** The corrosion process is conditional on an anodic reaction and a cathodic reaction taking place simultaneously. The anodic reaction causes the metal to dissolve. An oxidizing agent must be present for the cathodic reaction, and the most common agents are dissolved oxygen or hydrogen ions. If the availability of oxidizing agents is restricted, the corrosion process will be inhibited or will cease entirely.
2. **The electric conductivity of the electrolyte:** Corrosion involves electrochemical reactions, and an increase in the electrical conductivity of the electrolyte will therefore increase the corrosion rate. In sea water the chloride content causes rapidly increased conductivity.
3. **Temperature:** An increase in temperature will generally cause an increase in the corrosion rate. A rule of thumb is that temperature increases of 10ºC will double the corrosion rate.
4. **Concentration:** An increased concentration will normally increase the corrosion rate up to a maximum level. Higher concentration above this will not give higher corrosion rate. e.g. a chloride concentration above approximately 1500 ppm will not increase the corrosion rate.

ELLINGHAM DIAGRAM – THERMODYNAMICS OF METALLURGY

Gibbs equation helps us to predict the spontaneity of a reaction on the basis of enthalpy and entropy values directly. H.G.T Ellingham proposed the Ellingham diagram to predict the spontaneity of reduction of various metal oxides. Ellingham diagram was basically a curve which related the Gibbs energy value with the temperature.

Construction of an Ellingham Diagram

- An Ellingham diagram is a plot of ΔG versus temperature.
- Since ΔH and ΔS are essentially constant with temperature unless a phase change occurs, the free energy versus temperature plot can be drawn as a series of straight lines, where ΔS is the slope and ΔH is the y-intercept.
- The slope of the line changes when any of the materials involved melt or vaporize.
- Free energy of formation is negative for most metal oxides, and so the diagram is drawn with $\Delta G = 0$ at the top of the diagram, and the values of ΔG shown are all negative numbers.
- Temperatures where either the metal or oxide melt or vaporize are marked on the diagram. The Ellingham diagram shown is for metals reacting to form oxides (similar diagrams can also be drawn for metals reacting with sulfur, chlorine, etc., but the oxide form of the diagram is most common).
- The oxygen partial pressure is taken as 1 atmosphere, and all of the reactions are normalized to consume one mole of O_2.

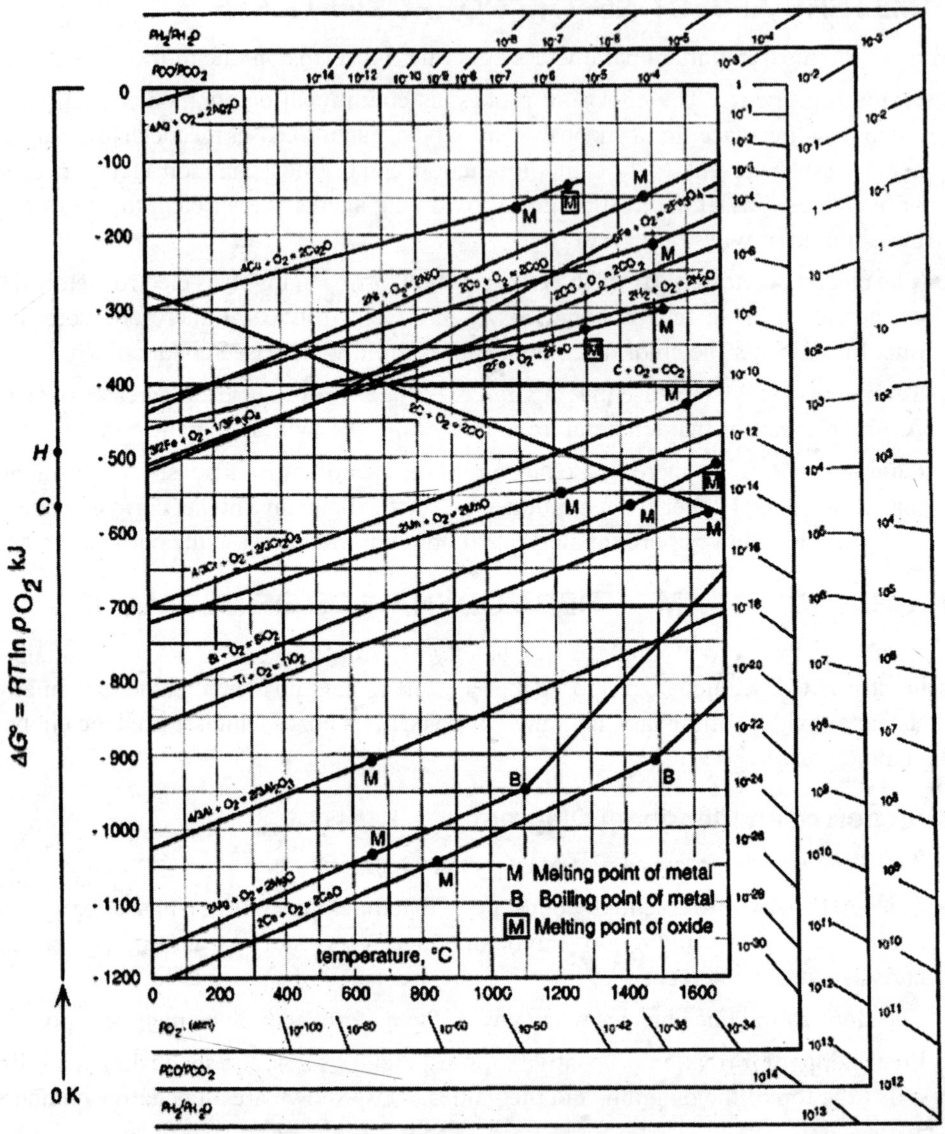

Fig.4.15: Ellingham Diagram

- The majority of the lines slope upwards, because both the metal and the oxide are present as condensed phases (solid or liquid). The reactions are therefore reacting a gas with a condensed phase to make another condensed phase, which reduces the entropy. A notable exception to this is the oxidation of solid carbon. The line for the reaction

$$C + O_2 \rightarrow CO_2$$

- is a solid reacting with a mole of gas to produce a mole of gas, and so there is little change in entropy and the line is nearly horizontal.For the reaction

$$2C + O_2 \rightarrow 2CO$$

we have a solid reacting with a gas to produce two moles of gas, and so there is a substantial increase in entropy and the line slopes rather sharply downward. Similar behavior can be seen in parts of the lines for lead and lithium, both of which have oxides that boil at slightly lower temperatures than the metal does.

There are three main uses of the Ellingham diagram

1. Determine the relative ease of reducing a given metallic oxide to metal;
2. Determine the partial pressure of oxygen that is in equilibrium with a metal oxide at a given temperature; and
3. Determine the ratio of carbon monoxide to carbon dioxide that will be able to reduce the oxide to metal at a given temperature.

Ease of Reduction

The position of the line for a given reaction on the Ellingham diagram shows the stability of the oxide as a function of temperature. Reactions closer to the top of the diagram are the most "noble" metals (for example, gold and platinum), and their oxides are unstable and easily reduced. As we move down toward the bottom of the diagram, the metals become progressively more reactive and their oxides become harder to reduce.

A given metal can reduce the oxides of all other metals whose lines lie above theirs on the diagram. For example, the $2Mg + O_2 \Rightarrow 2MgO$ line lies below the $Ti + O_2 \Rightarrow TiO_2$ line, and so magnesium can reduce titanium oxide to metallic titanium.

Since the $2C + O_2 \Rightarrow 2CO$ line is downward-sloping, it cuts across the lines for many of the other metals. This makes carbon unusually useful as a reducing agent, because as soon as the carbon oxidation line goes below a metal oxidation line, the carbon can then reduce the metal oxide to metal. So, for example, solid carbon can reduce chromium oxide once the temperature exceeds approximately 1225ºC, and can even reduce highly-stable compounds like silicon dioxide and titanium dioxide at temperatures above about 1620ºC and 1650ºC, respectively. For less stable oxides, carbon monoxide is often an adequate reducing agent.

Equilibrium Partial Pressure of Oxygen

The scale on the right side of the diagram labelled "PO_2" is used to determine what partial pressure of oxygen will be in equilibrium with the metal and metal oxide at a given temperature. The significance of this is that, if the oxygen partial pressure is higher than the equilibrium value, the metal will be oxidized, and if it is lower than the equilibrium value then the oxide will be reduced.

To use this scale, you will need a straightedge. First, find the temperature you are interested in, and find the point where the oxidation line of interest crosses that temperature. Then, line up the straightedge with both that point, and with the point labelled "0" that is marked with short radiating lines (upper left corner of the diagram). Now, with the straightedge running through these two points, read off the oxygen partial pressure (in atmospheres) where the straightedge crosses the "PO_2" scale, and this is the equilibrium partial pressure.

It is possible to reach the equilibrium oxygen partial pressure by use of a hard vacuum, purging with an inert gas to displace the oxygen, or using a scavenger chemical to consume the oxygen.

Ratio of CO/CO_2 Needed for Reduction

When using carbon as a reducing agent, there will be a minimum ratio of CO to CO_2 that will be able to reduce a given oxide. The harder the oxide is to reduce, the greater the proportion of CO needed in the gases.

To determine the CO/CO_2 ratio to reduce a metal oxide at a particular temperature, use the same procedure as for determining the equilibrium pressure of oxygen, except line up the straightedge with the point marked "C" (centre of the left side of the diagram), and read the ratio off of the scale marked "CO/CO_2".

LONG ANSWER TYPE QUESTIONS

1. Discuss various acid-base concepts with examples.

   Or

   Write an essay on acid-base concept.
2. Describe in detail the theory of bronsted and Lowry.
3. What are conjugate acids and bases? How are the relative Strengths of Bronsted acids. Bases related with conjugate pair? Explain with examples.
4. What do You understand by Lewis acid and base? Describe with examples. Give classifyAnd the limitations.
5. Discuss Bronsted Lowry theory. In what manner is it different from Lewis Concept.
6. Explain with suitable examples Lux Flood acid base Concept.
7. Solvent system concept is wide form of Arrhenius Concept. Explain with examples.
8. Choose the conjugate acid base pair in the following table. (an ion or molecule may be more than one time):
9. Write short notes on the following.
   (a) Acid base theory on the basis of concept of solvent system.
   (b) Arrhenius hypothesis.
   (c) Conjugated acid base pair
   (d) Lux-Flud concept.
10. Write the difference between acids base concepts given by Bronsted Lowry and Lewis. The utility and limitation of both concepts.
11. Discuss Bronsted Lowry acid base theory. What are its shortcomings and utility?
12. Explain Lewis acid base concept.
13. Classify acids and bases describing lewis acid base theory.
14. Clarify following :
    (i) Bronsted Lowry concept
    (ii) Relative Strength of hydra acids.

15. How are acids and bases defined on the basis of Lweis and Bronsted Lowry concept Explain with examples.
16. Choose Lewis hard acid and hard base :
 Al^{3+}, Cu^+, Ag^+, Au^+, Hg^{2+}, Ca^{2+}, Sr^{2+}, Hg^+, Ti^+, H_3COO^-, NO_3^-, H^-, I^-, R^-, SCN, CN^-, $Cl0_4^-$
17. Write an essay on relative strength of acids and base.
18. Write notes on the following :
 (i) Bronsted Lowry Concept
 (ii) Lewis concept.
19. On the basis of Bronsted and Lowry Concept of acids and bases explain following
 (a) Definition of acid and base with on example each
 (b) Classification of acid and bases
 (c) Classification of solvents.
20. Describe Lewis theory of acids and bases. Explain with reason which of the following are acids and bases: H_2O, CO, CO_2, BF_3, C_2H_5N
21. CO_2 and CU^{2+} are either Lewis acid or Lewis base? Give the reason.
22. What is conjugated acid and base? Explain
23. Explain the following:
 (i) Conjugate bases of strong acids are weak
 (ii) Mg (II) is a weak base
 (iii) $[COF_6]^{3-}$ is more stable than $[Col_6]^{3-}$
 (iv) Arrange in increasing order of acid strength :
 $ClCH_2COOH$, C_6H_5COOH, HCl, CH_3COOH

SHORT ANSWER TYPE

1. Define Arrhenius acids.
2. According to Arrhenius what are bases?
3. Give two main limitations of Arrhenius concept.
4. How are acids classified on the basis of Bronsted and Lowry Concept?
5. What do you understand by Cojugate pair? Give two examples of such pairs.
6. Explain solvent system theory.
7. How is Bronsted Lowry concept more useful than Arrhenius concept?
8. Describe solvent system concept of acids and bases.
9. What is Lewis acid and base?
10. What is the acidic order of
 (i) HF, HCl, HBr, HI (II) CH_4, NH_3, H_2O, HF.
11. What are hard and soft acid – base theory? Describe its utility.

12. The conjugate base of [al $(H_2O)_6]^{3+}$ will be ....
13. Acid is proton donor and base is proton acceptor, This concept is of ....
14. Choose Lewis acid and base from the following giving reasons –$AlCl_3$, Cu^{2+}, Ag+, H_2O, $(CH_3)_3N$, blanks:
15. Water acts both as an acid and base.
16. Name conjugated acid and base :

    (i) $NH_3 + H_2O \rightleftharpoons NH_4^+ + OH-$ (II) $HCl + H_2O \rightleftharpoons H_2O + Cl^-$

OBJECTIVE TYPE QUESTIONS

1. [Agl2]- is stable as :

   (i) Ag is hard acid while I– is soft base (ii) Ag+ is soft acid and I– is soft base

   (iii) Ag+ is hard acid and I– is hard base (iv) Ag + is soft acid and I– is hard base

2. On the basis of Bronsted Lowry acid base concept, bases are the substance which:

   (i) Donate OH^- (ii) Accept OH^-

   (iii) Have a tendency to donate proton (iv) Have a tendency to accept proton

3. On the basis of Bronsted Lowry concept [Al $(H_2o)_6]^{3+}$ is :

   (i) Cationic bae (ii) Molecular base

   (iii) Anionic base (iv) Molecular acid.

4. Pyridine behaves as:

   (i) Molecular base (ii) Cationic base

   (iii) Cationic acid (iv) Anionic base.

5. [Al $(H_2O)_5OH]^{2+}$ behaves as :

   (i) Cationic bae (ii) Molecular base

   (iii) Anionic base (iv) Molecular acid.

6. According to Cady and Elsey acid is a substance:

   (i) Which increases the concentration of cation of the solvent.

   (ii) Which decreases the concentration of cation of the solvent.

   (iii) Which does not change the concentration of caption.

   (iv) None of these.

7. Weakest base is:

   (i) Cl^- (ii) Br^-

   (iii) I^- (iv) F^-

8. In the reaction H+ =: $NH_3 \rightarrow [H \leftarrow NH_3]+$, H+ is

   (i) Lewis base (ii) Lewis acid

   (iii) Electron pair donor (iv) None of these.

9. Which of the following is not a Lewis acid?
   (i) Ag^+ (ii) Li^+ (iii) BF_3 (iv) N_2H_4
10. Which of the following is Lewis acid?
    (i) $SiCl_4$ (ii) Pyridine (iii) Hydrazine (iv) SCN^-
11. Strongest Bronsted base is:
    (i) ClO^- (ii) ClO_2^- (iii) ClO_3^- (iv) ClO_4^-
12. The Conjugate acid of NH_2 is:
    (i) NH_4^+ (ii) NH_3 (iii) N_2H_4 (iv) NH_2OH
13. Conjugated base of H_2SO_4
    (i) $H_2SO_4^-$ (ii) SO_3^{--} (iii) SO_4^{--} (iv) None of the above
14. Which one of the following acts both as an acid and a base?
    (i) SO_3^{2-} (ii) HCl (iii) HCO_3^- (iv) C_6H_6
15. The correct increasing order of acidic strength of acids $HClO_4$, $HClO_3$, $HClO_2$ and HClO is :
    (i) $HClO_4 < HClO_2 < HClO_3 < HClO_4$ (ii) $HClO_4 < HClO_3 < HClO_2 < HClO$
    (iii) $HClO_4 < HClO_2 < HClO < HClO_3$ (iv) None of these.

5 Periodic Properties

Effective Nuclear Charge, Penetration of orbitals, Variations of s, p, d and f orbital Energies of atoms in the Periodic table, Electronic Configurations, Atomic and Ionic sizes, Ionization Energies, Electron affinity and Electronegativity, Polarizability, Oxidation states, Coordination numbers and Geometries, Hard Soft Acids and Bases, Molecular Geometries.

EFFECTIVE NUCLEAR CHARGES

If an atom or an ion has only a single electron, its potential energy can be calculate by considering only the electrostatic attraction between the positively charged nucleus and the negatively charged electron. But, for more than one electron, the total energy of the atom or the ion depends not only on attractive electron-nucleus interactions but also on repulsive electron-electron interactions. It is impossible to exactly calculate the repulsive interactions. Consequently, we must use approximate methods to deal with the effect of electron-electron repulsions on orbital energies.

Definition – The effective nuclear charge is the net charge an electron experiences in an atom with multiple electrons.

Effective nuclear charge refers to the charge felt by the outermost (valence) electrons of a multi-electron atom after taking into account the number of shielding electrons that surround the nucleus.

The formula for calculating the effective nuclear charge for a single electron is $\mathbf{Z_{eff} = Z - S}$

Where, Z_{eff} is the effective nuclear charge, Z is the number of protons in the nucleus, and

S is the average amount of electron density between the nucleus and the electron for which you are solving.

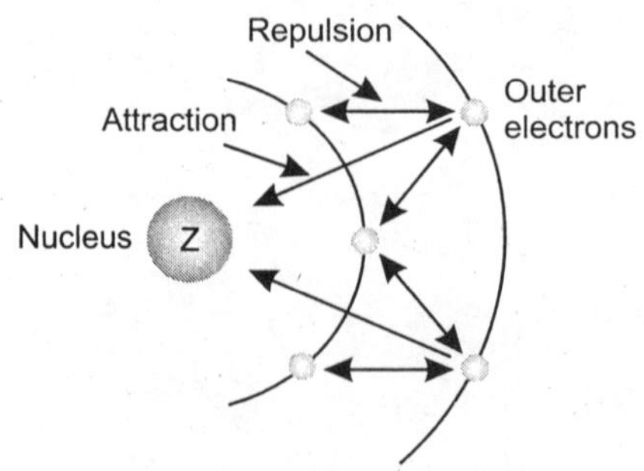

Fig.5.1: Shielding Effect due to interaction between electron and nucleus

Shielding Effect – If an electron is far from the nucleus (*i.e.*, if the distance r between the nucleus and the electron is large), then at any given moment, most of the other electrons will be between that electron and the nucleus. Hence the electrons will cancel a portion of the positive charge of the nucleus and thereby decrease the attractive interaction between it and the electron farther away. As a result, the electron farther away experiences an effective nuclear charge (Z_{eff}) that is less than the actual nuclear charge Z. This effect is called **electron shielding.**

The shielding effect is the name given to the balance between the attraction between valence electrons and protons and the repulsion between valence and inner electrons. The shielding effect explains the trend in atomic size on the periodic table and also why valence electrons are readily removed from an atom.

If the distance between an electron and the nucleus is–

(a) **Approaches infinity,** Z_{eff} approaches a value of 1 because all the other (Z – 1) electrons in the neutral atom are, on the average, between it and the nucleus.

(b) **Very close to the nucleus,** then at any given moment most of the other electrons are farther from the nucleus and do not shield the nuclear charge. At $r \approx 0$, the positive charge experienced by an electron is approximately the full nuclear charge, or $Z_{eff} \approx Z$.

(c) **At intermediate** values of r, the effective nuclear charge is somewhere between 1 and Z: $1 \leq Z_{eff} \leq Z$. Thus the actual Z_{eff} experienced by an electron in a given orbital depends not only on the spatial distribution of the electron in that orbital but also on the distribution of all the other electrons present.

For the elements of the first three rows of the periodic table. Notice that only for hydrogen does $Z_{eff} = Z$, and only for helium are Z_{eff} and Z comparable in magnitude.

PENETRATION OF ORBITALS

Penetration refers to how effectively electrons can get close to the nucleus. The electron probability density for s-orbitals is highest in the center of the orbital, or at the nucleus. Different orbitals have greater nuclear **penetration** than others.

Electrons are negatively charged and are pulled pretty close to each other by their attraction to the positive charge of a nucleus. The electrons are attracted to the nucleus at the same time as electrons repel each other. The balance between attractive and repulsive forces results in **shielding**.

Penetration and shielding are two underlying principles in determining the physical and chemical properties of elements. We can predict basic properties of elements by using shielding and penetration characteristics to assess basic trends. The orbital (n) and subshell (m_l) define how close an electron can approach the nucleus. The ability of an electron to get close to the nucleus is penetration.

Coulomb's Law (an analogy with classical physics) can be used to describe the attraction and repulsion between atomic particles:

Coulomb's Law is a law stating that, like charges repel and opposite charges attract, with a force proportional to the product of the charges and inversely proportional to the square of the distance between them.

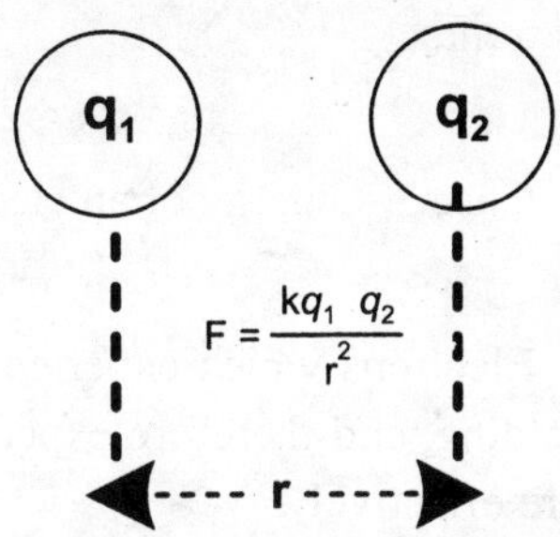

Fig.5.2: Coulomb's Law

$$F = k\,q_1\,q_2/r^2 \qquad ...(5.1)$$

Where q_1 and q_2 = Two charges

r = distance between charges

The force that an electron feels is dependent on the distance from the nearest charge (*i.e.*, an electron, usually with bigger atoms and on the outer shells) and the amount of charge. More distance between the charges will result in less force, and more charge will have more force of attraction or repulsion.

In the simplest case, or simply the electrons closest to the nucleus decrease the amount of nuclear charge affecting the outer electrons. Shielding is caused by the combination of partial neutralization of nuclear charge by core electrons, and by electron-electron repulsion.

The amount of charge felt by an electron depends on its distance from the nucleus. The closer an electron comes to the nucleus, or the more it penetrates, the stronger its attraction to the nucleus. Core electrons penetrate more and feel more of the nucleus than the other electrons.

$$F_{electron} - \text{nucleus} = kZe2r^2 \qquad ...(5.2)$$

$$F_{electron} - \text{nucleus} = kZe2r^2 \qquad ...(5.3)$$

with

- Z is the charge of the nucleus (*i.e.*, number of protons)
- e is the charge of an electron or proton
- r is the radius, or distance between the proton and the electron

Penetration and shielding result in an Effective force (F_{eff}) that holds the outer electrons to the atom For example, we see that since a 2s electron has more electron density near the nucleus than a 2p electron, it is penetrating the nucleus of the atom more than the 2p electron.

The penetration power of an electron, in a multi-electron atom, is dependent on the values of both the shell and subshell. We can learn by taking below cases-

Case 1 – Within the same shell value (n), the penetrating power of an electron follows this trend in subshells (m_l):

$$s > p > d > f$$

Case 2 – for different values of shell (n) and subshell (l), penetrating power of an electron follows this trend:

$$1s > 2s > 2p > 3s > 3p > 4s > 3d > 4p > 5s > 4d > 5p > 6s > 4f...$$

Case 3 – energy of an electron for each shell and subshell goes as follows...

$$1s < 2s < 2p < 3s < 3p < 4s < 3d < 4p...$$

Electrons which experience greater penetration experience stronger attraction to the nucleus, less shielding, and therefore experience a larger Effective Nuclear Charge (Z_{eff}), but shield other electrons more effectively.

VARIATION OF s, p, d, f ORBITALS ENERGIES OF ATOMS IN PERIODIC TABLE

Table 5.1: Periodic Table

1	2	3	4	5	6	7	8	9	10	11	12	13	14	15	16	17	18
1 H 1.008																	2 He 4.0026
3 Li 6.94	4 Be 9.0122											5 B 10.81	6 C 12.011	7 N 14.007	8 O 15.999	9 F 18.998	10 Ne 20.180
11 Na 22.990	12 Mg 24.305											13 Al 26.982	14 SI 28.085	15 P 30.974	16 S 32.06	17 Ci 35.45	18 Ar 39.948
19 K 39.098	20 Ca 40.078	21 Sc 44.956	22 Ti 47.867	23 V 50.242	24 Cr 51.996	25 Mn 54.938	26 Fe 55.845	27 Co 58.933	28 Ni 58.693	29 Cu 63.546	30 Zn 65.38	31 Ga 69.723	32 Ge 72.630	33 As 74.922	34 Se 78.97	35 Br 79.904	36 Kr 83.798
37 Rb 85.468	38 Sr 87.62	39 Y 88.906	40 Zr 91.224	41 Nb 92.906	42 Mo 95.95	43 Tc (98)	44 Ru 101.07	45 Rh 102.91	46 Pd 106.42	47 Ag 107.87	48 Cd 112.41	49 In 114.82	50 Sn 118.71	51 Sb 221.76	52 Te 127.60	53 I 126.90	54 Xe 131.29
55 Cs 132.91	56 Ba 137.33	57-71 *	72 Hf 178.39	73 Ta 180.95	74 W 183.84	75 Rc 186.21	76 Os 190.23	77 Ir 192.22	78 Pt 195.0	79 Au 196.97	80 Hg 200.59	81 Ti 204.38	82 Pb 207.2	83 Bi 208.98	84 Po (209)	85 At (210)	86 Rn (222)
87 Fr (223)	88 Ra (226)	89-103 #	104 Rf (265)	105 Db (268)	106 Sg (271)	107 Bh (270)	108 Hs (277)	109 Mt (276)	110 Ds (281)	111 Rg (280)	112 Cn (285)	113 Nh (286)	114 Fi (289)	115 Nc (289)	116 Lv (293)	117 Ts (294)	118 Og (294)

* Lanthanide series	57 La 138.91	58 Ce 140.12	59 Pr 140.91	60 Nd 144.24	61 Pm (145)	62 Sm 150.36	63 Eu 151.96	64 Gd 157.25	65 Tb 158.93	66 Dy 162.50	67 Ho 164.93	68 Er 167.26	69 Tm 168.93	70 Yb 173.05	71 Lu 174.97
* Actinide series	89 Ac (227)	90 Th 232.04	91 Pa 231.04	92 U 238.03	93 Np (237)	94 Pu (244)	95 Am (243)	96 Cm (247)	97 Bk (247)	98 Cr (251)	99 Es (252)	100 Fm (257)	101 Md (258)	102 No (259)	103 Lr (262)

s, p, d, f and so on are the names given to the orbitals that hold the electrons in atoms. These orbitals have different shapes (*e.g.* electron density distributions in space) and energies (*e.g.* 1s is lower energy than 2s which is lower energy than 3s; 2s is lower energy than 2p).

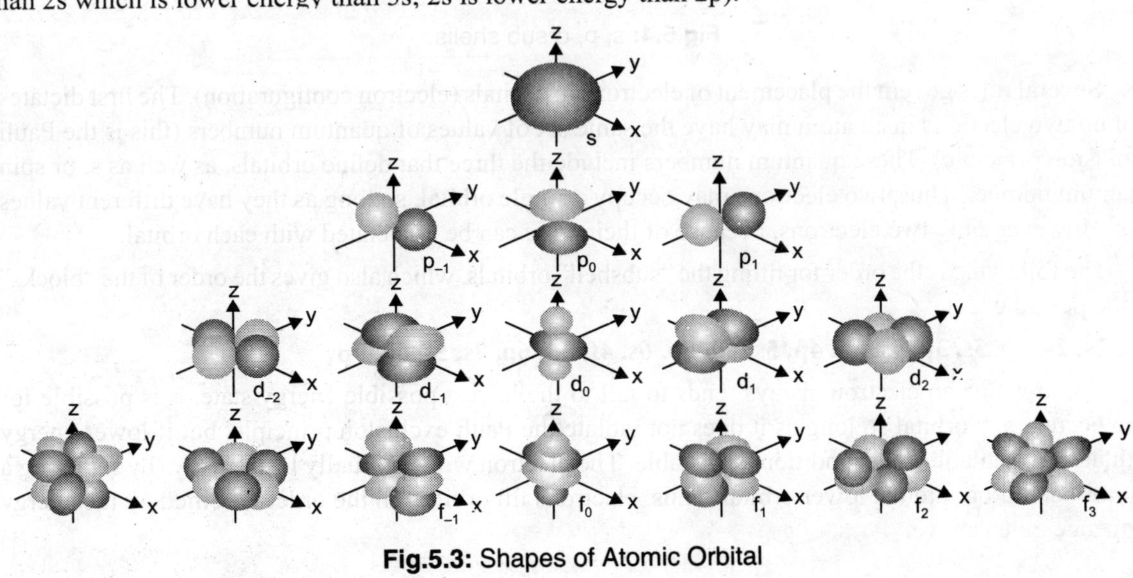

Fig.5.3: Shapes of Atomic Orbital

Where the finding of an electron is maximum is known as **orbital.**

The first shell contain s orbital in which two electron can be filled The second shell contain s & p orbital, the p orbital can have maximum 6 electrons. The third shell contain s, p & d orbital,d orbital can have maximum 10 electrons.The forth shell contain s, p, d & f orbital, f can have maximum 14 electrons.

- The shape of s orbital is round or spherical in shape.
- The p orbital have dumbbell shape, it contain subshell known as p_x which lies on x axis, py which lies on y axis and p_z which lies on z axis.
- The d orbital have double dumbbell shape. it contain subshells known as dxy lies between x and y axis, d_{yz} lies between y and z axis, d_{zx} lies between z and x axis, $d_{x^2-y^2}$ lies on the x and y axis & finally dz^2 which lies on z axis.

Here are the images of subshell

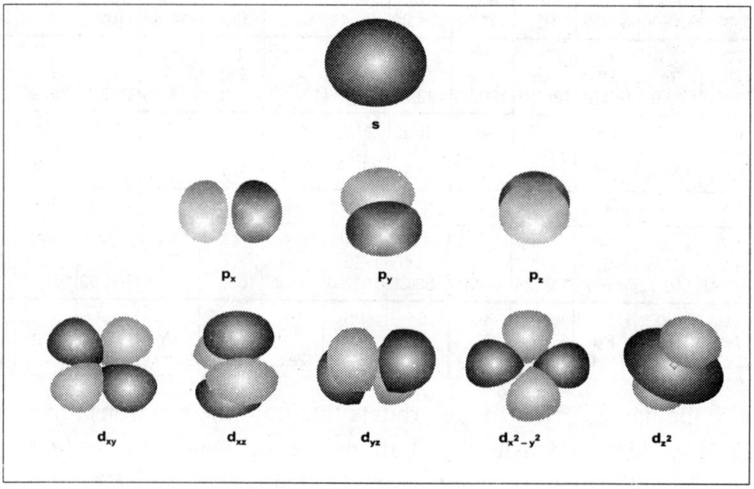

Fig.5.4: s, p, d sub shells

Several rules govern the placement of electrons in orbitals (electron configuration). The first dictates that no two electrons in an atom may have the same set of values of quantum numbers (this is the Pauli exclusion principle). These quantum numbers include the three that define orbitals, as well as s, or spin quantum number. Thus, two electrons may occupy a single orbital, so long as they have different values of s. However, only two electrons, because of their spin, can be associated with each orbital.

The following is the order for filling the "subshell" orbitals, which also gives the order of the "blocks" in the periodic table:

1s, 2s, 2p, 3s, 3p, 4s, 3d, 4p, 5s, 4d, 5p, 6s, 4f, 5d, 6p, 7s, 5f, 6d, 7p

Additionally, an electron always tends to fall to the lowest possible energy state. It is possible for it to occupy any orbital so long as it does not violate the Pauli exclusion principle, but if lower-energy orbitals are available, this condition is unstable. The electron will eventually lose energy (by releasing a photon) and drop into the lower orbital. Thus, electrons fill orbitals in the order specified by the energy sequence given above.

This behavior is responsible for the structure of the periodic table.

Table 5.2: Ionization energy of atomic orbital in (ev)

Atom	1s	2s	2p	3s	3p
H	13.6				
He	24.5				
Li		5.45			
Be		9.30			
B		14.0	8.30		
C		19.5	10.7		
N		25.5	13.1		
O		32.3	15.9		
F		46.4	18.7		
Ne		48.5	21.5		
Ne				5.21	
Na				7.68	
Al				11.3	5.95
Si				15.0	7.81
P				18.7	10.2
S				20.7	11.7
Cl				25.3	13.8
Ar				29.2	15.9

ELECTRON CONFIGURATION

Electron configuration is the distribution of electrons of an atom or molecule in the atomic or molecular orbitals. Commonly, the electron configuration is used to describe the orbitals of an atom in its ground state, but it can also be used to represent an atom that has ionized into a cation or anion by compensating for the loss of or gain of electrons in their subsequent orbitals.

In atomic physics and quantum chemistry, the electron configuration is the distribution of electrons of an atom or molecule (or other physical structure) in atomic or molecular orbitals. For example, the electron configuration of the neon atom is $1s^2\ 2s^2\ 2p^6$.

Importance of Electron Configuration

- To understand the structure of the periodic table of elements.
- This is also useful for describing the chemical bonds that hold atoms together.
- In bulk materials, this same idea helps explain the peculiar properties of lasers and semiconductors

Expression of Electron Configuration

- An electron configuration describes the distribution of electrons among the various orbital sin the atom.
- Electrons configurationis represented in two ways.

- The *spdf* notation uses numbers to designate a principal shell and letters (*s, p, d, f*) to identify a subshell; a superscript indicates the number of electrons in a designatedsubshell.

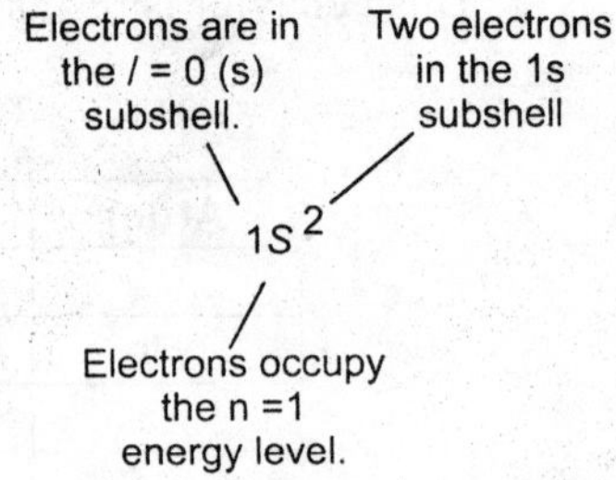

How to Write Electronic Configuration?

Basically, the distribution of electrons over various shells (energy levels) is such that, the overall energy of all electrons of an atom remains minimum for the stability of the atom. The electron distribution over various energy level is governed by the following rules:

- Electrons fill orbitals in a way to minimize the energy of the atom. Therefore, the electrons in an atom fill the principal energy levels in order of increasing energy (the electrons are getting farther from the nucleus). The order of levels filled looks like this:

  1s, 2s, 2p, 3s, 3p, 4s, 3d, 4p, 5s, 4d, 5p, 6s, 4f, 5d, 6p, 7s, 5f, 6d, and **7p** To help memorize this sequence check out this arrangement,

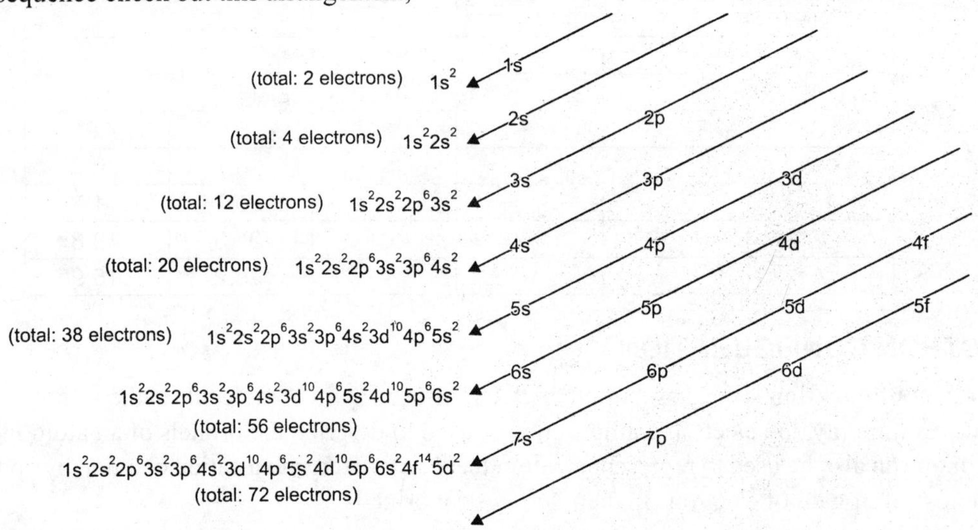

Fig.5.5: Electronic configuration

- **Pauli Exclusion Principle:** According to this principle, an orbital can have maximum two electrons and these must have opposite spins. In other words, no two electrons in an atom can have all the four quantum numbers exactly the same. As we know that fourth quantum number is m, 1/2 is spin up and –1/2 is spin down and hence if the first three quantum numbers are same for two electrons then the electrons must have opposite spins in an orbital.
- **Hund's Rule:** When assigning electrons in orbitals, each electron will first fill all the orbitals with similar energy (also referred to as degenerate) before pairing with another electron in a half-filled orbital.
- **Aufbau principle:** According to this principle, the electrons will first occupy the orbitals that have the lowest energy.
- Although the Aufbau rule accurately predicts the electron configuration of most elements, there are notable exceptions among the transition metals and heavier elements. The reason these

exceptions occur is that some elements are more stable with fewer electrons in some subshells and more electrons in others.

Table 5.3: Aufbau law

Aufbau Diagram Boxes

Orbitals are hard to draw and rthe Aufbau Diagram is an easier way to show electron location ,y using sets of boxes

Each box can hold up to 2 e$^-$

- s orbital – 1 box – total no. of 2e$^-$
- p orbital – 3 boxes – total no. of 6e$^-$
- d orbital – 5 boxes – total no. of 10e$^-$
- f orbital – 7 boxes – total no. of 14 e$^-$

Element	**Should be**	**Actually is**
Copper	$1s^22s^22p^63s^23p^6\mathbf{3d^44s^2}$	$1s^22s^22p^63s^23p^6\mathbf{3d^54s^1}$
Chromium	$1s^22s^22p^63s^23p^6\mathbf{3d^94s^2}$	$1s^22s^22p^63s^23p^6\mathbf{3d^{10}4s^1}$

When writing an electron configuration, first write the energy level (the period), then the subshell to be filled and the superscript, which is the number of electrons in that subshell. The total number of electrons is the atomic number, Z. The rules above allow one to write the electron configurations for all the elements in the periodic table.

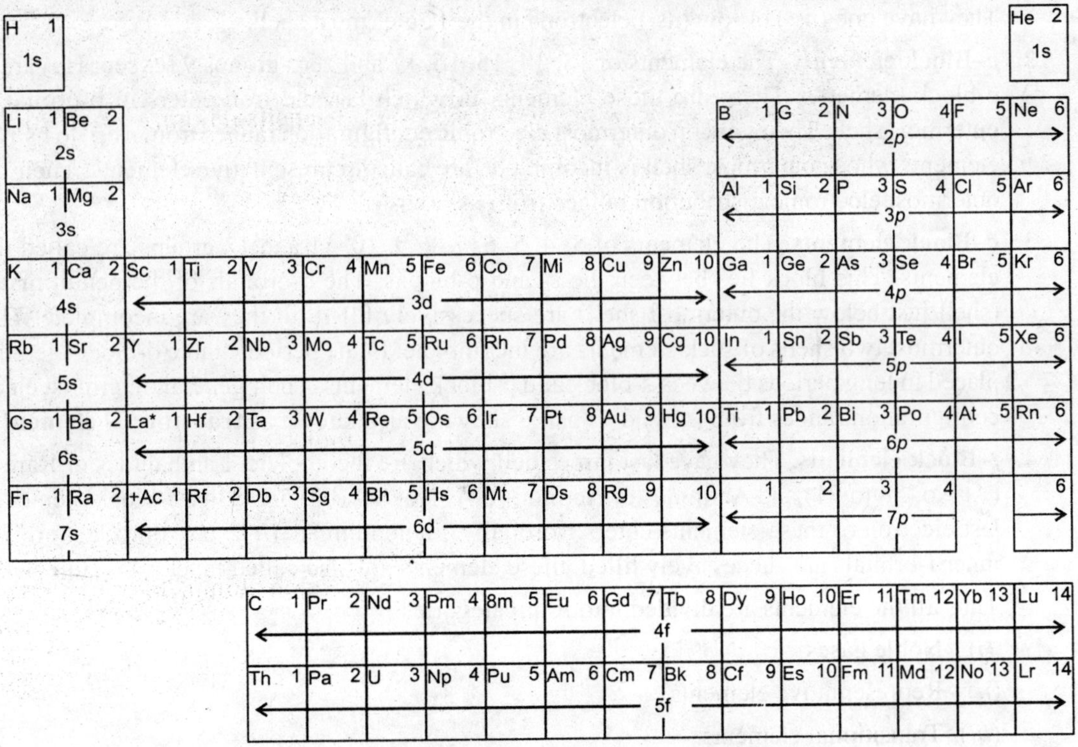

Fig.5.6: Periodic Table Electronic Configuration

For example, the Electronic configuration of iron can be written as:

$$1s^2\ 2s^2\ 2p^6\ 3s^2\ 3p^6\ 3d^6\ 4s^2 = [Ar]\ 3d^6 4s^2$$

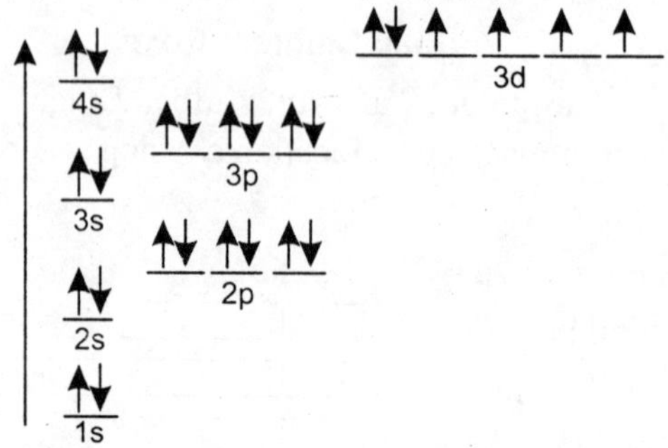

Fig.5.7: Electronic configuration – Iron

CLASSIFICATION OF ELEMENTS BASED ON ELECTRONIC CONFIGURATION OR [s, p, d and f-BLOCK ELEMENTS]

According to 'Long From' of periodic table, all elements are broadly divided into four blocks *i.e.* s, p, d and f blocks. This classification of elements is based on the entry of last electron in various subsheels or orbitals.

1. ***s*–Block elements.** The elements of 1 and 2 groups, hydrogen and helium belong to this group. They have one (ns^1) or tow (ns^2) electrons in their outer most shell.
2. ***p*–Block elements.** The elements of 13, 14, 15, 16, 17 and zero groups 9 (except He) are called p-block elements. These are those elements in which last electron enters in p-orbital of the outermost shell. Their general outermost electronic configuration ranges from ns^1p^1 to ns^2p^6. Only elements whose outermost shell is incomplete, are called representative elements. Their general outermost electronic distribution ranges from ns^1 to ns^2p^5.
3. ***d*–Block elements.** The elements of 3, 4, 5, 6, 7, 8, 9, 10, 11 and 12 groups are called d-block elements. This block lies between the s- and p-blocks. The d-orbitals of the penultimate shell (shell just below the outermost shell) are successively filled till they are incomplete. Thus the outermost two shells of such elements are incomplete. In the periodic table d-block elements are placed in long periods between s-block and p-block elements. Their general outermost electronic configuration ranges from $(n-1)\ d^{1-10}\ ns^{1-2}$. They are also known as **transitional elements.**
4. ***f*–Block elements.** They have last three shell which are incomplete. Lanthanides or Rare-earths (At No. 58 to 71) and Actinides or Actinons 9 At No. 89-103) are called f-block elements. The last electron of these elements enters f-orbitals of antepulimate (n–2) *i.e.* third last orbit. Since inner f-orbitals are successively filled, these elements are also called inner transition elements.

   Thus all the elements are divided into four categories.

   (*i*) Noble gases

   (*ii*) Representative elements.

   (*iii*) Transitional elements.

   (*iv*) Inner-transition elements.

ATOMIC RADII

The size of an atom is one of the most important properties of atoms that affects their chemical character. In dealing with atomic sizes, it is assumed that atoms are spherical. The atomic size is, therefore, generally expressed as the radius of the sphere or the distance between the centre of the nucleus and the outermost shell of electrons. However, the radius of an atoms cannot be determined precisely due to the following difficulties:

(*i*) Atoms are incapable of independent existence and it is not possible to isolate a single atom so as to determine its radius.

(*ii*) The positions of the electrons in the extra-nuclear portion are not definite and are only probable. Hence, the exact distance cannot be measured.

(*iii*) The probability of finding the electron around the nucleus never becomes zero even at very large distances. Therefore, there is no question of measuring the exact value of these distances.

(*iv*) Atoms are always surrounded by other atoms. The presence of these neighbouring atoms also influences the size of the aom.

Factors on which atomic size depends

(*i*) Nuclear charge. With increase in nuclear charge, the force of attraction between the nucleus and the electron cloud increases. Hence, the electron cloud moves closer to the nucleus and atomic size decreases.

(*ii*) Number of orbits. With increase in number of orbits, the distance between the nucleus and last orbit increases. Hence, atomic size increases. Similarly, with dercrease in number of orbits, the atomic size decreases.

Measurement. Atomic radii are measured by electron diffraction method in angstrom units (1Å = 10–8cm).

Study of atomic radii. Different types of atomic radii (which depend upon the environment of an atom) are as follows:

(*i*) Covalent radius (*ii*) Metallic radius
(*iii*) van cer Waal's radius (*iv*) Ionic radius

(1) Covalent Radius

The covalent radius may be defined as one-half of the distance between the centres of nuclear of two similar atoms held together by purely covalent single bond.

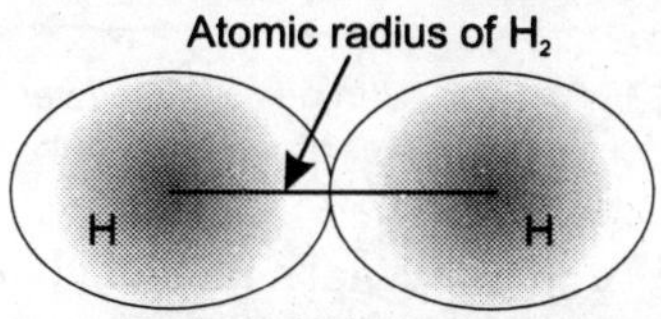

Fig.5.8: Covalent radius of hydrogen atom

1. The inter-nuclear distance of the bonded atoms can be **determined by X-ray diffraction of spectroscopic techniques.** Half of this distance gives the covalent radius.

$$\text{Covalent radius} = \frac{\text{Inter-nuclear distance between two covalentlyh bonded atoms}}{2}$$

e.g., (a) Inter-nuclear distance in a H_2-molecule = 0.74 Å

$$\text{Covalent radius of H-atom} = \frac{0.74}{2} = 0.37\ \text{Å}$$

(2) Metallic Radii (Crystal Radii)

The closed packing of the metal atoms is known as the crystal lattice and this packing is different in case of different elements. Half of the inter-nuclear distance between two adjacent atoms in a metallic bond represents the crystal radius or metallic radius. This is generally determined by X-ray diffraction method. *E.g.* In solid sodium inter-nuclear distance between two adjacent sodium atoms is 3.72 Å. Therefore.

Metallic radius of Na = 1.86Å (half of 3.72Å)

Similarly Metallic radius of K = 2.34Å

Since there are only weak bonding forces in the formation of the metallic bond and there is no actual overlap of the atomic orbitals, the value of the metallic radius is always higher than that of the covalent radius.

e.g., Metallic radius of Na = 1.86Å

Covalent radius of Na = 1.54Å

(3) Van der Waal's Radius

It is defined as **"Half of the distance between the nuclei of two similar adjacent atoms belonging to two neighbouring molecules in the solid state."**

There are some forces of attraction, known as van der Waal's forces operating between neutral atoms of inert gases and also between non-polar molecules of various substances. Due to these forces of attraction, the neutral atoms of inert elements and non-polar molecules of other elements can come quite close together in the solid state. One half of the inter-nuclear distance between the nearest non-bonded atoms in the solid state gives the van der Waal's radius of the atom. For instance, the van der Waal's radius of chlorine atom is 1.86Å and that of argon, an inert element existing in the mono-atomic form is 1.91Å. This is quite clear from Fig. 5.9 gives below.

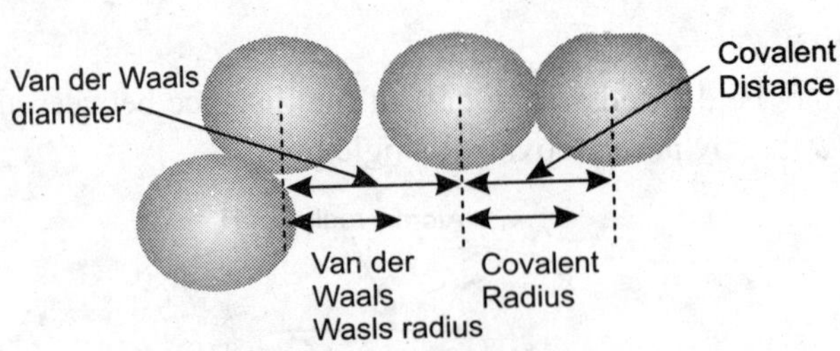

Fig.5.9: Van der Waal's radii of chlorine and argon

It is obvious from Fig. 5.9 that van der Waal's radii must be larger than covalent radii since the inter-nuclear distance between non-bonded atoms must be more than that between bonded atoms. The following figures for the covalent and van der Waal's radii of some elements clearly illustrate this fact.

Element	H	N	O	F	Cl	Br	I
van der Waal's radius	1.2	1.59	1.4	1.35	1.80	1.95	2.15
Covalent radius, Å	0.37	0.74	0.74	0.72	0.99	1.14	1.33

Variation of atomic radii in a period. There is a progressive decrease in size of the atoms along each series of horizontal rows (periods) from left to right as the atomic number increases. For example, sodium of group 1 is the first member is period 3, it has an atomic radius of 1.57Å; its size is larger than that of magnesium in group 2 in the same row which is equal to 1.36 Å; and the size of atoms in this row decreases progressively through chlorine, which has an atomic radius of 0.99 Å. The inert gas at the end of the period is argon; it has longer atomic radius which is 1.91 Å as indicated in Table.

As we move along the period from left to right, the atomic number increases. *i.e.*, the nuclear (positive) charge increases. This increased nuclear charge attracts the electrons in the shells more strongly due to increased force of electrostatic attraction between the nucleus and orbital electrons. This results in pulling the orbits closer to the nucleus and shrinking the atom in size. Thus, in theory, it explains the decrease in atomic radii of elements across the period.

Variations of atomic radii a group. Down the groups, the size of atoms (atomic radii) gradually increases as the atomic number increases. The atomic radius of lithium in Group–1 is 1.23Å which increases to 1.57 Å for sodium and to 2.35 Å for caesium.

(4) Ionic Radii

The ionic radii can be defined as **the distance from the nucleus of an ion upto which it has influence of its electron cloud.**

Inter-nuclear distance in any ionic compound is determined from X-ray studies. If the ions are takes as spheres, the inter-nuclear distance may be considered to be made up of the sum of the ionic radii of the two ions (Fig. 5.10). For example, the inter-nuclear distance of Na–Cl is 2.76Å. this is equal to the sum of the ionic radii of Na^+ and Cl^- ions. Therefore, if we know the radius of one ion, that of other ion can be calculated.

Ionic radii across a period

- The radii of positive ions decrease from Na^+ to Al^{3+}
- The radii of positive ions decrease from P^{3-} (phosphide ion) to Cl
- The ionic radii increase from the Al^{3+} to P^3.

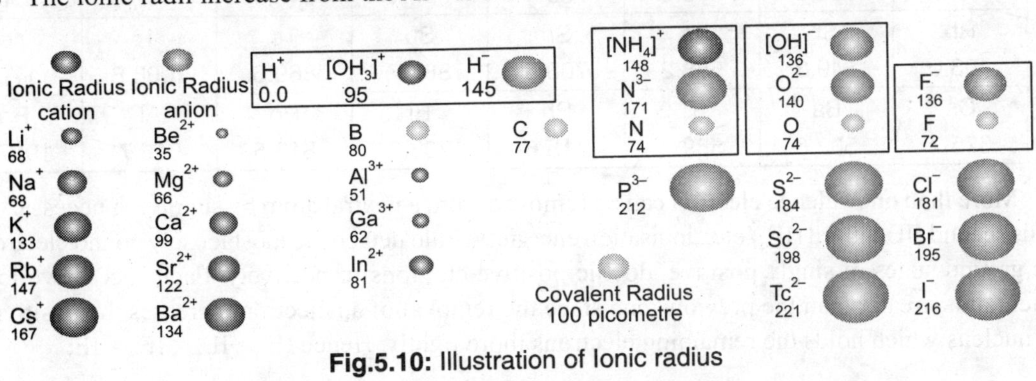

Fig.5.10: Illustration of Ionic radius

IONISATION ENERGY (*i.e.*) OR IONISATION POTENTIAL (I.P.)

The electrons in an atom are attracted by the positively charged nucleus. In order to remove an electron from an atom, energy has to be supplied.

First ionisation energy (IE). First ionisation energy or simply ionisation energy (*I.e.*) is defined as the amount of energy required to remove one valence electron from an isolated neutral gaseous atom resulting in the formation of a monovalent positive ion.

The process may be represented as :

Atom(g) + Energy $\rightarrow$ Positive ion(g) + Electron(g)

i.e., Li(g) + Energy (520.1kJMol^{-1}) Li^+(g) + e^-(g)

Unit. It is expressed in either electron-volts/atom or kcal/mole or kJ/mole/

1 eV/atom = 23.06 kcal/mole = 96.5 kJ/mole.

Ionisation energy is also called ionisation potential because it is measured as the amount of potential needed to remove the most loosely held electrons from the gaseous atom. The experimental values of ionisation energies are determined either by spectroscopic methods or by passing an electric current or gradually increasing intensity through the vapours of the element in a discharge tube.

Thus, the ionisation energy gives the ease with which an electron can be removed from an atom. Evidently, smaller the value of ionisation energy, easier it is to remove the electron from the atom.

The first ionisation energies for the various elements belonging to different groups are given in Table 5.4.

Table 5.4: First Ionization Energies of Various Elements (kJ mol^{-1})

Group 1	Group 2	Group 13	Group 14	Group 15	Group 16	Group 17	Group Zero (18)
H 1312.0							He 2372.1
Li 520.1	Be 899.3	B 800.6	C 1086.2	N 1402.1	O 1313.7	F 1680.8	Ne 2080.4
Na 495.2	Mg 737.6	Al 577.4	Si 786.4	P 1011.7	S 999.4	Cl 1255.0	Ar 1520.3
K 418.7	Ca 589.8	Ga 579.0	Ge 760.4	As 946.1	Se 940.9	Br 1142.8	Kr 1350.6
Rb 403.0	Sr 549.3	In 558.2	Sn 708.5	Sb 88.37	Te 869.6	I 1008.8	Xe 1170.2
Cs 375.7	Ba 502.6	Ti 589.2	Pb 715.6	Bi 703.2	Po 813.5	At 916.7	Rn 1037.0

More than one valence electron can be removed from a neutral atom by supplying necessary energy. Thus second (IE_2), third (IE_3) etc., ionisation energies would detach the most loosely bound electrons from the ground states of singly positive, double positive etc. Ions respectively. Each successive ionisation potential is greater than the previous one since the removal of an electron increases the positive charge on nucleus which holds the remaining electrons more tightly. Hence $IE_4 > IE_3 > IE_2 > IE_1$

$$M_{(g)} \rightarrow M^{+}_{(g)} + e^{-} \qquad IE_1 \text{ (First ionisation energy)}$$
$$M^{+}_{(g)} \rightarrow M^{2+}_{(g)} + e^{-} \qquad IE_2 \text{ (Second ionisation energy)}$$

MEASUREMENT OF IONISATION ENERGIES

The amount of energy required to detach an electron from an atom can be measured by supplying the required energy as thermal energy, electrical energy, or radiant energy. Thus ionisation energies can be determined from the spectrum of the element or by any of the two methods outlined as:

(1) The Electrical Method

(2) Photo-ionisation Method

(3) Spectral Method

Factors on which Ionisation Energy Depends

1. **Nuclear charge.** With the increase in nuclear charge, the force of attraction between the nucleus and valence electrons increases. Consequently, more energy is required to remove a valence electron. Hence, ionisation energy increases with increase in nuclear charge. *E.g.* ionisation energy of Ca (At. No. 20) is more that K (At. No. 19).
2. **Atomic size.** With increase in atomic size, the distance between the nucleus and the valence electrons increases. Thus, the force of attraction $\left(\frac{Z_2 xe}{r^2}\right)$ between the nucleus and the valence electron decreases. Hence, energy required to move valence electrons to ionisation energy decreases with increase in atomic size.

   When size of an atom and nuclear charge increase simultaneously (*e.g.*, down a group) the size of the atom dominates the magnitude of nuclear charge. As a result, the ionisation energy generally decreases.
3. **The screening or shielding effect of the inner electronic orbitals.** The electrons that lie between the nucleus and the valence shell exert a repulsive force on electrons in outermost orbitals away from the nucleus. In other words, the inner orbitals act as a shield on screen between the nucleus and the outermost orbit. This will reduce the forms of attraction on the outermost elements by the nucleus and hence the ionisation energy with decrease. Higher the number of intervening electrons, greater will be the shielding effect; smaller the attraction of the nucleus, lower is the ionisation energy.

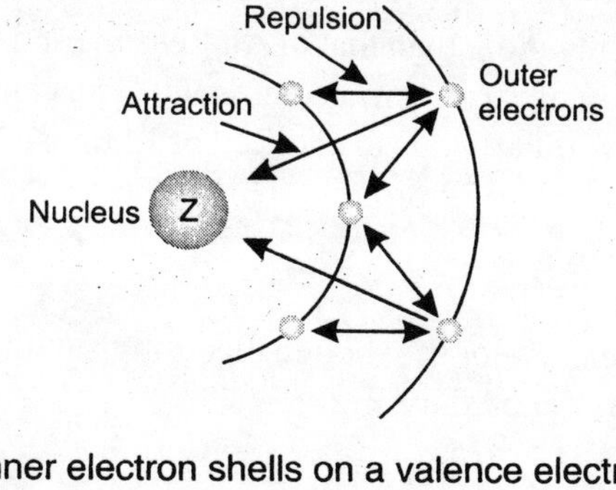

Fig.5.11: Effect of inner electron shells on a valence electron (shielding effect).

4. **Penetration of sub-shells.** Value of ionisation energy also depends on the azimuthal quantum number (l) of the outermost electrons. The order of increase in energy of different sub-shells is s < p < d < f. Thus, the order of penetration of different sub-shells towards is s > p > f, s-subshell is more penetrating towards nucleus that p. Thus, greter energy (*I.e.*) will be required to remove an electron from s-subsheel than p-subshell. Similarly, it can be explained that lesser energy will be required to remove an electron from f-subshell than in case of d-subshell.

   Example. Ionisation energy of Al is less than that of Mg.

Reason. In case of Al ($1s^2, 2s^2, 2p^6, 3s^2, 3p_x^{\ 1}$), the electron is to be removed from p-subshell. In case of Mg ($1s^2, 2s^2, 2p^6, 3s^2$), the electron is to be removed from s-subshell. Since s-subshell is more penetrating towards nucleus (and strongly attracted) than p, thus ionisation energy of Mg is more than that of Al.

Table 5.5: Ionisation Energies of Group 1 Elements

Element	At. No.	Electronic configuration	Size of atom (Å)	*I.e.* (E_1) kJ mole^{-1}
H	1	1	0.37	1312
Li	3	2, 1	1.23	520
Na	11	2, 8, 1	1.57	495
K	19	2, 8, 8, 1	2.03	419
Rb	37	2, 8, 18, 8, 1	2.16	403
Cs	35	2, 8, 18, 18, 8, 1	2.35	376

But this decreases in values of ionisation energies is not as regular in group 13 (III A) as in group I and II (Table 5.5). The main reason of this irregularity is excessive increases in nuclear charge in comparison to the size of atom.

Table 5.6: Ionisation Energies of Group 13 (IIIA) Elements

Element	IE_1	IE_2	IE_3
B (5)	801	2427	3659
Al (13)	577	1816	2744
Ga (31)	579	1979	2962
In (49)	558	1820	2704
Tl (81)	589	1971	2877

Ionisation energy of Ga is slightly more than that of Al. Reason for this is that the difference of atomic number of Ga and Al is 18 (addition of 10 transition elements). Thus large increase of nuclear charge of Ga increases its ionisation energy. Similarly due to addition of 14 transition elements between In (49) and Tl (81), nuclear charge of Tl increases by 32 protons. This is the reason of high ionisation energy of Tl.

Successive Ionisation Energies

The energy required to remove the most loosely bound electron from the isolated gaseous atom is called its **first ionisation energy** and it is denoted by IE_1.

$$M(g) + IE_1 \rightarrow M^+(g) + e^-$$

Similarly, the energies required to knock out second and third electrons are called **second and third ionisation energies** respectively, *e.g.*,

$$M^+(g) + IE_2 \rightarrow M^{2+}(g) + e^-$$

$$M^{2+}(g) + IE_1 \rightarrow M^{3+}(g) + e^-$$

Where IE_2 and IE_3 are the second and the third ionisation energies respectively.

When one electron has been removed form the neutral gaseous atom, the positively charged ion formed has one electron less than the number of protons in the nucleus. As a result, the electrostatic attraction between the nucleus and the remaining electrons in the cation increases due to increased effective nuclear charge. In other words, the positive ion holds its remaining electrons more firmly. Therefore, the energy required to remove another electron from this positively charged ion or second electron from the neutral atom must be higher than the first, *i.e.*, IE_2 must be higher than IE_1. Further, the removal of two electrons from the neutral atom gives a doubly positively charged ion which will hold its remaining electrons even more tightly. As a result, the energy required to remove the third electron from the gaseous atom should be even more than that required for second electron, *i.e.*, IE_3 must be greater than IE_2. Thus, IE_2 is always greater than IE_1 and IE_3 is always greater than IE_2, *i.e.*,

$$IE_3 > IE_2 > IE_1$$

For example, first second and third ionisation energies of Al are 138, 434 and 646 kcal/mole respectively.

The ionisation energies required to remove first, second and third electrons from a gaseous atom are called **successive ionisation energies.**

ELECTRON AFFINITY (E.A.)

Definition.The amount of energy released when an electron is added to a neutral gaseous atom of an element to form an anion is called first electron affinity. For example,

$A(g) + e^-$	$\rightarrow$	$A^-_{(g)}$	+	Energy (called electron affinity)
Neutral Atom		Anion		

Unit. It is measured in kcal/mole or electron volt (eV). 1 eV = 23.06 kcl/mole.

Example. When one mole of F atom is converted into one mole of F^- ions, 83 kcals of energy is released.

$$F_{(g)} + e^- \rightarrow F^-_{(g)} + 83 \text{ kcal/mole} (= \text{E.A.})$$

Higher the energy released, greater the value of electron affinity of an atom. Similarly higher the value of E.A. easier the formation of anion from atom and more strong the oxidising nature of element.

Successive Electron Affinities

When the first electron is added to the gaseous atom, it forms a uninegative ion and the energy released during the process is called **first electron affinity**. For example,

$$O^-_{(g)} + e^- \rightarrow O^-_{(g)} + E.A_1 (= 34.0 \text{ kcal/mole})$$

Now, if an electron is added to the uninegative ion, it experiences a repulsive force from the anion, As a result, the energy has to be supplied to overcome the repulsive force. Thus, in order to force the second electron the energy is required rather than its release and this is called **second electron affinity**. For example,

$$O^-_{(g)} + e^- + E.A_2 (= 168.3 \text{ kcal/mole}) \rightarrow O^{2-}_{(g)}$$

Thus, the value of second electron affinity is negative.

From the above discussion, it is clear that the electrons get added to gaseous atoms one after the other and not simultaneously. The phenomenon of adding the electrons one after the other *i.e.*, in succession is called **successive electron affinity.**

Determination of Electron Affinity of an Element

Electron affinity can be measured directly by the spectroscopic methods. But it can also be measured indirectly by applying the Born-Haber cycle (M. Born & F. Haber, 1919). This cycle relates the lattice energy of a crystal to other thermochemical data. For example, consider the formation of an alkali metal halide say, NaCl from elementary solid sodium metal and gaseous Cl_2. Heat of formation of NaCl crystal Q is –98.2 kcal/mole.

$$Na_{(g)} + \frac{1}{2}Cl_{2(g)} \xrightarrow{-98.2} NaCl_{(g)} \text{Crystal}$$

It may be assumed that the solid Na is vaporised to $Na_{(g)}$ on receiving its sublimation energy S and the gaseous $Cl_{2(g)}$ is dissociated into Cl atoms by supplying dissociation energy D.

$$Na_{(s)} + S \rightarrow Na_{(g)}$$

$$\frac{1}{2}Cl_{2(g)} + \frac{1}{2}D \rightarrow Cl_{(g)}$$

The $Na_{(g)}$ is converted in to $Na^+_{(g)}$ by giving ionisation potential (IE) and Cl(g) into $Cl^-_{(g)}$ by releasing electron affinity (EA)

$$Na_{(g)} + IE \rightarrow Na^+_{(g)}$$

$$Cl_{(g)} \rightarrow Cl^-_{(g)} + E.A.$$

These gaseous ions are brought closer to give the product $NaCl_{(s)}$ and in this process of crystal formation lattice energy U_0 is lost.

$$Na^+_{(g)} + Cl^-_{(g)} \rightarrow NaCl_{(s)} + U_0$$

According to Hess law of heat summation.

$$Q = S\frac{1}{2}D + IE + EA + U_0$$

$$-98.2 = 26.0 + 28.9 + 118.0 + EA - 180.4$$

$$EA = -90.7 \text{ kcal/g atom of chlorine}$$

Obviously, electron affinity of Cl, $(EA)_{Cl}$ can be determined, if other energy terms, namely heat of formation of NaCl, heat of sublimation of Na, ionisation potential of Na, heat of dissociation of Cl_2 and lattice energy of NaCl are known.

Factors on which Electron Affinity Depends

There are many factors which govern the electron affinity but the following are some important factors on which it mostly depends:

(*i*) **Nuclear charge.** The electron affinity increases as the nuclear charge increases. This is due to greater attraction for the incoming electron if nuclear charge is high.

(*ii*) **Size of the atom.** With the increase in size of the atom, the distance between the nucleus and the incoming electrons increase and this results in lesser attraction. Consequently, the electron affinity value will decrease.

(*iii*) **Electronic configuration.** The elements having stable electronic configuration of half and completely filled valence sub-shell show very small tendency to accept additional electron and thus, electron affinities are low or almost zero in certain cases.

(*iv*) **Screening and Penetration Effects.** Value of electron affinity also depends on that orbital in which electron enters. Value of electron affinity for the electron entering in the s-orbital is less than the p, d and f orbitals. This is due to more penetration power of s-electrons.

Periodic Trends

Due to the lack of sufficient data, the changing trends in electron affinities on moving along the periodic table are not well-defined as those for ionisation energies. However, it has been observed that the electron affinity in general increases from left to right in a period and decreases from top to bottom in a group. These variations are discussed below.

Variation down a group. Electron affinity generally decreases on moving down the group. This is expected on account of increases in the size of atoms, the effective nuclear attraction for electrons decreases and hence there is a lesser tendency to attract additional electrons with increase in atomic number.

Variation along a period. Electron affinity generally increases as we move from left to right in a period. This is due to increase in the nuclear charge which results in greater attraction for electrons. In the second period, fluorine has a maximum value of electron affinity. As the electron affinity increases, the element becomes less metallic and more non-metallic.

ELECTRONEGATIVITY

Electronegativity is defined as a measure of the ability of an atom in a covalent molecule to attract the shared electron pair towards itself. It is denoted by Greek letter χ (chi).

Consider a covalent bond formed by two atoms (say A and B) of equal electronegativity. The bonded electron pair is equally shared between the atoms and situated between the centre of two atoms. *e.g.* H_2, N_2, Cl_2, NCl_3, Cl_4, SI_2 etc.

However, if atoms of dissimilar electronegativites form a covalent bond, the shared pair of electrons is not exactly in the middle but shifted towards the side of greater electronegativity. Consider two atoms which are not identical, *e.g.*, H and Cl, forming the covalent compound H : Cl. The polarity of the molecule arises due to a diffence between the electronegativity of hydrogen and that of chlorine. Since the electronegativity of chlorine is greater than that of hydrogen, the former has the larger share in the electron pair than the letter; the chlorine end of the molecule is the negative end of the dipole, *i.e.*,

$$\overset{\delta^+}{H}-\overset{\delta^-}{Cl} \quad \text{or} \quad \overset{\delta^+}{H}:\overset{\delta^-}{Cl}$$

Measurement of Electronegativity

There scales of electonegativity are prevalent and used for measuring the percentage ionic character of a compound and heat of formation in a reaction.

(a) Mulliken's Scale, (b) Pauling's Scale, (c) Allred and Rochow method.

Periodic Trend

In general when we move across the period from left to right in the periodic table, the size of the atom gradually decreases because of gradual increase in the effective nublear charge, so the electonegativity will also increase gradually when we move across the period from left to right in the periodic table.

Whereas, when we move down the group, the size of the atom goes on increasing gradually, the result is that there is a gradual decrease in the electonegativity when we move down the group.

Application of electronegativity. Electronegativity has a very wide range of application. Knowledge of electronegativity helps in analyzing various reactions, *e.g.* understanding mechanism of some organic reactions where polarity is involved very much in activating some group, some pair or electrons, types of bond formation either ionic, covalent or co-ordinate and so on. Some of the important applications are given below:

1. Electronegativity values and nature of bonds. The nature of the bond formed between any two atoms may also be predicted by making use of electronegative which give relative tendencies of atoms to attract electronegativity from an electrovalent bond by the complete transfer of an electron from one atom to the other. For example, Na(0.9) and Cl(3.0) form Na^+, Cl^-, an electrovalent compound. When two identical atoms combine (*i.e.*, the difference in electronegativity values is not appreciable, they form non-polar covalent bond. A form non-polar bond is one in which the electron pair is shared equally; there is no charge distribution in the bond. For example, formation of hydrogen and chlorine molecules from atoms.Cl(3.0) and Cl(3.0), similarly H(2.1) and H(2.1). When there is an appreciable electronegativity difference between two elements, their atoms still form covalent bond, but the more electronegative atom pulls the shared electron pair towards itself. The electrons are still shared; the bbond is still covalent, but due to the displacement of the electrons towards one atom and away from the other, the more electronegative atom acquires a slight negative charge and the comparative electronegative atom acquires positive charge. The resulting bond is called polar covalent bond. For example, in HCl, the bond formed will be covalent but polar.

2. Electronegativity and hydrolysis mechanism. Examples of the hydrolysis of chlorine dioxide, Cl_2O, and boron trichloride, BCl_3, will clarify the effect electronegativity on hydrolysis.

In chlorine dioxide molecule the oxygen atom will have a small negative charge (say δ^-) and the two chlorine atom will have a small positive charge (say δ^+), because oxygen is more electronegative than chlorine. When water attacks Cl_2O, the positive hydrogen of water will attack at the oxygen of Cl_2O and the products of hydrolysis, will be hypochlorous acid (HOCl) and water.

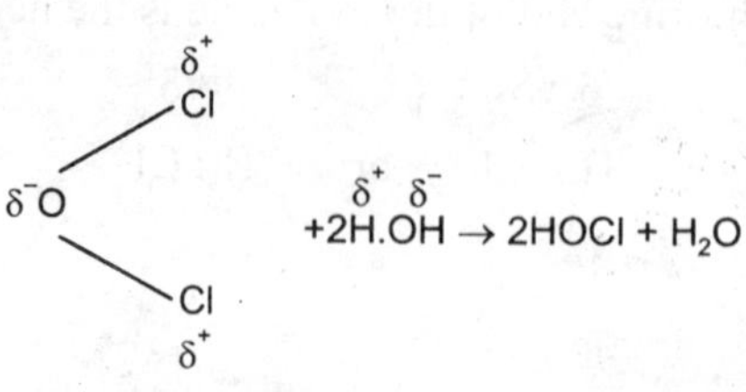

Boron trichloride hydrolyses in a different way because boron is less electronegative than chlorine. When water attacks BCl_3, boron is attacked by more electronegative oxygen.

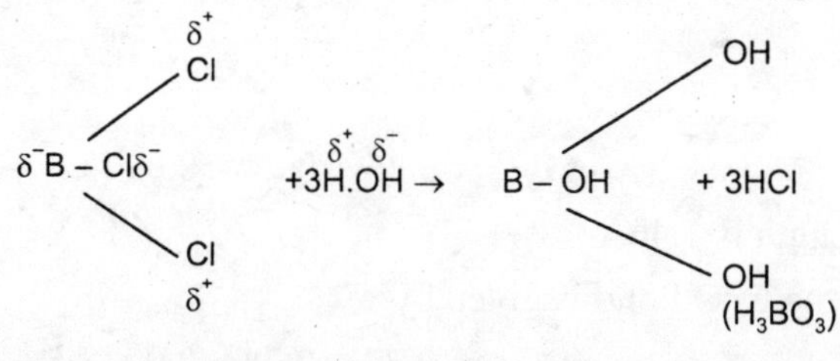

3. Electronegativity and nomenclature of compounds. In binary compounds, which may be regarded as derivatives of more electronegative elements, it has become customary to put the name of the less electronegative first. For example, the binary compound of iodine and chlorine, an interhalogen compound, is called iodine chloride, ICl. Similarly, The binary compound between oxygen and fluorine is writer as OF_2; that between oxygen and chlorine is written as Cl_2O. Some names in use since long have however, been retained although the electronegativity of the first element is more *e.g.* NH_3.

4. Calculation of partial ionic character of a covalent bond. The development of ionic character in a covalent bond between tow atoms, say, A and B, is due to the difference in the electronegativities, greater would be the development of ionic character and as a result, higher would be the stability of the resulting bond.

Pauling proposed the following correlation between the electronegativity differences $\chi_A - \chi_B$ and the percent ionic character of the covalent bond:

Percent ionic character = $[1 - e^{-0.25(\chi A - \chi B)}] \times 100$

Hanry & Smyth (1946) revised the Pauling equation as follows:

% ionic character = $16\,(\chi_A - \chi_B) + 3.5\,(\chi_A - \chi_B)^2$

They have shown that bond in hydrogen fluoride is 43% ionic rather than 60% as assumed by Pauling.

Example. The electronegativity difference of the elements in HF is 1.9. Calucatte the percent ionic character of the H–F bond.

Solution : According to Pauling's equation,

Percent ionic character of H-F bond = $[1 - e^{-0.25(1.9)}] \times 100$

= **37.77**

5. Calculation of bond length. If the two atoms A and B bonded together through a covalent bond vary in their electronegativities; then the covalent bond would acquire some ionic character, as illustrated above. In other words, the bond acquires polarity. The greater the polarity, the shorter would be the length of the bond formed between A and B Shoemake and Stevenson proposed the following empirical equation for calculating the bond lengthin such cases.

$$D_{A-B} = r_A + r_B - 0.09\,(\chi_A - \chi_B)$$

Where D_{A-B} is the actual bond distance between A and B, r_A and r_B are the covalent radii of A and B and $\chi_A - \chi_B$ is the electronegativity difference. The normal length of the covalent bond between A and B would have been equal to $r_A + r_B$. Evidently, the shortening of the bond between A and B due to electronegativity difference is equal to $0.09\,(\chi_A - \chi_B)$. Due to decrease in length, the bond would be more stable.

Let us calculate the bond distance N–O with the help of above equation. Covalent radii of nitrogen and oxygen atoms are, respectively, 0.75Å and 0.74Å and the electronegativities of N and O are 3.0 and 3.5, respectively.

Thus,

$$D_{N-O} = 0.75 + 0.74 - 0.09\,(3.5 - 3.0) = \mathbf{1.445Å}$$

The actual N–O bond length is 1.46Å

Similarly, the length of the C–O bond is given by

$$D_{C-O} = 0.77 + 0.74 - 0.09\,(3.5 - 2.5) = \mathbf{1.42Å}$$

The experimental value is exactly the same.

6. Explanation for bond angles. Lower the electronegativity of the central atom in a polyatomic molecule, lesser would be the magnitude of the bond angle. This may be explained polyatomic molecule, lasser would be the magnitude of the bond angle. This may be explained as follows: If the electronegativity of the central atom is less, it could not be able to hold the bonding electron pairs towards itself. Therefore, the bonding electron pairs would shift more towards the other atoms attached to the central atom. This would lead to the reduction in the bond pair-bond pair repulsion and hence a decrease in the bond angle.

For illustration, the bond angles and electronegativities of trifluorides of the fifth group elements are given below. As expected, the decrease in electronegativity of the central atom results in a decrease of the bond angle.

Table 5.7: Electronegativity of central atom and Bond angle of compound

Compound	Electronegativity of central atom	Bond angle
NF_3	3.0	102º
PF_3	2.1	97º
AsF_3	2.0	96º
SbF_3	1.9	88º

7. Explanation of diagonal relationship. The electronegativity increases as we go from Li (1.0) to Be (1.5) (variation of electronegativity in a period) but it decreases as we move from Be to Mg (1.2) (variation of electronegativity in a group). Consequently, these two opposite changes (one along the period and the other down the group), as we move diagonally, both these effects partly cancel each other and there is no marked change in electronegativity. This is the reason why Li (1.0) and Mg (1.2) have close values of electronegativities.

Because of similar values of electronegativities and turn of electron-attracting powers, Li and Mg have similar chemical properties.

8. Percent ionic character of bond. The electronegativity scale provides a ready method of expressing the percent ionic character. This was first done by pauling in an empirical manner by establishing a relation between the amount of ionic character of a single bond between two atoms A and B and their electronegativity difference $\chi_A - \chi_B$.

According to Pauling. "The bond between two atoms is 50% ionic, if the difference in electronegativity of two atoms is 2.1 id the difference in the electronegative value is more or less than 2.1, the bond is correspondingly more or less ionic".

A curve is obtained for halogen halides by plotting electronegativity difference against percent of ionic character. The percent ionic character for this purpose has been obtained from the work of Henry and Smyth who developed the following relationships:

Table 5.8: Differences between Electronegativity and Electron Affinity

No.	Electronegativity	No.	Electron affinity
1.	It is the tendency of an atom in a molecule to attract covalent electrons toward itself.	1.	It is the tendency of an isolated atom to attract an electron.
2.	It is the property of atoms in a bonded state.	2.	It is the property of an isolated atom.
3.	It has no unit.	3.	It is measured in kcal/mole or kJ/mole.
4.	It is a comparative value.	4.	Its values are actual.
5.	It is a periodic property.	5.	It is not a periodic property.

POLARIZABILITY

Anions formed due to gain of electrons and cations formed due to loss of electron. Anions are larger in size than cations, therefore, electron clouds are less tightly held around anions. Therefore a small highly charged cation shall be able to destroy the electron cloud of the large anions in a manner that it increases the electron density between the nuclei. For example, the large iodide ion, the anion is pulled towards the positive lithium ion. The iodide ion is said to be **polarized.** The polarization effect introduces covalence in lithium iodide, LiI molecule.

The power of an ion to destroy the other ion is known as its polarizing power and the tendency of the ion to the ion to distort is known as its polarizability. The polarizing power of a cation is proportional to its charge radius ratio. This is illustrated in where a large and soft anion comes under the influence of a small cation. The cation is able to polarize the anion. This result in a deformation of electron cloud on anion, called the polarization of anion. Due to small size of the cation, polarization of the cation is not much. Extensive polarization would be witnessed when cation penetrates the anionic electron cloud giving a covalent bond. In a compound consisting of large negative ions and small positive ions he polarization tends to partial covalent character in ionic compounds. Examples of such ionic-covalent compounds are $FeCI_3$, $AICI_3$, LiBr etc.

Table 5.9: Polarizing Power of Cations

polarizing power of cations				
Li^+	Be^{2+}	Al^{3+}	O^{2-}	F^-
Na^+	Mg^{2+}	high	S^{2-}	Cl^-
K^+	Ca^{2+}		Se^{2-}	Br^-
Rb^+	Sr^{2+}		high	I^-
Cs^+	Ba^{2+}	polarizability of anions		

For polarisation (covalent character) to be significant for a bond, the cation must be polarising and the anion must be polarisable.if the cation is strongly polarising, the anion must be polarisable if the cation is moderately polarising, the anion must be very polarisable

Bonds between polarising cations and polarizable anions have significant covalent character. The polarizability of the anion depends both on its size and charge.

Anions of lower charge density are more polarizable.

For anions having the same charge, the larger anion is polarised to a greater extent $F^- < Cl^- < Br^- < I^-$.

For anions of similar size, the more polarizable anion has a greater negative charge $S^{2-} - > Cl^-$

This is because a larger charge indicates a larger excess of electrons over protons.

FAJANS' RULES

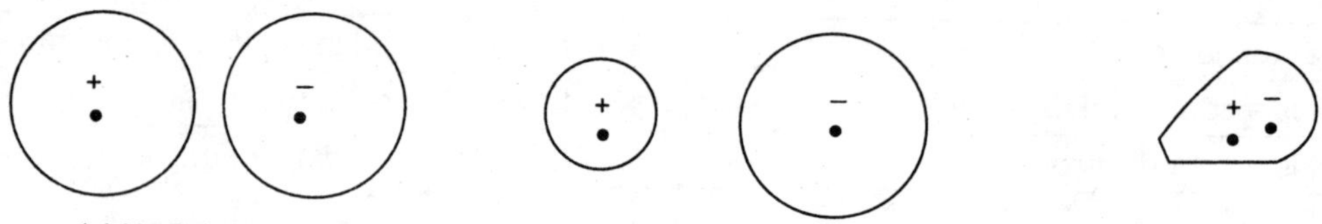

(a) No Polarization (b) Polarized anion (c) Polarization feading to a covalent bond

Fig.5.12: Interaction between cation and anion

Chemical bonds are broadly classed as covalent bond and ionic bond. In both cases the energy of the bonded atoms is lower than the energy of the separated atoms due to an electrostatic force of attraction between the bonded atoms. In covalent bonds the electrostatic attraction is between the bonding electrons and the nuclei and core electrons of both atoms. In ionic bonds it is the attraction of the positive ion for the negative ion. In fact, very few bonds are purely covalent or purely ionic. For example the covalent OH bond in water is polarised due to the difference in electronegativity of hydrogen and oxygen.

Ionic **bonds** may have considerable **covalent character** if the **cation** is **small** and/or **highly charged**. These cations are said to be **polarising** (or have high polarising power) because they distort the electron cloud of the anion, and the electron density between the nuclei is higher than in the corresponding ionic bond.

Fajans' rules are formulated by Kazimierz Fajans in 1923. Rules are used to predict whether a chemical bond will be covalent or ionic, and depend on the charge on the cation and the relative sizes of the cation and anion. They can be summarized in the following table:

Table 5.10: Chart illustrating the relationship between atomic and ionic radius

Ionic	**Covalent**
Low positive charge	High positive charge
Large cation	Small cation
Small anion	Large anion

Example 1. Why Nacl is ionic in nature but and AlI_3 is Covalent.

Reason– sodium chloride, NaCl has a low positive charge (+1), a fairly large cation (~1 Å) and relatively small anion (0.2 Å) so, it is ionic in nature.But aluminium iodide AlI_3 has a high positive charge (+3) and a large anion so it is covalent in nature.

Polarization will be increased by:

i. high charge and small size of the cation.

ii. Ionic potential Å Z^+/r^+ (= polarizing power)

iii. High charge and large size of the anion

iv. The polarizability of an anion is related to the deformability of its electron cloud.

v. An incomplete valence shell electron configuration

vi. Noble gas configuration of the cation produces better shielding and less polarizing power *e.g.* Hg^{2+} (r^+ = 102 pm) is more polarizing than Ca^{2+} (r^+ = 100 pm)

Factors that Influence Polarizability

The relationship between polarizability and the factors of electron density, atomic radii, and molecular orientation is as follows:

1. The greater the number of electrons, the less control the nuclear charge has on charge distribution, and thus the increased polarizability of the atom.
2. The greater the distance of electrons from nuclear charge, the less control the nuclear charge has on the charge distribution, and thus the increased polarizability of the atom.
3. Molecular orientation with respect to an electric field can affect polarizibility (labeled Orientation-dependent), except for molecules that are: tetrahedral, octahedral or icosahedral (labeled Orientation-independent). This factor is more important for unsaturated molecules that contain areas of electron dense regions, such as 2, 4-hexadiene. Greatest polarizability in these molecules is achieved when the electric field is applied parallel to the molecule rather than perpendicular to the molecule.

OXIDATION STATE

An oxidation state is a number that is assigned to an element in a chemical combination. This number represents the number of electrons that an atom can gain, lose, or share when chemically bonding with an atom of another element. The terms "oxidation state" and "oxidation number" are often used interchangeably. The transfer of electrons is described by the oxidation state of the molecule. One might mistaken formal charge for oxidationstate but they are different. Oxidation state is commonly used to determine the changes in redox reactions and is numerically similar to valence electrons, but different from formal charge. Formal charge determines the arrangement of atoms and the likelihood of the molecule existing.

IUPAC definition of oxidation state is "Oxidation state of an atom is the charge of this atom after ionic approximation of its heteronuclear bonds,and the term oxidation number is nearly synonymous".

The oxidation state, sometimes referred to as oxidation number, describes degree of oxidation (loss of electrons) of an atom in a chemical compound. Oxidation states are typically represented by integers which may be positive, zero, or negative. In some cases, the average oxidation state of an element is a fraction, such as 8⁄3 for iron in magnetite.

- The **highest oxidationstate** is +9 in the iridium tetroxide cation (IrO_4^+).It is predicted that even a +10 oxidation state may be achievable by platinum in the platinum tetroxide dication (PtO_4^{2+}).
- The **lowest oxidation state** is −4, as for carbon in methane or for chromium in $[Cr(CO)_4]^{4-}$.

- The increase in **oxidation** state of an atom, through a chemical reaction, is known as an oxidation; a decrease in oxidation state is known as a **reduction.**
- In inorganic nomenclature, the **oxidation state is represented by a Roman numeral** placed after the element name inside a parenthesis or as a superscript after the element symbol.

Oxidation State Rules

To effectively assign oxidation states to a compound, the seven basic rules must be followed in order.

RULE 1: Any individual atom uncombined with other elements has the oxidation state of 0 (zero).

Ex. The OS for Ag is 0. The oxidation state for O (oxygen) or O_2 is 0 as long as it is uncombined with any other element.

RULE 2: The total sum of the oxidation state of all atoms in any given species is equal to the net charge on that species.

(a) In neutral species, the total sum of the oxidation state of all atoms is 0. **Ex.** The sum of OS for NaCl is 0 since the OS of Na = +1 and the OS of Cl = –1, therefore NaCl total OS = 0.

(b) In ions, the total sum of the oxidation state is the charge of the ion. **Ex.** The OS of Ca^{2+} (Calcium) is = +2. The total sum of the OS of all atoms in CrO_4^{2-},Chromate ion is –2. The total sum of the oxidation states of all atoms in CH_3COO^-, Acetate ion is –1.

RULE 3: In a compound, the Oxidation state for Group 1metal is +1 and for Group 2metal, the oxidation state is +2.

Ex. In NaCl, Na has the oxidation state of +1 since it is a Group 1 Alkali metal. Cl would have an oxidation number of –1 to make the sum of the oxidation states 0 (Rule 2). In $MgCl_2$, Mg has the oxidation state of +2, since it is a Group 2 Alkaline Earth metal. Cl would have an oxidation state of –1, and since there are 2 Cl atoms, the overall charge of the species would again be 0 (Rule 2).

RULE 4: The oxidation state of FLOURINE is –1 in a compound.

Ex. OS of F is –1 in HF, SF6.

RULE 5: The oxidation state of HYDROGEN is +1 in a compound.

Ex. OS of H is +1 in HI, CH_4, NH_4^+.

RULE 6: The oxidation state of OXYGEN is –2 in a compound.

Ex. OS of Oxyen is –2 in OH^-, H_2O, CO_3^{2-}.

RULE 7: In two-element compounds with metals, Group 15(3A) elements have the oxidation state of –3, Group 16(6A) elements have the oxidation state of –2, and Group 17(7A) elements have the oxidation state of –1.

Ex. In $MgBr_2$, Br has the oxidation state of –1, since it is a Group 17 element. In Li_2S, S has the oxidation state of –2, since it is a Group 16.

Another method for Determination of Oxidation numbers

1. The oxidation number of an atom is zero in a neutral substance that contains atoms of only one element. Thus, the atoms in O_2, O_3, P_4, S_8 and aluminum metal all have an oxidation number of 0.

$[AlCl_4]^-$	Al^{3+}	4
$[AlF_6]^{3-}$	Al^{3+}	6

The Lewis acid in coordination complexes, often called a **central metal** ion (or atom), is often a transition metal or inner transition metal, although main group elements can also form coordination compounds. The Lewis base donors, called **ligands**, can be a wide variety of chemicals—atoms, molecules, or ions. The only requirement is that they have one or more electron pairs, which can be donated to the central metal. Example -The coordination number for the silver ion in $[Ag(NH_3)_2]^+$ is two

How To Calculate Coordination Number

Here are the steps for identifying the coordination number of a coordination compound.

- Identify the central atom in the chemical formula. Usually, this is a transition metal.
- Locate the atom, molecule, or ion nearest the central metal atom. To do this, find the molecule or ion directly beside the metal symbol in the chemical formula of the coordination compound. If the central atom is in the middle of the formula, there will be neighboring atoms/molecules/ ions on both sides.
- Add the number of atoms of the nearest atom/molecule/ions. The central atom may only be bonded to one other element, but you still need to note the number of atoms of that element in the formula. If the central atom is in the middle of the formula, you'll need to add up the atoms in the entire molecule.
- Find the total number of nearest atoms. If the metal has two bonded atoms, add together both numbers,

Coordination Number Geometry

There are multiple possible geometric configurations for most coordination numbers.

Coordination Number 2 – linear

Coordination Number 3 – trigonal planar (*e.g.*, CO_3^{2-}), trigonal pyramid, T-shaped

Coordination Number 4 – tetrahedral, square planar

Coordination Number 5 – square pyramid (*e.g.*, oxovanadium salts, vanadyl VO^{2+}), trigonal bipyramid,

Coordination Number 6 – hexagonal planar, trigonal prism, octahedral

Coordination Number 7 – capped octahedron, capped trigonal prism, pentagonal bipyramid

Coordination Number 8 – dodecahedron, cube, square antiprism, hexagonal bipyramid

Coordination Number 9 – three-face centered trigonal prism

Coordination Number 10 – bicapped square antiprism

Coordination Number 11 – all-faced capped trigonal prism

Coordination Number 12 – cuboctahedron (*e.g.*, Ceric ammonium nitrate –$(NH_4)_2Ce(NO_3)_6$)

2. The oxidation number of simple ions is equal to the charge on the ion. The oxidation number of sodium in the Na^+ ion is +1, for example, and the oxidation number of chlorine in the Cl^- ion is –1.
3. The oxidation number of hydrogen is +1 when it is combined with a nonmetal as in CH_4, NH_3, H_2O, and HCl.
4. The oxidation number of hydrogen is –1 when it is combined with a metal as in. LiH, NaH, CaH_2, and $LiAlH_4$.
5. The metals in Group IA form compounds (such as Li_3N and Na_2S) in which the metal atom has an oxidation number of +1.
6. The elements in Group IIA form compounds,such as Mg_3N_2 and $CaCO_3$ in which the metal atom has a +2 oxidation number.
7. Oxygen usually has an oxidation number of –2. Exceptions include molecules and polyatomic ions that contain O–O bonds, such as O_2, O_3, H_2O_2, and the O_2^- ion.
8. The elements in Group VIIA often form compounds such as AlF_3, HCl, and $ZnBr_2$, in which the nonmetal has a –1 oxidation number.
9. The sum of the oxidation numbers in a neutral compound is zero.
   $H_2O : 2(+1) + (-2) = 0$
10. The sum of the oxidation numbers in a polyatomic ion is equal to the charge on the ion. The oxidation number of the sulfur atom in the SO_4^{2-} ion must be +6, for example, because the sum of the oxidation numbers of the atoms in this ion must equal –2.
    $SO_4^{2-} : (+6) + 4(-2) = -2$
11. Elements toward the bottom left corner of the periodic table are more likely to have positive oxidation numbers than those toward the upper right corner of the table. Sulfur has a positive oxidation number in SO_2, for example, because it is below oxygen in the periodic table.
    $SO_2 : (+4) + 2(-2) = 0$

EXERCISE

Example 1. Find the oxidation state of Cr in CrO_4^{-2}.

Steps:

1. From Rule –2, we know that the sum of OS for this compound is –2.
2. From Rule –6, we know that Oxygen in a compound is –2. Since we have four oxygen, –2 × 4 = –8.
3. With those information, set up an short equation $Cr^+(-8) = -2Cr^+(-8) = -2$
   So the oxidation state of **Cr is +6.**

Example 2. Find the oxidation state of S in $S_2O_3^{2-}$

Steps:

1. From Rule –2, we know that the sum of OS for this compound is –2
2. From Rule –6, we know that Oxygen in a compound is –2. Since we have three oxygen, –2 × 3 = –6
3. With those information, set up an equation $2S + (-6) = -2$

Formula will be

$2S - 6 = -2, 2S = 4$

$S = 4/2 = 2$

The oxidation state for S is 2.

Example 3. Find the oxidation state of C in $C_2H_3O_2^-$

Steps:

1. From Rule –2, we know that the sum of OS for this compound is –1
2. From Rule –5, we know that Hydrogen in a compound is +1. Since we have three Hydrogen, $+1 \times 3 = +3$
3. From Rule –6, we know that Oxygen in a compound is –2. Since we have two oxygen, $-2 \times 2 = -4$

Formula will be

$2C + (+3) + (-4) = -12C + (+3) + (-4) = -1$

$2C = 0$

$C = 0/2 = 0$

The oxidation state for C is 0.

Example 4. Fractional Values for Oxidation States The oxidation states are usually in whole numbers, but in some cases, they are in fractional numbers.

Steps: Consider Fe_3O_4.

1. Using rule –2, the sum of the oxidation state of all atoms is 0
2. Using rule –6, the O has the oxidation state of –2. $4 \times (-2) = -8$
3. Equation: $3Fe + (-8) = 0$
4. $3Fe = 8$
5. $Fe = 8/3$

Confidence Building Questions

- **What is the oxidation state of Mo in MoO_4^{2-}?**

  Skill:

  Note that $(+6) + 4*(-2) = -2$, the charge of the ion MoO_4^{2-}. An oxidation state is a signed number, in this case +6.

- **What is the oxidation state of S in SO_3^{2-}?**

  Discussion:

  Even though S is in group VIA: O, S, Se, Te, Bi, the sulfur in sulfite ions has an oxidation number +4.

- **What is the oxidation state of As in AsO_4^{3-}?**

  Discussion:

  This is the normal oxidation state of group VA: N, P, As, Sb, Bi.

- **What is the oxidation state of Br in BrO_3^-?**

  Discussion:

  Br has many oxidation states, *e.g.* +7 in BrO_4^-.

- **What is the oxidation state of As in As_2S_3?**

  Discussion:

  The oxidation for S is –2, since it is in the group VIA: O, S, Se, Te, Po. Super! S being in the same group as O, has an oxidation state of –2.

- **What is the oxidation state of Cr in $Cr(OH)_3$?**

  Discussion:

  There are three OH^- groups in the compound. Chromium hydroxide, $Cr(OH)_3$, is a greenish gelatinous solid.

- **In the reaction, $P_4(s) + 5O_2(g) = P_4O_{10}(s)$ which element is reduced?**

Write as others

What are the oxidation states of O in O_2 and P_4O_{10}? This equation represents the burning of phosphorus in air. P_4O_{10} dissolves in water to form phosphoric acid.

- Assign the oxidation numbers of the atoms Al, Xe and Cr in the following compounds.

  (a) Al_2O_3 (b) XeF_4 (c) $K_2Cr_2O_7$

- Arrange the following compounds in order of increasing oxidation state for the carbon atom.

  (a) CO (b) CO_2 (c) H_2CO (d) CH_3OH (e) CH_4

COORDINATION NUMBERS AND GEOMETRY

Coordination Number–Definition

1. "A coordination number is the number of ions and atoms around a central neutron, or the total number of attachment locations to the central element in a molecule".
2. "The coordination number is the sum of the atoms in a given molecule bonded to the central metal atom of a compound".

Coordination numbers typically ranges from 2 to 16 but is usually 6. Coordination number, also called Ligancy, the number of atoms, ions, or molecules that a central atom or ion holds as its nearest neighbours in a complex or coordination compound or in a crystal.

Table 5.11: Co-ordination numbers of metal ions

Example	metal atom	Coordination number
$[Mo(CN)_8]_4^-$ and $[Sr(H_2O)_8]_2^+$	Mo, Sr	8
$[ZrF_7]_3^-$	Zr	7
$Fe(CN)_6^{4-}$, $Co(NH_3)_6^{3+}$ and $Ni(NH_3)_6^{2+}$	Fe^{2+}, Co^{3+}, Ni^{2+}	6
$[Zn(CN)_4]^{2-}$, $[Cu(CN)_4]^{3-}$, and $[Ni(CN)_4]^{4-}$	Zn, Cu, Ni	4
$[Ag(NH_3)_2]^+$, $[AuCl_2]^-$, and $[HgCl_2]$	Ag, Au, Hg	2

2. The oxidation number of simple ions is equal to the charge on the ion. The oxidation number of sodium in the Na^+ ion is +1, for example, and the oxidation number of chlorine in the Cl^- ion is –1.
3. The oxidation number of hydrogen is +1 when it is combined with a nonmetal as in CH_4, NH_3, H_2O, and HCl.
4. The oxidation number of hydrogen is –1 when it is combined with a metal as in. LiH, NaH, CaH_2, and $LiAlH_4$.
5. The metals in Group IA form compounds (such as Li_3N and Na_2S) in which the metal atom has an oxidation number of +1.
6. The elements in Group IIA form compounds,such as Mg_3N_2 and $CaCO_3$ in which the metal atom has a +2 oxidation number.
7. Oxygen usually has an oxidation number of –2. Exceptions include molecules and polyatomic ions that contain O–O bonds, such as O_2, O_3, H_2O_2, and the O_2^- ion.
8. The elements in Group VIIA often form compounds such as AlF_3, HCl, and $ZnBr_2$, in which the nonmetal has a –1 oxidation number.
9. The sum of the oxidation numbers in a neutral compound is zero.

   $H_2O : 2(+1) + (-2) = 0$
10. The sum of the oxidation numbers in a polyatomic ion is equal to the charge on the ion. The oxidation number of the sulfur atom in the SO_4^{2-} ion must be +6, for example, because the sum of the oxidation numbers of the atoms in this ion must equal –2.

    $SO_4^{2-} : (+6) + 4(-2) = -2$
11. Elements toward the bottom left corner of the periodic table are more likely to have positive oxidation numbers than those toward the upper right corner of the table. Sulfur has a positive oxidation number in SO_2, for example, because it is below oxygen in the periodic table.

    $SO_2 : (+4) + 2(-2) = 0$

EXERCISE

Example 1. Find the oxidation state of Cr in CrO_4^{-2}.

Steps:

1. From Rule –2, we know that the sum of OS for this compound is –2.
2. From Rule –6, we know that Oxygen in a compound is –2. Since we have four oxygen, $-2 \times 4 = -8$.
3. With those information, set up an short equation $Cr^+(-8) = -2Cr^+(-8) = -2$

   So the oxidation state of **Cr is +6.**

Example 2. Find the oxidation state of S in $S_2O_3^{2-}$

Steps:

1. From Rule –2, we know that the sum of OS for this compound is –2
2. From Rule –6, we know that Oxygen in a compound is –2. Since we have three oxygen, $-2 \times 3 = -6$
3. With those information, set up an equation $2S + (-6) = -2$

Formula will be

$2S - 6 = -2, 2S = 4$

$S = 4/2 = 2$

The oxidation state for S is 2.

Example 3. Find the oxidation state of C in $C_2H_3O_2^-$

Steps:

1. From Rule –2, we know that the sum of OS for this compound is –1
2. From Rule –5, we know that Hydrogen in a compound is +1. Since we have three Hydrogen, $+1 \times 3 = +3$
3. From Rule –6, we know that Oxygen in a compound is –2. Since we have two oxygen, $-2 \times 2 = -4$

Formula will be

$2C + (+3) + (-4) = -12C + (+3) + (-4) = -1$

$2C = 0$

$C = 0/2 = 0$

The oxidation state for C is 0.

Example 4. Fractional Values for Oxidation States The oxidation states are usually in whole numbers, but in some cases, they are in fractional numbers.

Steps: Consider Fe_3O_4.

1. Using rule –2, the sum of the oxidation state of all atoms is 0
2. Using rule –6, the O has the oxidation state of –2. $4 \times (-2) = -8$
3. Equation: $3Fe + (-8) = 0$
4. $3Fe = 8$
5. $Fe = 8/3$

Confidence Building Questions

- **What is the oxidation state of Mo in MoO_4^{2-}?**

  Skill:

  Note that $(+6) + 4*(-2) = -2$, the charge of the ion MoO_4^{2-}. An oxidation state is a signed number, in this case +6.

- **What is the oxidation state of S in SO_3^{2-}?**

  Discussion:

  Even though S is in group VIA: O, S, Se, Te, Bi, the sulfur in sulfite ions has an oxidation number +4.

- **What is the oxidation state of As in AsO_4^{3-}?**

  Discussion:

  This is the normal oxidation state of group VA: N, P, As, Sb, Bi.

- **What is the oxidation state of Br in BrO_3^-?**

  Discussion:

  Br has many oxidation states, *e.g.* +7 in BrO_4^-.

- **What is the oxidation state of As in As_2S_3?**

  Discussion:

  The oxidation for S is –2, since it is in the group VIA: O, S, Se, Te, Po. Super! S being in the same group as O, has an oxidation state of –2.

- **What is the oxidation state of Cr in $Cr(OH)_3$?**

  Discussion:

  There are three OH^- groups in the compound. Chromium hydroxide, $Cr(OH)_3$, is a greenish gelatinous solid.

- **In the reaction, $P_4(s) + 5O_2(g) = P_4O_{10}(s)$ which element is reduced?**

Write as others

What are the oxidation states of O in O_2 and P_4O_{10}? This equation represents the burning of phosphorus in air. P_4O_{10} dissolves in water to form phosphoric acid.

- Assign the oxidation numbers of the atoms Al, Xe and Cr in the following compounds.
  (a) Al_2O_3 (b) XeF_4 (c) $K_2Cr_2O_7$
- Arrange the following compounds in order of increasing oxidation state for the carbon atom.
  (a) CO (b) CO_2 (c) H_2CO (d) CH_3OH (e) CH_4

COORDINATION NUMBERS AND GEOMETRY

Coordination Number–Definition

1. "A coordination number is the number of ions and atoms around a central neutron, or the total number of attachment locations to the central element in a molecule".
2. "The coordination number is the sum of the atoms in a given molecule bonded to the central metal atom of a compound".

Coordination numbers typically ranges from 2 to 16 but is usually 6. Coordination number, also called Ligancy, the number of atoms, ions, or molecules that a central atom or ion holds as its nearest neighbours in a complex or coordination compound or in a crystal.

Table 5.11: Co-ordination numbers of metal ions

Example	metal atom	Coordination number
$[Mo(CN)_8]_4^-$ and $[Sr(H_2O)_8]_2^+$	Mo, Sr	8
$[ZrF_7]_3^-$	Zr	7
$Fe(CN)_6^{4-}$, $Co(NH_3)_6^{3+}$ and $Ni(NH_3)_6^{2+}$	Fe^{2+}, Co^{3+}, Ni^{2+}	6
$[Zn(CN)_4]^{2-}$, $[Cu(CN)_4]^{3-}$, and $[Ni(CN)_4]^{4-}$	Zn, Cu, Ni	4
$[Ag(NH_3)_2]^+$, $[AuCl_2]^-$, and $[HgCl_2]$	Ag, Au, Hg	2

$[AlCl_4]^-$	Al^{3+}	4
$[AlF_6]^{3-}$	Al^{3+}	6

The Lewis acid in coordination complexes, often called a **central metal** ion (or atom), is often a transition metal or inner transition metal, although main group elements can also form coordination compounds. The Lewis base donors, called **ligands**, can be a wide variety of chemicals—atoms, molecules, or ions. The only requirement is that they have one or more electron pairs, which can be donated to the central metal. Example -The coordination number for the silver ion in $[Ag(NH_3)_2]^+$ is two

How To Calculate Coordination Number

Here are the steps for identifying the coordination number of a coordination compound.

- Identify the central atom in the chemical formula. Usually, this is a transition metal.
- Locate the atom, molecule, or ion nearest the central metal atom. To do this, find the molecule or ion directly beside the metal symbol in the chemical formula of the coordination compound. If the central atom is in the middle of the formula, there will be neighboring atoms/molecules/ions on both sides.
- Add the number of atoms of the nearest atom/molecule/ions. The central atom may only be bonded to one other element, but you still need to note the number of atoms of that element in the formula. If the central atom is in the middle of the formula, you'll need to add up the atoms in the entire molecule.
- Find the total number of nearest atoms. If the metal has two bonded atoms, add together both numbers,

Coordination Number Geometry

There are multiple possible geometric configurations for most coordination numbers.

Coordination Number 2 – linear

Coordination Number 3 – trigonal planar (*e.g.*, CO_3^{2-}), trigonal pyramid, T-shaped

Coordination Number 4 – tetrahedral, square planar

Coordination Number 5 – square pyramid (*e.g.*, oxovanadium salts, vanadyl VO^{2+}), trigonal bipyramid,

Coordination Number 6 – hexagonal planar, trigonal prism, octahedral

Coordination Number 7 – capped octahedron, capped trigonal prism, pentagonal bipyramid

Coordination Number 8 – dodecahedron, cube, square antiprism, hexagonal bipyramid

Coordination Number 9 – three-face centered trigonal prism

Coordination Number 10 – bicapped square antiprism

Coordination Number 11 – all-faced capped trigonal prism

Coordination Number 12 – cuboctahedron (*e.g.*, Ceric ammonium nitrate –$(NH_4)_2Ce(NO_3)_6$)

Table 5.12: Co-ordination numbers and shapes of metal ions

Coordination numbers and shapes of some complex ions			
Coordination Number	**Shape**		**Examples**
2	Linear	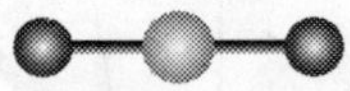	$[CuCl_2]^-$, $[Ag(NH_3)_2]^+$, $[AuCl_2]^-$
4	Square planar	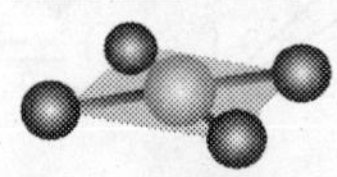	$[Ni(CN)_4]^{2-}$, $[PdCl_4]^{2-}$, $[Pt(NH_3)_4]^{2+}$, $[Cu(NH_3)_4]^{2+}$
4	Tetrahedral	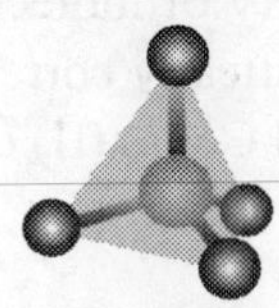	$[Cu(CN)_4]^{3-}$, $[Zn(NH_3)_4]^{2+}$, $[CdCl_4]^{2-}$, $[MnCl_4]^{2-}$
6	Octahedral	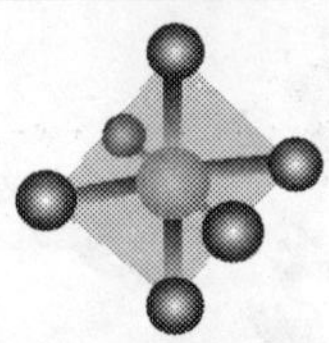	$[Ti(H_2O)_6]^{3+}$, $[V(CN)_6]^{4-}$, $[Cr(NH_3)_4Cl_2]^+$, $[Mn(H_2O)_6]^{2+}$, $[FeCl_6]^{3-}$, $[Co(en)_3]^{3+}$

Complex with a specific coordination number will have a choice of several possible geometries (*i.e.*, there is not a one-to-one correspondence between coordination number and geometry).

Coordination Number 2

This arrangement is not very common for first row transition metal ion complexes and some of the best known examples are for Silver(I). In this case we have a low charge and an ion at the right hand side of the d-block indicating smaller size. Bond angle = 180°.

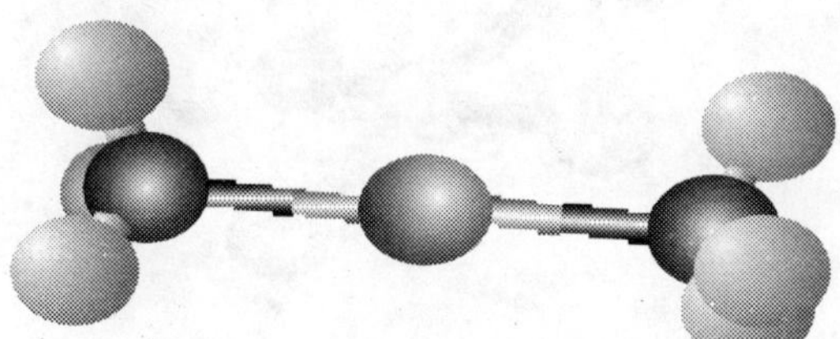

Fig.5.13: The linear $[Ag(NH_3)_2]^+$ ion

Coordination Number 4

Two different geometries are possible tetrahedron and square planar. The tetrahedron is the more common while the square planar is found almost exclusively with metal ions having a d^8 electronic configuration.

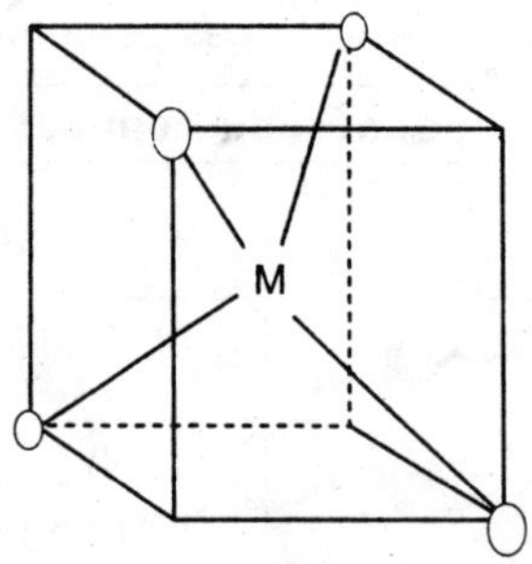

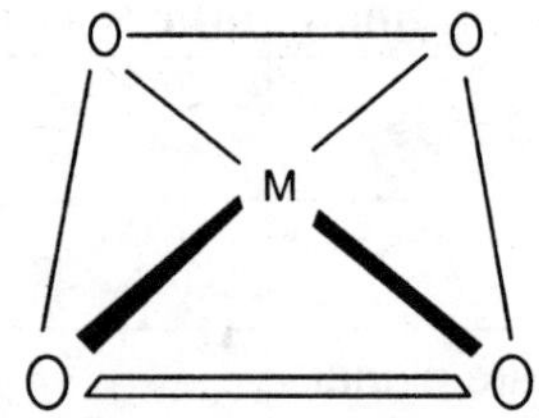

- **Tetrahedral Geometry:** The chemistry of molecules centered around a tetrahedral C atom is covered in organic courses. To be politically correct, please change all occurrences of C to Co. There are large numbers of tetrahedral Cobalt(II) complexes known.

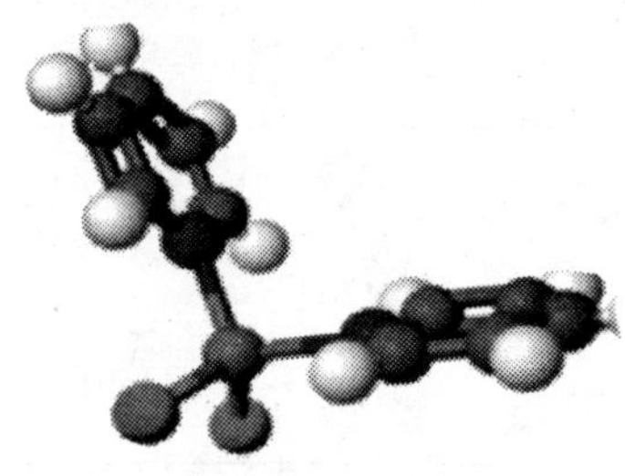

Fig.5.14: $CoCl_2pyr_2$

- **Square Planar Geometry:** This is fairly rare and is included only because some extremely important molecules exist with this shape.

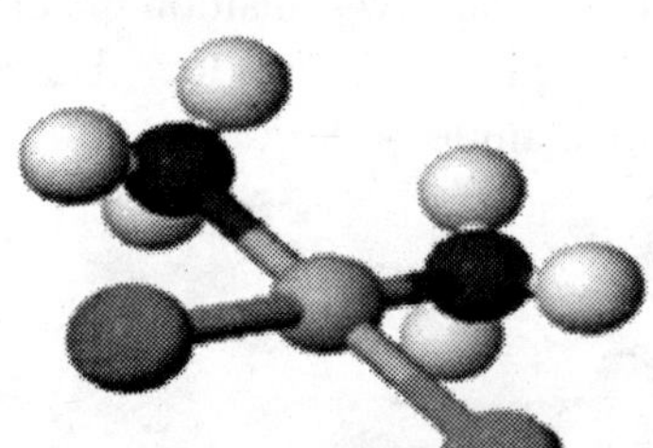

Fig.5.15: cisplatin –cis–$PtCl_2(NH_3)_2$

Coordination Number 6

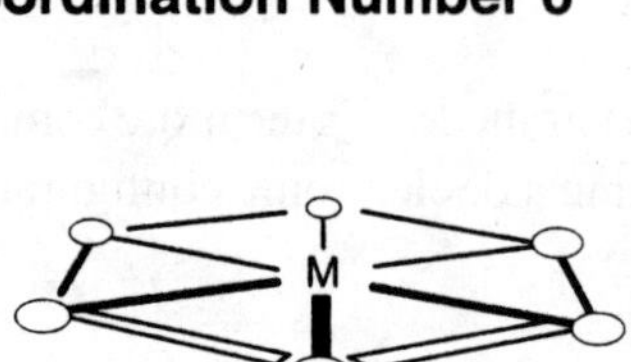

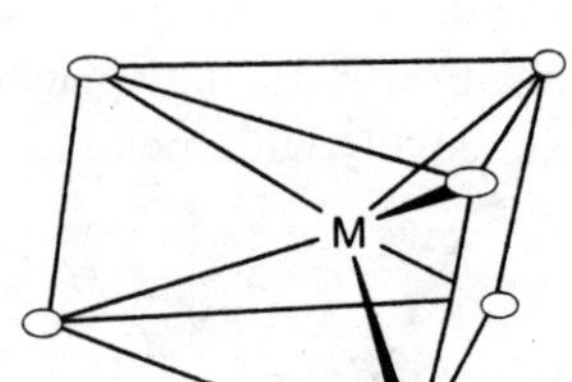

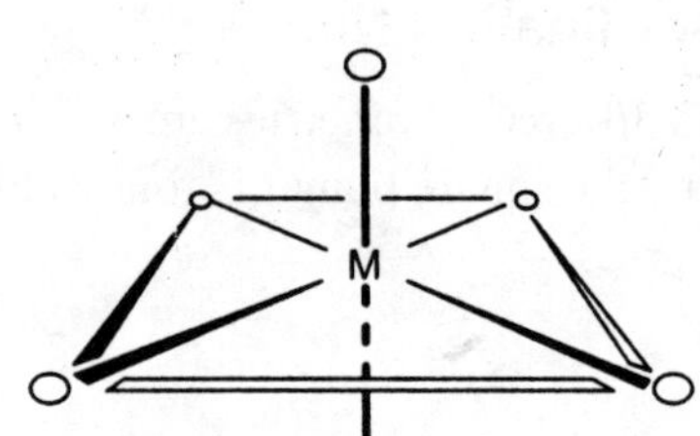

- **Hexagonal planar Geometry:** Unknown for first row transition metal ions, although the arrangement of six groups in a plane is found in some higher coordination number geometries.
- **Trigonal prism Geometry:** Most trigonal prismatic compounds have three bidentate ligands such as dithiolates or oxalates and few are known for first row transition metal ions.
- **Octahedral Geometry (Oh):** The most common geometry found for first row transition metal ions, including all aqua ions. In some cases distortions are observed and these can sometimes be explained in terms of the Jahn-Teller Theorem.

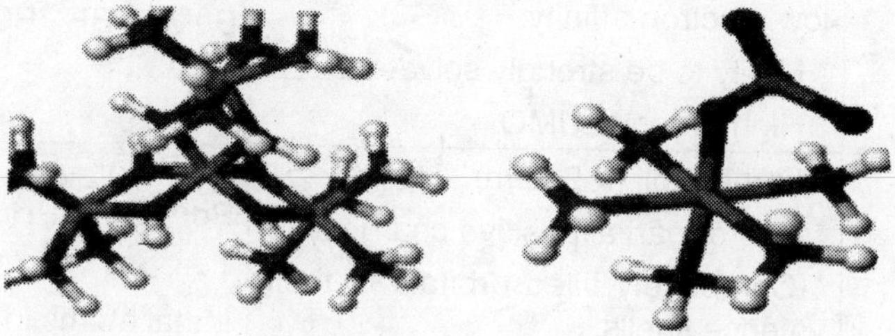

Fig.5.16: $[Co(en)_3]Cl_3$(left) and $[CoCO_3(NH3)_5]^+$ (right)

HARD AND SOFT ACIDS AND BASES

Hard and Soft Acids and Bases (HSAB) Theory is a qualitative concept introduced by Ralph Pearson to explain the stability of metal complexes and the mechanisms of their reactions.

HSAB Principle: According to HSAB concept, hard acids prefer binding to the hard bases to give ionic complexes, whereas the soft acids prefer binding to soft bases to give covalent complexes.

- The large electronegativity differences between hard acids and hard bases give rise to strong ionic interactions.
- The electronegativities of soft acids and soft bases are almost same and hence have less ionic interactions. *i.e.*, the interactions between them are more covalent.
- The interactions between hard acid - soft base or soft acid - hard base are mostly polar covalent and tend to be more reactive or less stable. The polar covalent compounds readily form either more ionic or more covalent compounds if they are allowed to react.

The Lewis acids and bases can be further divided into hard or soft or border line types.

In short, Hard acids and bases are small and non-polarizable, whereas Soft acids and bases are larger and more polarizable.

Characteristics of hard and soft acids, bases and borderline acids & bases are summarized in table.

Table 5.13: Characteristics of Hard, Soft & Borderline Acids & Bases

Type of Acid/Base	CHARACTERISTICS	EXAMPLES
Hard acids	* Atomic centres of small ionic radii (<90 pm). * High positive charge. * Empty orbitals in their valence shells. * Low electronegativity (0.7–1.6) and low electron affinity. * Likely to be strongly solvated. * High energy LUMO.	H^+, Li^+, Na^+, K^+, Be^{2+}, Mg^{2+}, Ca^{2+}, Sr^{2+}, Sn^{2+} Al^{3+}, Ga^{3+}, In^{3+}, Cr^{3+}, Co^{3+}, Fe^{3+}, Ir^{3+}, La^{3+}, Si^{4+}, Ti^{4+}, Zr^{4+}, Th^{4+}, VO^{2+}, UO_2^{2+} $BeMe_2$, BF_3, BCl_3, $B(OR)_3$, $AlMe_3$
Soft acids	* Large radii (> 90 pm). * Low or partial positive charge. * Completely filled orbitals in their valence shells. * Intermediate electronegativities (1.9–2.5) * Low energy LUMO's with large magnitude of LUMO coefficients.	Cu^+, Ag^+, Au^+, Hg^+, Cs^+, Tl^+, Hg^{2+}, Pd^{2+}, Cd^{2+}, Pt^{2+} Metal atoms in zero oxidation states
Border line acids		Fe^{2+}, Co^{2+}, Ni^{2+}, Cu^{2+}, Zn^{2+}, Pb^{2+}, $B(CH_3)_3$, SO_2, NO^+
Hard bases	* Small radii (around 120pm) & highly solvated. * electronegative atomic centres (3.0-4.0). * Weakly polarizable. * Difficult to be oxidized. * High energy HOMO.	H_2O, OH^-, F^-, Cl^-, $CH_3CO_{2^-}$, PO_4^{3-}, SO_4^{2-}, CO_3^{2-}, NO_3^-, ClO_4^-, ROH, RO^-, R_2O, NH_3, RNH_2, N_2H_4
Soft bases	* Large atoms (> 170 pm) with intermediate electronegativity (2.5–3.0). * High polarizability * Easily undergo oxidation. * Low energy HOMO's but large magnitude HOMO coefficients.	RSH, RS^-, R_2S, I^-, CN^-, SCN^-, $S_2O_3^-$, R_3P, R_3As $(RO)_3P$, RNC, CO, C_2H_4, C_6H_6, R^-, H^-
Border line bases		Aniline, pyridine, N_3^-, Br^-, NO_2^-, SO_3^{2-}, N_2

APPLICATIONS OF HSAB PRINCIPLE

1. **Application of in hydrogen bonding:** The strong hydrogen bond is possible in cases of H_2O, NH_3 and HF, since the donor atoms (F, O & N) are hard lewi bases and their interactions with partially positively charged H, which is a hard acid, are stronger.

2. **Linkage of ambidentate ligands to metal atoms:** The ambidentate ligand, SCN can bind either by S end or N end. The bonding mode can be determined by using HSAB principle.

   It bonds through sulfur atom (soft base) when bonded to Pt^{2+}, a soft acid.

   However it bonds through nitrogen atom (a hard base) when linked to Cr^{3+}, a hard acid.

3. **Site preference in organic reactions:** RCOX is a hard acid and reacts with the nitrogen end of SCN^- ion to form an acyl isothiocyanate.

$$CH_3COX + SCN^- \xrightarrow{-X^-} RC(O)NCS$$

   Whereas the softer methyl group bonds to the Sulfur atom and forms methyl thiocyanate.

$$CH_3X + SCN^- \xrightarrow{-X^-} CH_3SCN$$

4. **Inorganic reactions:** HSAB principle is used to predict the outcome of some of the reactions.

   (1) The following reaction is possible because As is softer than P and I^- is softer than F^-.

$$AsF_3 + PI_3 \longrightarrow AsI_3 + PF_3$$

   Remember that both As and P are soft but relatively As is softer.

   (2) The following reaction is possible since Mg^{2+} is harder acid than Ba^{2+} and O^{2-} is harder base than S^{2-}.

$$MgS + BaO \longrightarrow MgO + BaS$$

5. **Precipitation reactions:** The softer acids like Ag^+, Hg^+, Hg^{2+} etc., and border line acids like Fe^{2+}, Ni^{2+}, Cu^{2+}, Zn^{2+}, Pb^{2+} etc., can be precipitated as sulfides from their aqueous solutions since S^{2-} ion is a softer base.

MOLECULAR GEOMETRIES

Molecular geometry is the three-dimensional arrangement of the atoms that constitute a molecule. It influences several properties of a substance including its reactivity, polarity, phase of matter, color, magnetism and biological activity. The angles between bonds that an atom forms depend only weakly on the rest of molecule, *i.e.* they can be understood as approximately local and hence transferable properties.

Determination of molecular Geometry

- The molecular geometry can be determined by various spectroscopic methods and diffraction methods.
- IR, microwave and Raman spectroscopy can give information about the molecule geometry from the details of the vibrational and rotational absorbance detected by these techniques.
- X-ray crystallography, neutron diffraction and electron diffraction can give molecular structure for crystalline solids based on the distance between nuclei and concentration of electron density.
- Gas electron diffraction can be used for small molecules in the gas phase. NMR and FRET methods can be used to determine complementary information including relative distances dihedral angles, angles, and connectivity.
- Molecular geometries are best determined at low temperature because at higher temperatures the molecular structure is averaged over more accessible geometries.

Types of molecular structure

A bond angle is the geometric angle between two adjacent bonds. Some common shapes of simple molecules include:

Linear: In a linear model, atoms are connected in a straight line. The bond angles are set at 180°. For example, carbon dioxide and nitric oxide have a linear molecular shape.

Trigonal planar: Molecules with the trigonal planar shape are somewhat triangular and in one plane (flat). Consequently, the bond angles are set at 120°. For example, boron trifluoride.

Bent: Bent or angular molecules have a non-linear shape. For example, water (H_2O), which has an angle of about 105°. A water molecule has two pairs of bonded electrons and two unshared lone pairs.

Tetrahedral: Tetra- signifies four, and -hedral relates to a face of a solid, so "tetrahedral" literally means "having four faces". This shape is found when there are four bonds all on one central atom, with no extra unshared electron pairs. In accordance with the VSEPR (valence-shell electron pair repulsion theory), the bond angles between the electron bonds are $\arccos(-1/3) = 109.47°$. For example, methane (CH_4) is a tetrahedral molecule.

Octahedral: Octa- signifies eight, and -hedral relates to a face of a solid, so "octahedral" means "having eight faces". The bond angle is 90 degrees. For example, sulfur hexafluoride (SF_6) is an octahedral molecule.

Trigonal pyramidal: A trigonal pyramidal molecule has a pyramid-like shape with a triangular base. Unlike the linear and trigonal planar shapes but similar to the tetrahedral orientation, pyramidal shapes require three dimensions in order to fully separate the electrons. Here, there are only three pairs of bonded electrons, leaving one unshared lone pair. Lone pair – bond pair repulsions change the bond angle from the tetrahedral angle to a slightly lower value.[9] For example, ammonia (NH_3).

Table 5.14: The Table Below Summarizes The Molecular And Electron-Pair Geometries For Different Combinations Of Bonding Groups And Nonbonding Pairs Of Electrons On The Central Atom.

No. of bonding groups/domains on 'central' atom	No. of lone pair electrons on 'central' atom	Electron-pair Geometry	Molecular Geometry	Bond Angle in degree
2	0	linear	linear	180
3	0	trigonal planar	trigonal planar	120
2	1	trigonal planar	bent	less than 120
4	0	tetrahedral	tetrahedral	109.5
3	1	tetrahedral	trigonal pyramidal	less than 109.5
2	2	tetrahedral	bent	less than 109.5
5	0	trigonal bipyramidal	trigonal bipyramidal	90, 120 and 180
4	1	trigonal bipyramidal	seesaw	90, 120 and 180
3	2	trigonal bipyramidal	T-shaped	90 and 180

2	3	trigonal bipyramidal	linear	180
6	0	octahedral	octrahedral	90 and 180
5	1	octahedral	square pyramidal	90 and 180
4	2	octahedral	square planar	90 and 180

Molecular Geometry Summary Sheet

Shaded squares represent geometries which give non-polar molecules when all substitents X are identical

Table 5.15: The Table Below Summarizes The Molecular And Electron-Pair Geometries And Shapes

Total No. of Groups of e–	Electron Pair Geometry (Hybridization)	Approximate Bond Angle	No. of Bonding Directions (No. of X)	No. of Lone Pairs (No. of E)	Geometry Name (VSEPR class)	Shape	Examples
2	linear (sp)	180º	2	0	**linear** (AX_2)		BeH_2, CO_2
3	trigonal planar (sp^2)	120º	3	0	**trigonal planar** (AX_3)		BF3, NO_3^-
			2	1	**bent** (AX_2E)		
4	tetrahedral (sp^3)	109.5º	4	0	**tetrahedral** (AX_4)	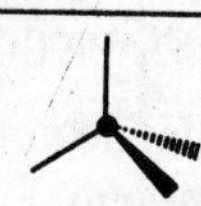	CH_4
			3	1	**trigonal pyramidal** (AX_3E)		NH_3
			2	2	**bent** (AX_2E_2)		H_2O
5	trigonal bipyramidal (sp^3d)	120º (in plane) & 90º (above & below)	5	0	**trigonal bipyramidal** (AX_5)	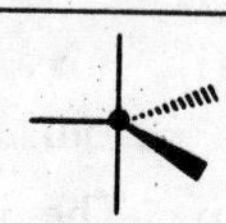	PCl_5
			4	1	**seesaw** (AX_4E)		SF_3
			3	2	**T-shaped** (AX_3E_2)	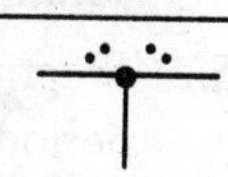	ClF_3
			2	3	**linear** (AX_2E_3)	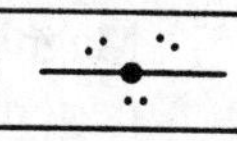	XeF_2

6	octahedral (sp^3d^2)	90º	6	0	**octahedral** (AX_6)	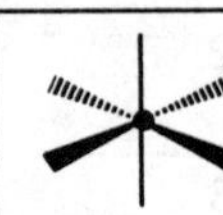	SF_6
			5	1	**spuare pyramidal** (AX_5E)		BrF_5
			4	2	**square planar** (AX_4E_2)		XeF_4

Note: for bent molecular geometry when the electron-pair geometry is trigonal planar the bond angle is slightly less than 120 degrees, around 118 degrees. For trigonal pyramidal geometry the bond angle is slightly less than 109.5 degrees, around 107 degrees. For bent molecular geometry when the electron-pair geometry is tetrahedral the bond angle is around 105 degrees.

LONG ANSWER TYPE QUESTIONS

1. What do you understand by periodic properties? Write the name of four periodic properties and explain in detail the factors which affect these properties.
2. What do you understand by ionisation potential? Describe its periodicity and the factors which affect it. Describe electrical method for its measurement.
3. Give a brief description of the factors which affect the ionisation potential.
4. What do you understand by electron affinity? Describe its periodicity in the periodic table. Describe the factors which influence the electron affinity.
5. What do you understand by electronegativity? Show its periodicity in the periodic table. Describe the factors which affect it.
6. What is the main difference between electronegativity and electron affinity? Give application of electronegativity.
7. What do you understand by iso-electronic ions?
8. Explain the following giving reasons:
   (i) For a metal, why is the value of second ionisation potential higher than the value of first ionisation potential?
   (ii) The first ionisation potential for nitrogen is higher than that for oxygen.
   (iii) In size, chloride ion is bigger than chlorine atom, whereas sodium ion is smaller than sodium atom.
   (iv) Magnitude of electronegativity increases when we move from left to right in a period of periodic table.
   (v) Magnitude of electron affinity decreases when we move from top to bottom in a group of periodic table generally.
9. (a) Why is the first ionisation energy of Mg greater than that of Na and Al?
   (b) Why is the electron affinity of fluorine is less than that of chlorine?
   (c) Why is the ionisation energy of boron lesser than beryllium?

10. How do the following properties vary in the periodic table?
    (i) Ionisation energy (ii) Atomic radii
    (iii) Electronegativity.
11. Explain the variation of atomic size in groups and periods.
12. Discuss the statement that the size of a neutral atom is less than that of anion and more than that of cation.
13. Define ionisation energy. Talking example of Aluminium, explain which ionization potential will be more and why? How ionization potential explains chmical behaviour of elements?
14. What do you understand by 'Ionisation potential'? Which factors affect its value and what are its trends in the periodic table?
15. Define electronegativity. Describe Pauling's method of its determination. How can you determine the nature of bond with the help of electronegativity?
16. Explain electronegativity. In the following pairs which will have the higher electronegativity?
    (i) F and Cl (ii) Li and K.
17. Explain, with examples, the difference between covalent radii and van der Waal's radii. Explain, in short, the iso-electronic ions.
18. Discuss the variation in (i) ionisation potential and (ii) atomic radius moving across a period in accordance with the increasing a atomic numbers.
19. What is meant by the size of an atom? Discuss briefly the various types of atomic radii.
20. Define atomic, ionic, covalent, metallic and van der Waal's radii for an elements.
21. Explain the following:
    (i) Pauling's electronegativity scale.
    (ii) Electronegativity increases as the s-character increases in the hybrid orbitals.
    (iii) The second ionisation energy of an element is much higher than the first one.
    (iv) There occurs a decrease in the first ionisation energy from Be to B and from N to O.
    (v) The firs ionisation energy of transition elements is reasonable constant.
    (vi) (a) The first ionisation energy values of Nb (4) and Ta (73) are very close.
    (b) The ionisation potential values of the isoelectronic species : H, He^+, Li^{2+} and Be^{2+} are 13.6, 54.4, 122.4 and 217.7 eV, respectively, Comment on these values.
    (vii) Why are the electron affinities of beryllium and magnesium almost zero?
    (viii) Radius of chloride ion is larger than that of chlorine atom. Explain.
    (ix) First ionisation energy of Al is less than that of Mg.
    (x) Electron affinities of noble gases are zero.
    (xi) Ionic radius of K^+ is smaller than that of Cl although they possess isoelectronic configuration.
    (xii) Change is size from Li to Be is much grater than that from K to Ca.
    (xiii) There is variation of ionisation energy in a period of periodic table.

(xiv) Electronegativity varies as F > Cl > Br > I.

(xv) First ionisation energy of nitrogen is greater than that of oxygen.

(xvi) van der Waal's radius of chlorine is greater than covalent radius.

(xvii) Number of electrons in Na^+ and Mg^{2+} ions are same but energy required to remove electron from Mg^{2+} is greater than Na^+ ion, why?

(xviii) Explain why F^- ion is more easily formed from F as compared to Cl^- ion from Cl.

SHORT ANSWER TYPE QUESTIONS

1. (a) What do you understand by the terms (i) Atomic radius (ii) Ionic radius?

   (b) The atomic radii increase on moving down a group but decrease in moving from left to right in the period. Comment on this statement.

2. (a) Define ionisation enrgy.

   (b) Describe the factors which affect the ionisation potential of the atoms.

   (c) Discuss the variation in ionisation energy in a period and in a group.

3. (a) Explain why van der Waal's radii is greater than covalent radii.

   (b) Ionisation energy of Al is lesser than Mg.

   (c) K^+ and Cl^- ions are isoelectronic, but K^+ is smaller than Cl^-.

   (d) Arrange the following in increasing order of their radii:

   (i) O^{2-} (ii) F^- (iii) Mg^{2+} (iv) Na^+

   (e) Why are lithium and beryllium placed in the same period in sequence?

   (f) The atomic radii of inert gases are greater.

4. State giving reasons which of the following have higher:

   (i) First ionisation energy, (i) F or Cl; (ii) B or Be.

   (ii) Electron affinity, O or O^-

   (iii) Ionic radius K^+ or Cl^-; and

   (iv) Polarising power, Na^+ or F^-?

5. Explain, why cations are smaller in size while anions are larger in size than corresponding atoms.

6. Arrange the F, Cl, Br and I ion order of their decreasing electron affinity.

7. Nitrogen has first ionisation potential more than that of oxygen. Explain, why?

8. Explain the following giving suitable reasons:

   (i) In a period left to right atomic radius decreases.

   (ii) Be, Mg, N and Ar have electron affinity zero.

   (iii) The radius of a cation is invariably smaller and that of an anioninvariably larger than that of the corresponding atom.

   (iv) The electronegativities of elements A, B and C are 1.0, 2.5 and 3.0 units respectively on Pauling's scale. Why is the tpe of bonding in the compounds AC and BC ionic and covalent respectively?

(v) Electron affinities of noble gases are zero and those of N and P very low.

(vi) Why there is not much change in atomic radius of transition elements? Distinguish between ionic and covalent radii.

9. Suggest the reason for decreasing order in first ionisation energy from N to O and from P to S.
10. Arrange the following in order of increasing atomic radii : Br, Cl, C and O.
11. Write short notes on the following:

(i) Ionisation potential (ii) Electron affinity

(iii) Electronegativity (iv) Atomic radii

(v) Ionic radii (vi) Isoelectronic ions

(vii) Periodicity (viii) Van der Waal's radius

(ix) Shielding effect (x) Covalent radii

(xi) Lanthanide contraction.

12. Discuss variation of electronegativity and electron affinity in a group of periodic table.
13. How are electron affinity and ionisation potential related to electronegativety?
14. Write the following in order of increasing ionisation potential?

Na, Be, Rb, K, Cs, Ca, Al, Ga.

OBJECTIVE TYPE QUESTIONS

1. Which of the following ions is the largest?

(a) Ap^{+} (b) Ba^{++} (c) Mg^{++} (d) Na^{+}

2. Which of the following ions has the smaller ionic radii?

(a) Ca^{+2} (b) Mg^{+2} (c) Zn^{+2} (d) Na^{+}

3. Which is the smallest ion from the following?

(a) Na^{+} (b) Mg^{2+} (c) Al^{3+} (d) Si^{4+}

4. The element with the highest ionisation potential is:

(a) Li (b) H (c) Mg (d) Cs

5. Which of the following elements has the minimum ionisation potential?

(a) N (b) F (c) O (d) Ne

6. On the following elements, which one has the maximum first ionisation potential?

(a) Li (b) Be (c) Na (d) Mg

7. From the following electronic configurations which atom has the minimum ionisation potential?

(a) $1s^2$ (b) $1s^2, 2s^2 2p^6$

(c) $1s^2, 2s^2 2p^6, 3s^1$ (d) $1s^2, 2s^2 2p^6, 3s^2$

8. In the periodic table, the elements of which group have the maximum ionisation potential?

(a) 0 group (b) I-A group (c) II-A group (d) III-A group

9. The first ionisation potential of magnesium atom is less than which one of the following.

(a) Li (b) Na (c) Ca (d) Be

10. The increasing order of the electron affinity of F, Cl, Br and I will be :
    (a) $F < Cl < Br < I$ (b) $Cl < F < Br < I$ (c) $I < Br < Cl < F$ (d) $I < Br < F < Cl$
11. Which one of these elements has zero electron affinity :
    (a) C (b) N (c) B (d) O
12. Which of the following elements has the highest electron affinity :
    (a) F (b) Br (c) Cl (d) I
13. The element of maximum ionisation potential is :
    (a) Be (b) Br (c) Cl (d) Li
14. The element of maximum electronegativity is :
    (a) $[Na]\ 3s^2\ 3p^1$ (b) $[Ne]\ 3s^2\ 3p^2$ (c) $[Ne]\ 3s^2\ 3p^3$ (d) $[Ar]\ 3d^{10}\ 4s^2\ 4p^3$
15. Out of the following electronic configurations, which atom will have the lowest value of ionisation potential :
    (a) $1s^2$ (b) $1s^2, 2s^2\ 2p^6$
    (c) $1s^2, 2s^2\ 2p^6, 3s^2$ (d) $1s^2, 2s^2\ 2p^6, 3s^{23}p^6n$
16. The sizes of I^+, I^- follows the order :
    (a) $I^+ > I^- > I$ (b) $I^- > I > I^+$ (c) $I^- > I^+ > I$ (d) $I > I^- > I^+$
17. Both of elements of 1st period contains valence electrons in
    (a) M shell (b) N shell
    (c) K shell (d) S shell
18. In periodic table, helium is placed at
    (a) top left corner (b) bottom right corner
    (c) bottom left corner (d) top right corner
19. Across period atomic size decreases due to
    (a) shielding effect (b) photoelectric effect
    (c) increase in nuclear force of attraction (d) decrease in nuclear force of attraction
20. First three periods are
    (a) long periods (b) short periods
    (c) moderate periods (d) none of above
21. On basis of electronic configuration group and period of ^9_5B is
    (a) 2 and IIIA (b) 3 and IIA (c) 4 and VIA (d) 5 and VIIA
22. Chemical properties depends upon
    (a) electronic configuration (b) valence shell electronic configuration
    (c) atomic mass (d) none of above
23. Nobel gases are present in
    (a) VIIA group (b) VIA group (c) VA group (d) VIIIA group

24. Energy required to remove an electron from outermost shell is called
    (a) potential energy (b) kinetic energy
    (c) mechanical energy (d) ionization energy
25. Which of the following has the largest size :
    (a) Al (b) Al^+ (c) Al^{2+} (d) Al^{3+}
26. The highest electron affinity among halogens is of :
    (a) F (b) Cl (c) Br (d) I
27. The alkali metals in each period have :
    (a) smallest size (b) lowest ionisation energy
    (c) Highest electron affinity (d) highest electonegativity
28. Which of the following elements has the highest ionisation energy :
    (a) N (b) O
    (c) P (d) S
29. Consider the iso-electonic series : K^+, S^{2-}, Cl^- and Ca^{2+}. The radii of the ions decrease as :
    (a) $Ca^{2+} > K^+ > Cl^- > S^{2-}$ (b) $Cl^- > S^{2-} > K^+ > Ca^{2+}$
    (c) $S^{2-} > Cl^- > K^+ > Ca^{2+}$ (d) $K^+ > Ca^{2+} > S^{2-} > Cl^-$
30. In which of the following pairs, the first atom or ion is not larger than the second :
    (a) Fe^{2+}, Fe^{3+} (b) O, S (c) N, O (d) Cl^-, Cl
31. Which metal is strongest reducing agent?
    (a) Ca (b) Sr (c) Ba (d) Mg
32. The ionisation potential of elements depends upon :
    (a) atomic size (b) nuclear charge of the atom
    (c) number of electrons in inner orbitals (d) all of the above.
33. As we go from top to bottom in a group shielding effect
    (a) doesn't change (b) decreases (c) increases (d) none of above
34. Group A elements are called
    (a) metals (b) alkali metals
    (c) representative elements (d) transition elements

ANSWERS

1. (a)	2. (b)	3. (d)	4. (b)	5. (c)	6. (b)	7. (c)	8. (a)	9. (d)	10. (b)
11. (b)	12. (c)	13. (c)	14. (c)	14. (c)	16. (b)	17. (c)	18. (d)	19. (c)	20. (b)
21. (a)	22. (b)	23. (d)	24. (d)	25. (a)	26. (b)	27. (b)	28. (a)	29. (c)	30. (b)
31. (b)	32. (d)	33. (c)	34. (c)						

6 Stereochemistry and Organic Reactions

Representations of 3 dimensional structures, structural isomers and stereoisomers, configurations and symmetry and chirality, enantiomers, diastereomers, optical activity, absolute configurations and conformational analysis. Isomerism in transitional metal compounds, Introduction to organic reactions

Introduction

Organic molecules are not normally presented in such a straight-forward way. A variety of notations can be used and compared, and molecules can be oriented in space such that their relationship is not obvious. There are two main categories of molecules, those that are in a straight-chain, or acyclic, form, and those that are in a ring, or cyclic, orientation. Information on the different structural notations are given below-

1. Straight-Chain Structures
   - Wedge-Dash Notation
   - Newman Projections
   - Sawhorse Projections
   - Fischer Projections
2. Cyclic Structures
   - Wedge-Dash Notation
   - Haworth Projections
   - Chair Conformations

THREE-DIMENSIONAL REPRESENTATIONS

Representation of three dimensional structures of Organic molecules are

(a) Wedge-Dash Notation

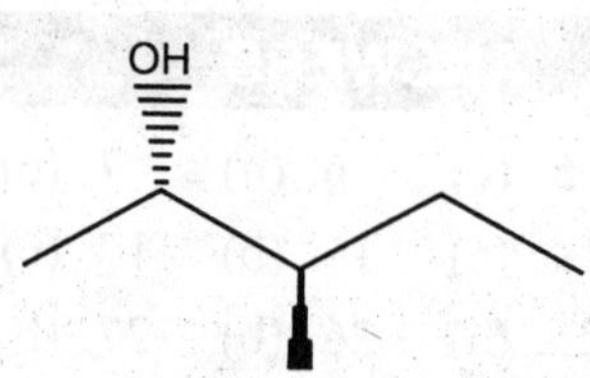

Fig.6.1: Representation of Wedge-Dash Notation

The most commonly-used notation for straight-chain molecules in organic chemistry is the **Wedge-Dash Notation.**

1. In this notation, two bonds are drawn in the plane of the page (sticks), one bond is drawn coming toward you, out of the page (wedged), and one bond is drawn going away from you, behind the page (dashed).
2. In the structure above, the alcohol on the second carbon is behind the page. Since there must be four bonds to each carbon, and two sticks and one dash have been drawn in, the remaining hydrogen must be wedged.
3. **Solid Wedge**- A bond represented by a solid wedge shows that the groups or substituents are facing out of the plane of paper towards the observer.
4. **Light Wedge** - A light wedge or dashed wedge shows that the substituents or groups are facing inside of the plane or below the plane.

(b) Newman Projections

Newman Projections are used mainly for determining conformational relationships. Recall that conformers are molecules that can be converted into one another by a rotation around a single bond. Newman Projections are useful when studying a reaction involving prochiral molecules that have a double bond, in which the addition of a new group creates a new stereo - centre.

The front carbon of this bond is represented by a dot, and the back carbon is represented by a large circle. The three remaining bonds are drawn as sticks coming off the dot (or circle), separated by one another by 120 degrees.

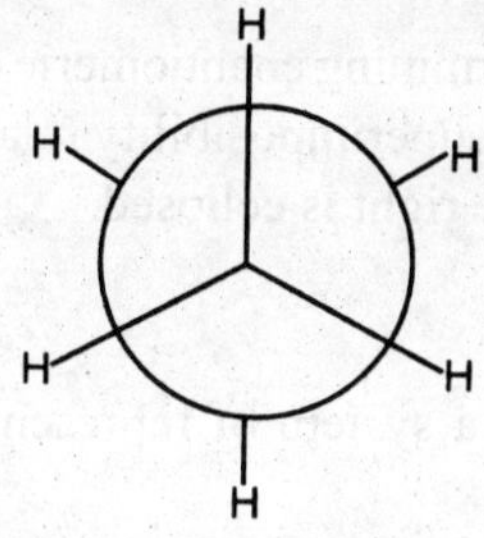

Fig.6.2: Representation of Newman projection

A Newman Projection can be drawn such that the groups on the front carbon are staggered (60 degrees apart) or eclipsed (directly overlapping) with the groups on the back carbon.

Below are two Newman Projections of ethane, C_2H_6. The structure on the left is staggered, and the structure on the right is eclipsed.

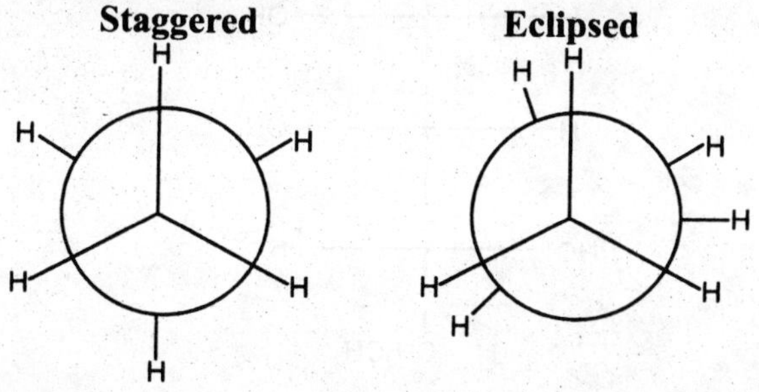

Fig.6.3: Newman projection staggered and eclipsed forms

(c) Sawhorse Projections

Sawhorse Projections are very similar to Newman Projections, but are used more often because the carbon-carbon bond that is compressed in a Newman Projection is fully drawn out in a Sawhorse Projection.

Like with Newman Projections, a Sawhorse Projection is a view of a molecule down a particular carbon-carbon bond, and groups connected to both the front and back carbons are drawn using sticks at 120 degree angles.

Sawhorse Projections can also be drawn so that the groups on the front carbon are staggered (60 degrees apart) or eclipsed (directly overlapping) with the groups on the back carbon.Below are two Sawhorse Projections of ethane.

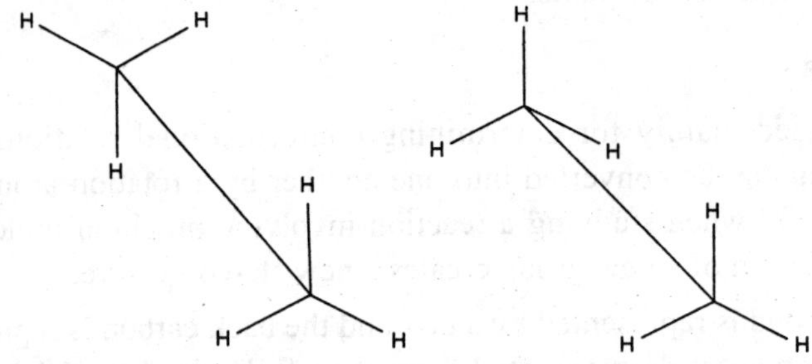

Fig.6.4: Sawhorse projection formulas of ethane

Sawhorse Projections are useful for determining enantiomeric or diasteromeric relationships between two molecules, because the mirror image or superimposibility relationships are clearer. The structure on the left is staggered, and the structure on the right is eclipsed.

(d) Fischer Projections

Fischera German chemist developed such a system of representing the structure in two dimensional known as Fischer Projections .

In a Fischer Projection, each place where the horizontal and vertical lines cross represents a carbon. The vertical lines are actually oriented away from you (similar to dashes in the Wedge-Dash Notation) and the horizontal lines are oriented toward you (similar to wedges in the Wedge-Dash Notation).

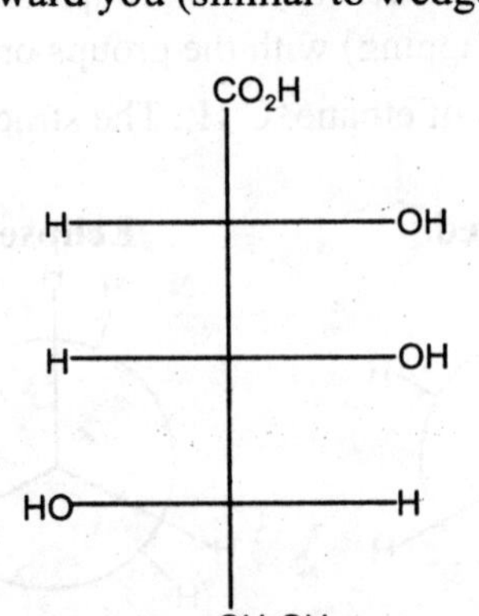

Fig.6.5: Representation of Fischer Projection

Cyclic Structures

Cyclic, or ring, structures can also be drawn in a variety of ways. Rings can have many different sizes, with the smallest ring being cyclopropane (three carbons). The most common ring types in organic chemistry are cyclopentane (five carbons) and cyclohexane (six carbons).

Cyclic Wedge-Dash Notation

Wedge-Dash Notation is also commonly used for cyclic systems. As with straight chains, cyclic Wedge-Dash Notation has two bonds drawn in the plane of the page (sticks), one bond drawn coming toward you, out of the page and above the ring (wedged), and one bond drawn going away from you, behind the page and below the ring (dashed). It is important to note that groups with the same orientation (both wedged or both dashed) are cis to one another, and groups with opposite orientations (one wedged and one dashed) are trans to one another.

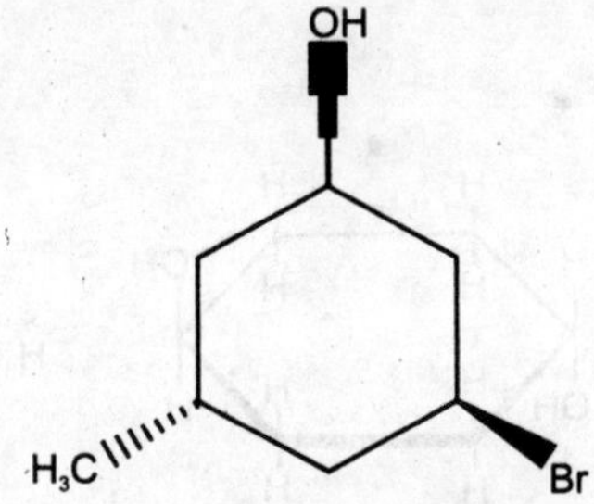

Fig.6.6: Representation of Cyclic Wedge-Dash Notation

In the structure above, the methyl group is behind the page. Since there must be four bonds to each carbon, and two sticks and one dash have already been drawn in, the remaining hydrogen must be wedged.

Haworth Projections

Haworth Projections are commonly used for cyclic sugars. Haworth Projections are useful for being able to clearly determine whether groups are above or below the ring.

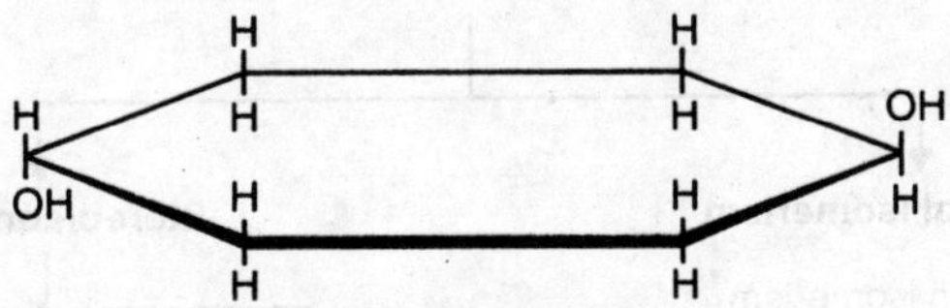

Fig.6.7: Representation of Haworth Projections

Haworth Projections are similar to the cyclic Wedge-Dash notation, in that the ring itself is drawn as planar. The groups on a Haworth Projection are simply oriented perpendicular to the plane of the ring, either up or down. It is important to note that two groups that are both oriented up or both oriented down are cis to one another. Two groups that are oriented one up and one down are trans to one another.

In the structure above, the alcohols on the right "end" carbon is oriented up, above the plane of the ring. The remaining three alcohols are oriented down, below the plane of the ring.

Chair Conformations

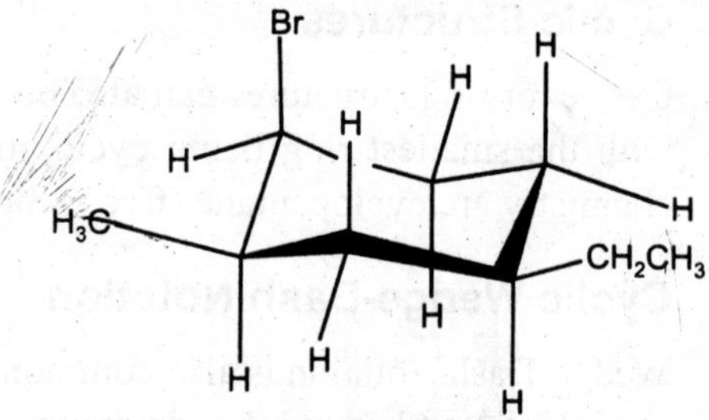

Fig.6.8: Representation of Chair Conformations

Chair Conformations are the most accurate representations of how cyclohexane rings are actually oriented in space. They appropriately reflect the angles between the carbons in the ring and the positions of the groups on each carbon in the ring. However, chairs are commonly harder to draw and interpret for students of organic chemistry.

Chairs show the cyclohexane ring in its true, non-planar, orientation. Chairs can be drawn on paper by drawing three sets of parallel lines. This drawing process is animated below.

From left to right, these structures are the Wedge-Dash Notation, the Haworth Projection, and the Chair Conformation.

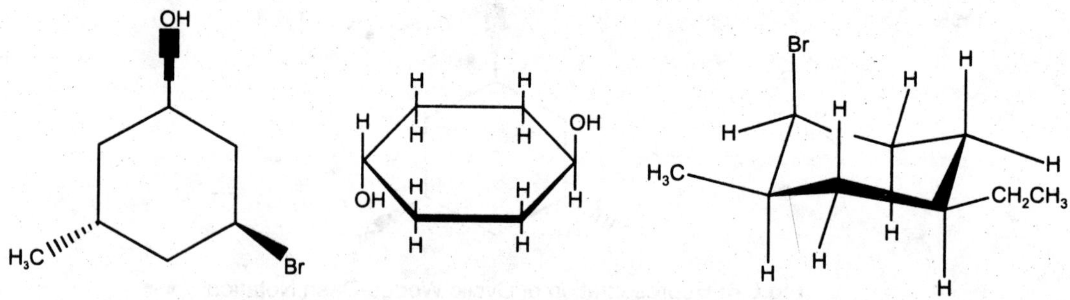

Fig.6.9: Representation of all three dimensional notations

ISOMERISM

In organic chemistry, isomers are molecules with the same molecular formula (i.e. the same number of atoms of each element), but different structural or spatial arrangements of the atoms within the molecule.

Isomerism

- **Structural Isomerism**
  - (1) Chain isomerism
  - (2) Position isomerism
  - (3) Functional isomerism
  - (4) Metamerism
  - (5) Tautomerism
  - (6) Ring-chain isomerism
- **Stereoisomerism**
  - Configurational
    - (1) Geometrical
      - (i) cis
      - (ii) Trans
    - (2) Optical
      - (i) Diesteromers
      - (ii) Enantiomers
  - Confirmational
    - (1) Eclipsed
    - (2) Staggerd
    - (3) Gauche

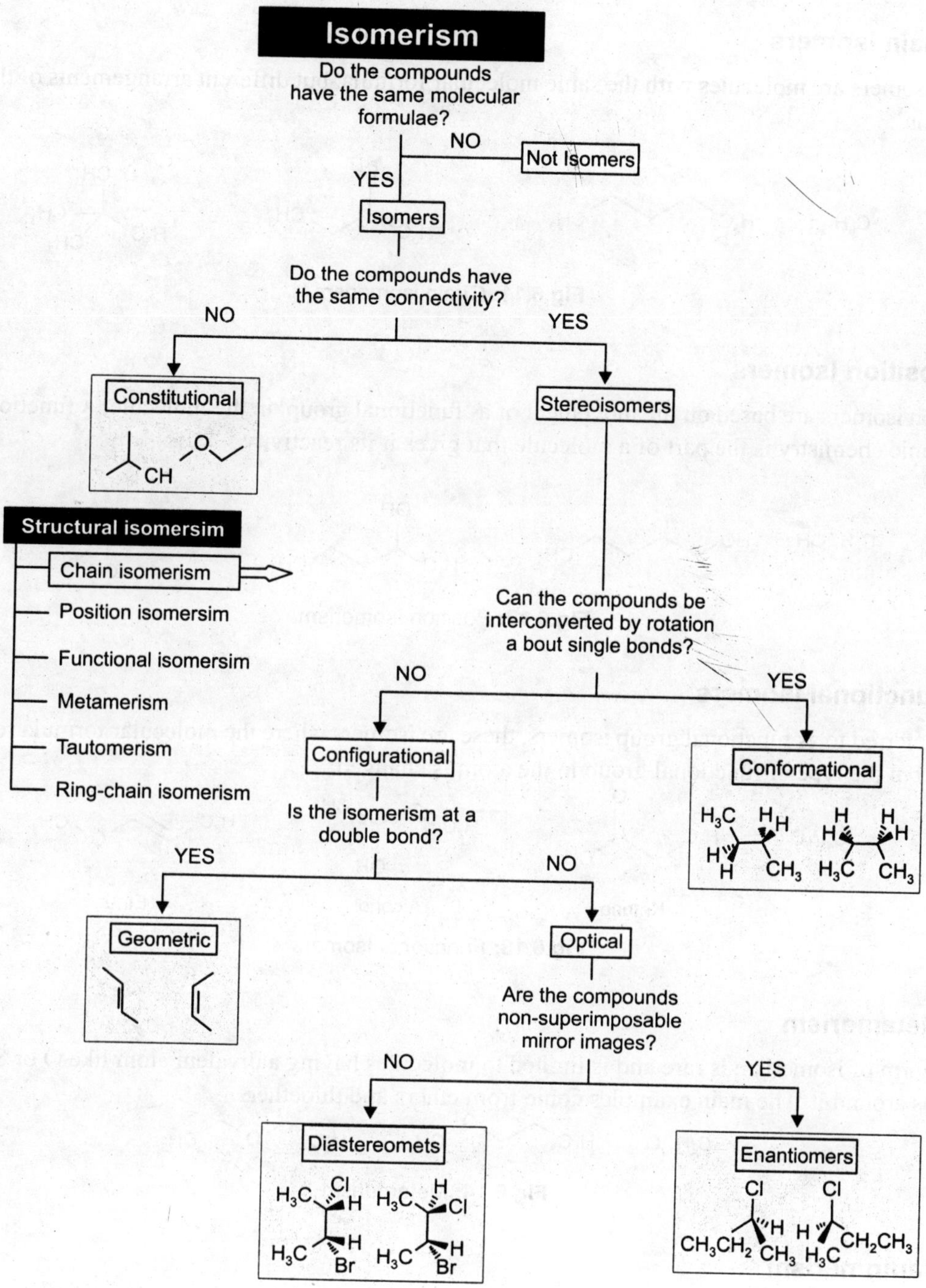

Fig.6.10: Flow chart representing summary of Isomerisation in molecules

STRUCTURAL ISOMERISM

Isomers can be split into two broad groups – structural (or constitutional) isomers, and stereoisomers.

(1) Chain Isomers

Chain isomers are molecules with the same molecular formula, but different arrangements of the carbon 'skeleton'.

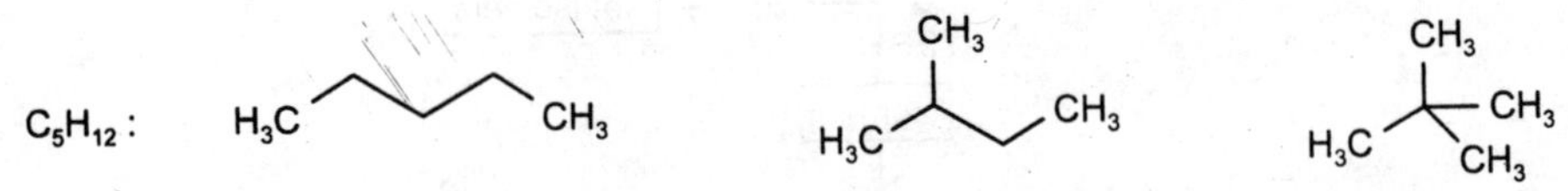

Fig.6.11: Chain isomerism

(2) Position Isomers

Position isomers are based on the movement of a 'functional group' in the molecule. A functional group in organic chemistry is the part of a molecule that gives it its reactivity.

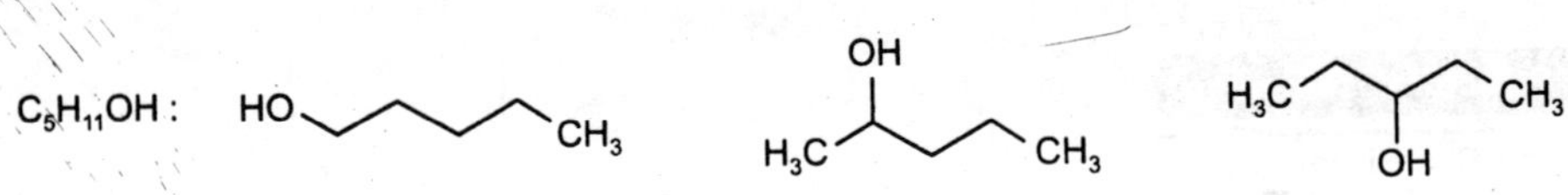

Fig.6.12: Position isomerism

(3) Functional Isomers

Also referred to as functional group isomers, these are isomers where the molecular formula remains the same, but the type of functional group in the atom is changed.

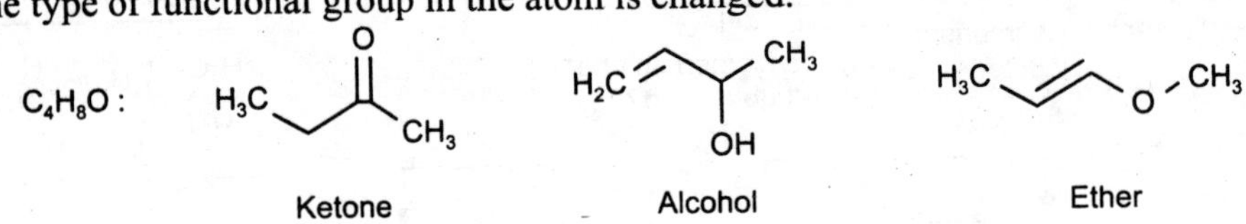

Fig.6.13: Functional Isomers

(4) Metamerism

This form of isomerism is rare and is limited to molecules having a divalent atom like O or S and alkyl groups around it. The main examples come from ethers and thioethers.

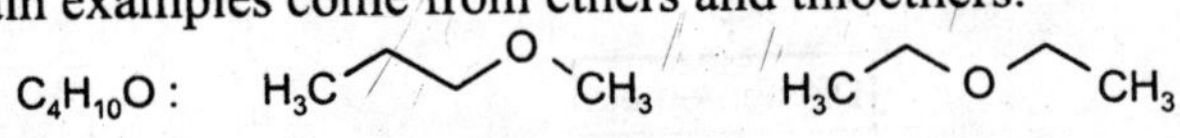

Fig.6.14: Metamerism

(5) Tautomerism

This isomerism is due to spontaneous inter conversion of two isomeric forms with different functional groups. The prerequisites for this is the presence of the C=O, C==N or N=O in the usual cases and an alpha H atom. The most usual is the 'keto-enol' tautomerism, but there can be others like nitro-aci and amine-imine forms.

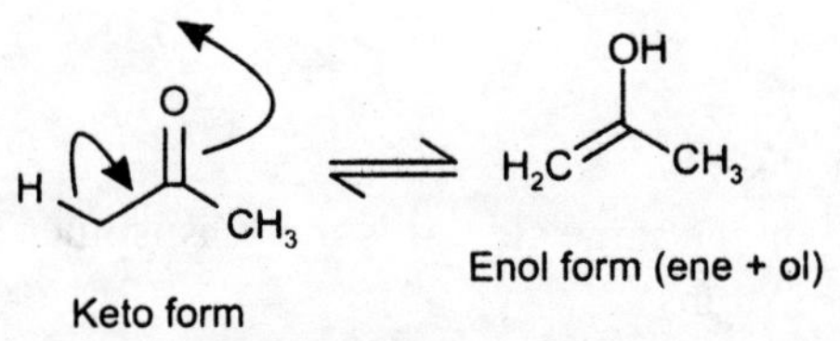

Fig.6.15: Tautomerism

(6) Ring-Chain isomerism

Here one isomer is an open chain molecule and the other a cyclic molecule.

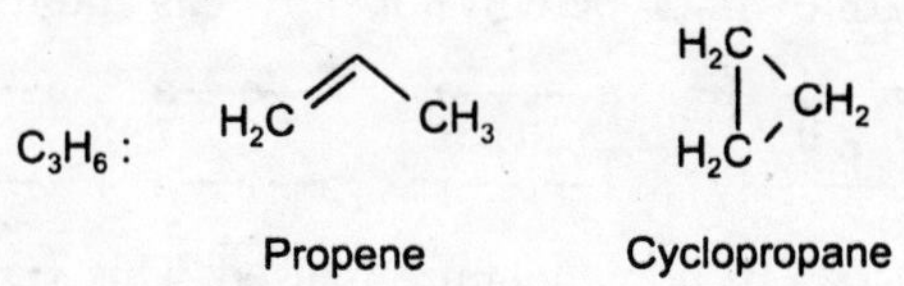

Fig.6.16: Ring-chain isomerism

1. STEREOISOMERISM

Molecules with the same connectivity but different arrangements of the atoms in space are called stereo isomers. It exists between compounds of identical structures and is due to the difference in their configuration.There are two main types of stereoisomerism –

- Geometric isomerism
- Optical isomerism.

Geometrical Isomerism

The isomers which possess the same structural formula but differ in spatial arrangement of groups or atoms around the double bond are known as geometrical isomers and the phenomenon is known as geometrical isomerism.

In this system, the two atoms or groups attached to each of doubly bonded carbon are, put in order of priority on the basis of sequence rules.

The symbol 'E' is assigned to an isomer in which the atoms or groups, of higher priority are on the opposite side, 'E' stands for German word 'Entgegen' which means opposite or ,' across.

The symbol 'Z' is assigned to an isomer in which the atoms or groups of higher priority are on, the same side, 'Z' stands for 'Zusammen' which means together.

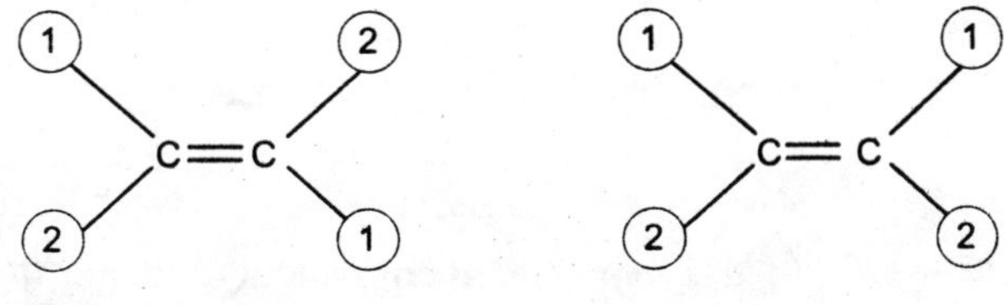

Fig.6.17: Geometrical Isomerism

1. for higher priority groups
2. for lower priority groups.

When the similar groups lie on the same side, it is the 'Cis isomer', while when the similar groups, lie on opposite sides, the isomer is trans.

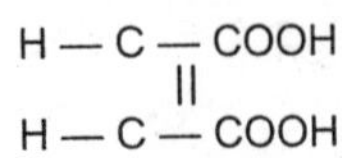

Maleic acid (Cis)

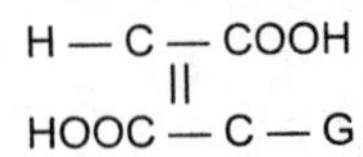

Fumaric acid (trans)

Geometrical isomerism is possible only when each of the doubly bonded Carbon atom has two diffeent groups. Thus the compound of following type will not show Cis and trans isomerism.

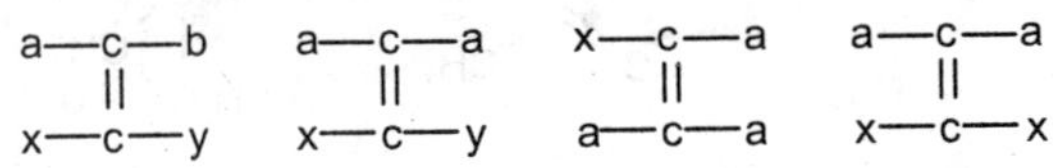

In such cases 'E and Z' system of nomenclature is used. This system is based on priority system given by Cahn, Ingold and Prelog.

Sequence Rules

(a) Higher priority is assigned to .atoms (directly attached to Carbon) of higher atomic number. For example, order of priority in following atoms H, Br, I, Cl will-be I (at no. 53) > Br (at no. 35) > Cl (at no. 17) > H (at no. 1)

(b) If isotopes of same element are attached, the isotope with higher mass number will have higher priority.

(c) In the groups, the order of priority is also decided on the basis of atomic number of first atom of group. For eg, in following

$$-Cl, -OH, -COOH, -NH_2, -SO_3H$$

Order of priority will be

$$\underset{\text{at no.: } 17}{-Cl} > \underset{16}{-SO_3H} > \underset{8}{-OH} > \underset{7}{-NH_2} > \underset{6}{-COOH}$$

When the order of priority of groups cannot be settled on first atom, second atom or subsequent atoms in groups are considered. Hence in following :

$$-CH_2, -CH_2-CH_3, -COOH$$

First atom is C in all groups hence 2nd atom will be considered.

Priority order will be

$$\underset{\text{at no. : 8}}{-COOH} > \underset{\text{at no. : 6}}{CH_2CH_3} > \underset{\text{at no. : 1}}{-CH_2}$$

(*d*) A doubly or triply bonded atom is considered equivalent to two or three such atoms.

For example :

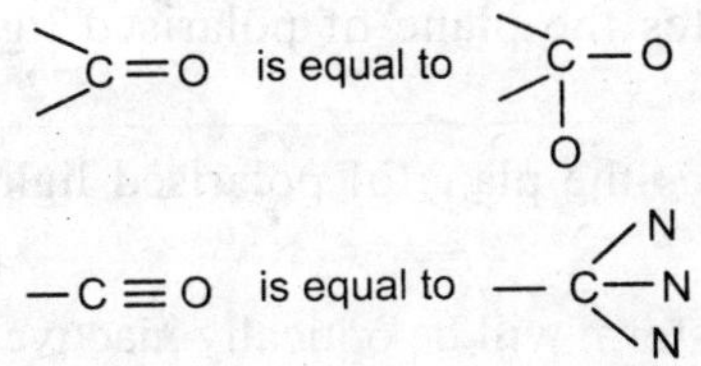

Examples:

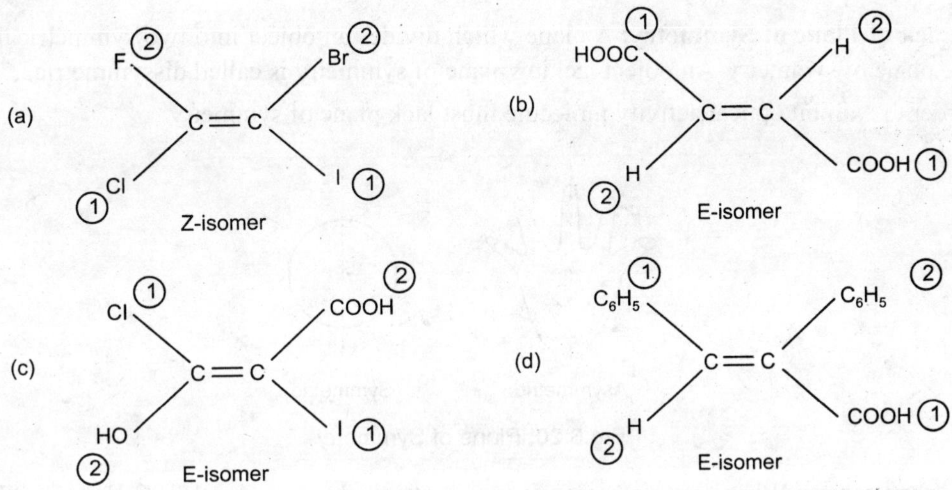

Fig.6.18: E-Z isomers

OPTICAL ISOMERISM

(2) Optical Activity

Ordinary light consists of rays of varying wavelengths vibrating in all possible directions perpendicular to the direction of propagation. When light is passed through Nicol prism, the vibrations are adjusted in a single plane only. The light whose vibrations occur only in one plane is termed as plane polarised light.

Some substances have the ability to rotate the plane of polarised light either to clockwise or to the anticlockwise. These substances are termed as optically active and their property is called optical activity.

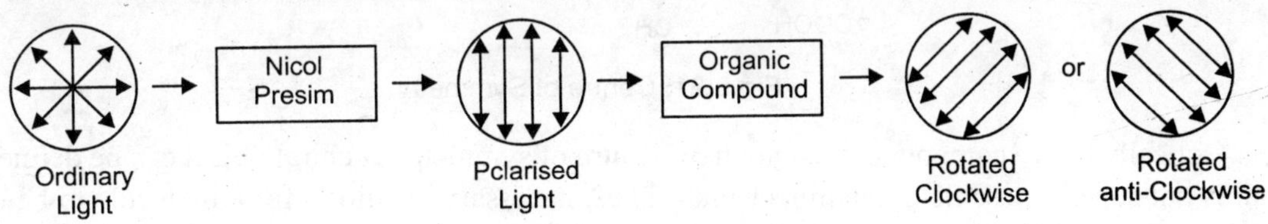

Fig.6.19: Optical Activity

OPTICAL ISOMERISM

Compounds having similar physical and chemical properties but differing only in their behaviour towards polarised light are called optical isomers and this phenomenon is known as optical isomerism.

Optical isomers which rotates the plane of polarised light to right (clockwise) is known as dextrorotatory isomer (*d*-form).

Optical isomers which rotates the plane of polarised light to left (anticlockwise) is known as laevorotatory (*l*-form).

Equimolar mixture of d and *l*-form will be optically inactive and is termed as racemic mixture.

Cause of optical activity

(a) **Lack of Plane of Symmetry:** A plane which divides an object into two symmetrical halves is said to be plane of symmetry. An object lacking plane of symmetry is called dissymmetric. .

In order to exihibit optical activity molecule must lack plane of symmetry.

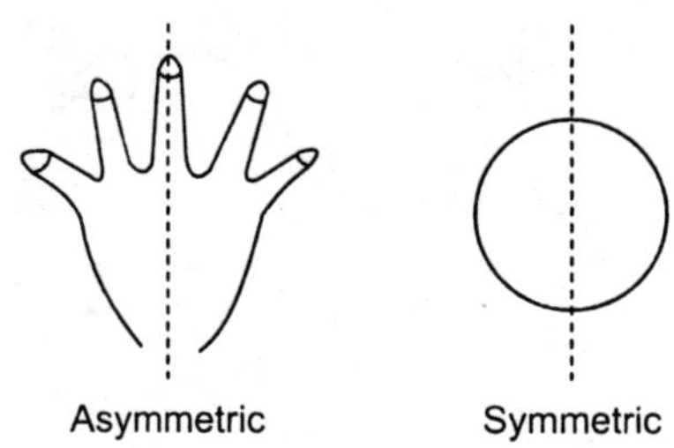

Fig.6.20: Plane of Symmetry

(b) **Chiral centre:** When four different groups are attached to Carbon, then Carbon is called chir'al carbon.

Optically active compounds must have a chiral centre.

(c) **Centre of Symmetry:** It is a point within a molecule through which if straight lines ; drawn, they meet identical atoms or groups in both direction from that point.

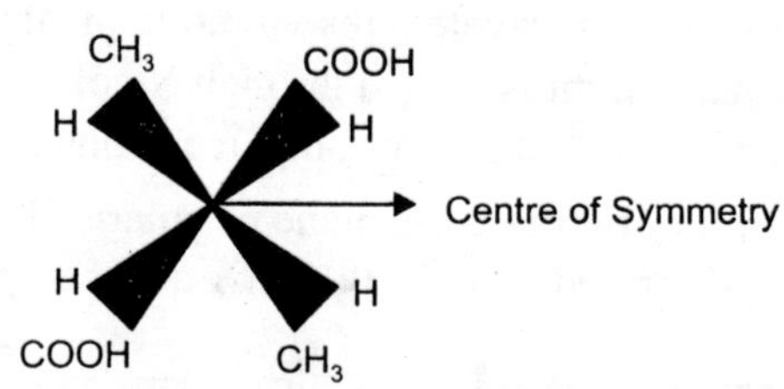

Fig.6.21: Centre of Symmetry

Optically active compound must not have centre of symmetry. A chiral object can be defined as one that is not superimposable on its mirror imag Thus, necessary condition for a molecule to be optically active is that geometric structure of molecule should not superimpose on its mirror image.

Optical isomerism of Lactic Acid

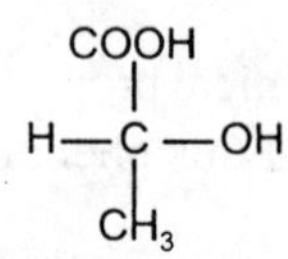

It contains one asymmetric carbon.

There are three forms of lactic acid :

***(i)* dextro-rotatory (d form) or (+) lactic acid:** It rotates plane of polarised light to right.

***(ii)* laevo rotatory (l form) or (-) lactic acid:** It rotates plane of polarised light to left.

***(iii)* Racemic (dl form) or (±) lactic acid:** It is equimolar mixture of d and *l*-forms. It is optically inactive. The inactivity is due to external compensation.

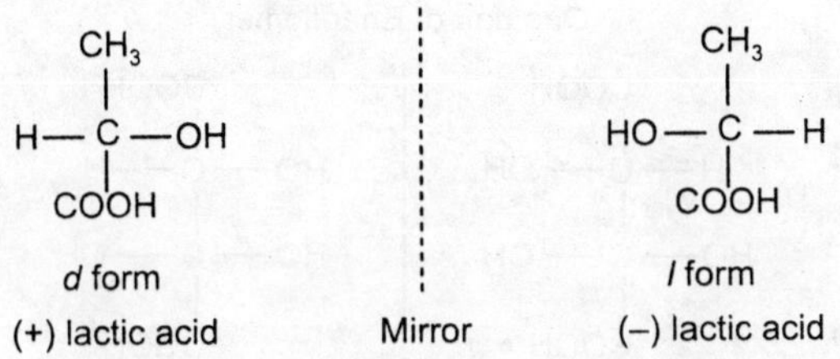

Fig.6.22: Two Enantiomers of Lactic Acid

Optical isomerism in tartaric acid: Tartaric acid has 2 asymmetric carbon.

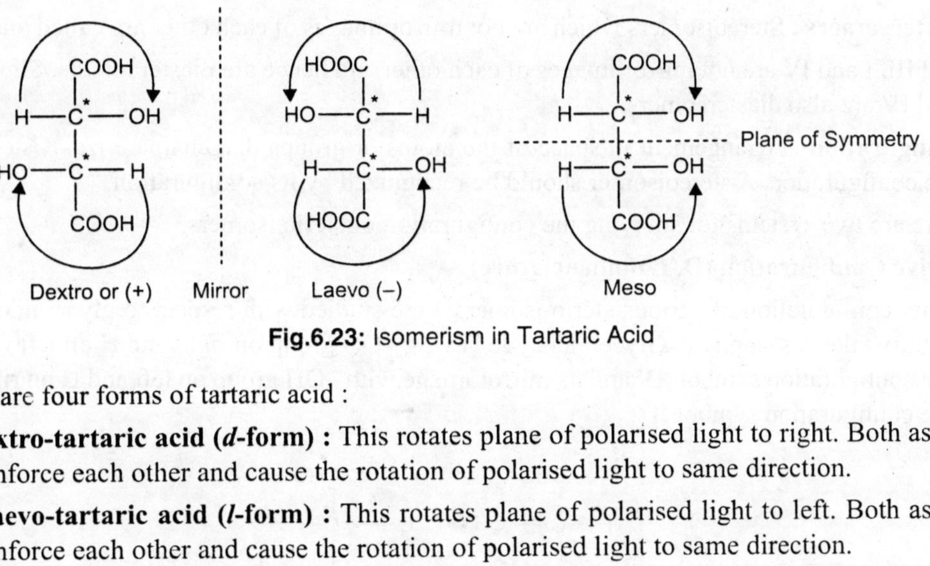

Fig.6.23: Isomerism in Tartaric Acid

There are four forms of tartaric acid :

***(i)* Dextro-tartaric acid (*d*-form) :** This rotates plane of polarised light to right. Both asymmetric carbon reinforce each other and cause the rotation of polarised light to same direction.

***(ii)* Laevo-tartaric acid (*l*-form) :** This rotates plane of polarised light to left. Both asymmetric carbon reinforce each other and cause the rotation of polarised light to same direction.

***(iii)* Meso-tartaric acid :** It possess a plane of symmetry and as a result it is optically inactive. One of the asymmetric carbon turns the plane of polarised light to right while other to left to the same extent so that rotation due to upper half is compensated by lower half.

***(iv)* Racemic tartaric acid :** It is equimolar mixture of d and *l*-forms. The rotation of d-form is compensated externally by *l*-form.

ENANTIOMERS AND DIASTEREOMERS

Enantiomers: Pair of stereoisomers which are non-super imposable mirror images of each other are called enantiomers.

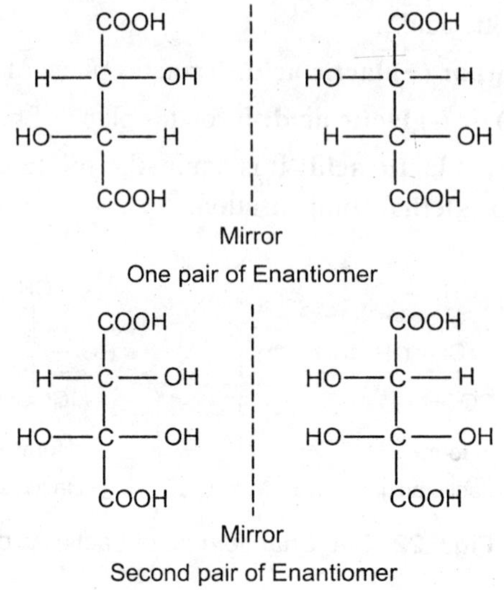

Fig.6.24: Enantiomers

Diastereomers : Stereoisomers which are not mirror images of each other are called diastereomers.

I and III, I and IV are not mirror images of each other and hence are diastereomers. Similarly II and III, II and IV are also diastereomers.

Configuration : Arrangement in space of the atoms or groups that characterizes a sterioisomer is called its configuration. A stereoisomer should be recognized by its configuration.

There are two systems for studying the configuration of sterio isomers.

(*i*) Relative Configuration (D, L nomenclature)

Earlier configuration of various sterio isomers were studied with respect to glyceraldehyde which was arbitrary taken as standard. Glyceraldehyde having –OH group on right and H on left was arbitrary given the configuration symbol 'D' and its mirror image with –OH group on left and H on right side was given the configuration symbol 'L'.

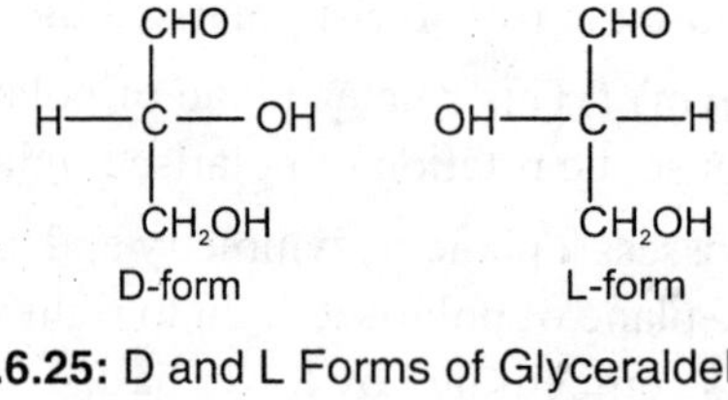

Fig.6.25: D and L Forms of Glyceraldehyde

Any sterio isomer which can be obtained from or converted into D-glyceraldehyde will belong to D-series. Similarly any sterioisomer which can be obtained from or converted into L-glyceraldehyde will belong to L-series. (+) or (–) sign added after D and L indicates direction of optical rotation.

Thus an optically active substance having D-configuration can be dextro or laevorotatory. Similarly an optically active substance with L-configuration can be dextro or laevo.

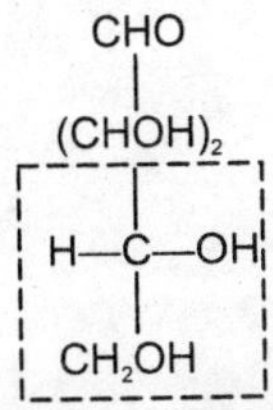

D(+) Glucose

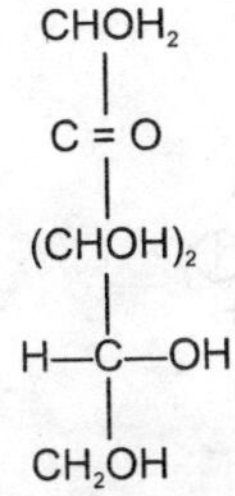

D(—) Fructose

(*ii*) Absolute Configuration : (R, S nomenclature)

This sytem was given by Cahn-Ingold-Prelog.

Important features of this system are :

(*i*) The four different groups attached to chiral carbon atom assigned priority order 1, 2, 3 or 4 on the basis of sequence rules (given earlier). 1 is the highest priority group while 4 is the lowest priority group.

(*ii*) The formula is now rotated so that the group with lowest priority (4) is directed away from viewer. The arrangement of the remaining three groups is viewed in decreasing order of their priorities. In looking so, if eye travels in clock wise direction, the configuration is designated as 'R' (Rectus means right), while if eye travels in anticlockwise direction, the configuration is designated as 'S' (Sinister means left).

Priority order is as follows :

I, Br, Cl, SO_3, H, F, OR, OH, NO_2, NR_2, NHR, NH_2, COOR, COOH, $CONH_2$, $COCH_3$, CHO, CH_2OH, CN, $-CH = CH_2$, $-C{\equiv}CH$, C, H_5, CR_3, CH_2R, CH_3, D, H

(*iii*) When molecule contains more than one chiral centre, the same procedure is applied to each.

Hint for assigning R, S configuration do not include 'H' in giving priority. If 'H' is on vertical plane :

(*a*) if eye travels in clockwise direction, then configuration is assigned as 'R'.

(*b*) if eye travels in anticlockwise direction, then configuration is assigned as 'S'.

If 'H' is on horizontal plane :

(*a*) if eye travels in clockwise direction, the configuration is specified as 'S'.

(*b*) for anticlockwise direction, the configuration is designated as 'R'.

Examples: Assign R or S configuration to each of the following:

(a) 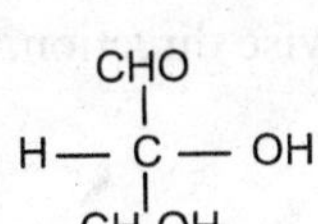

(b) 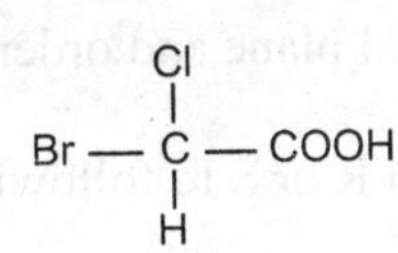

(c) 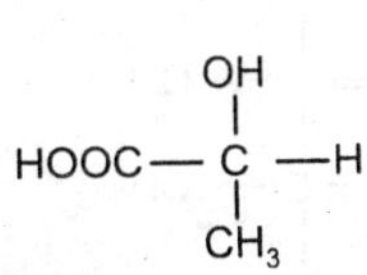

(d) 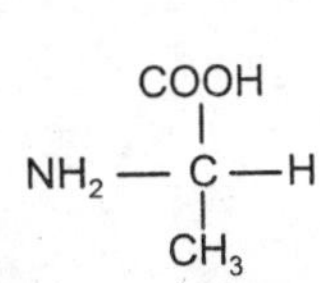

Solutions:

(*a*) 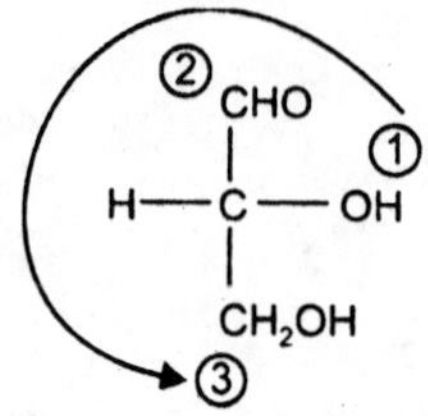

'H' is on horizontal plane and priority order is in anticlockwise direction so, it is specified as 'R'.

(*b*) 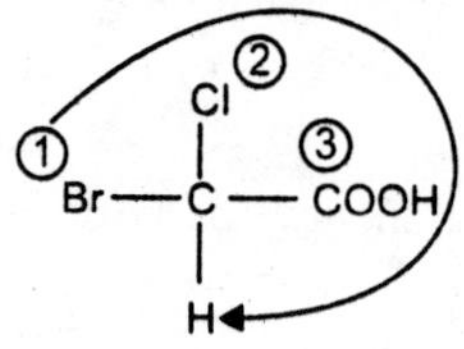

'H' is an vertical plane and priority order is in clockwise direction, so it is specified as

(*c*) 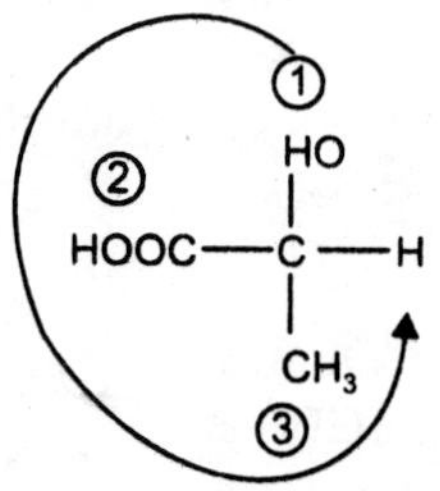

'H' is on horizontal plane and order of priority is in anticlockwise direction, so it is designated as 'R'.

(*d*) 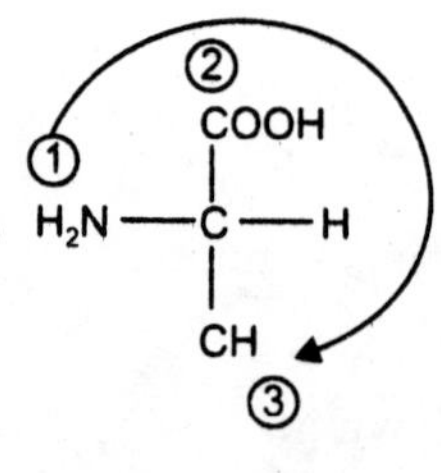

'H' is in horizontal plane and order of priority is in clockwise direction, so it is designated as 'S'.

Example: Assign R or S to following :

(*a*) 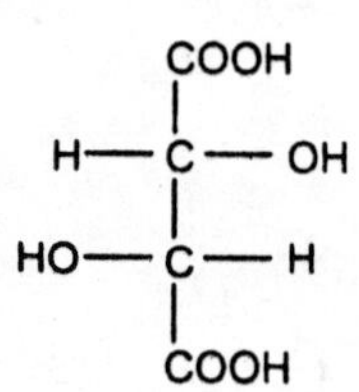

(*b*) 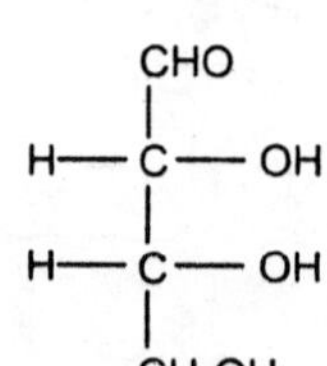

Sol.

(*a*) 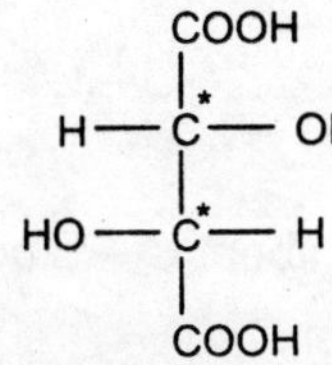

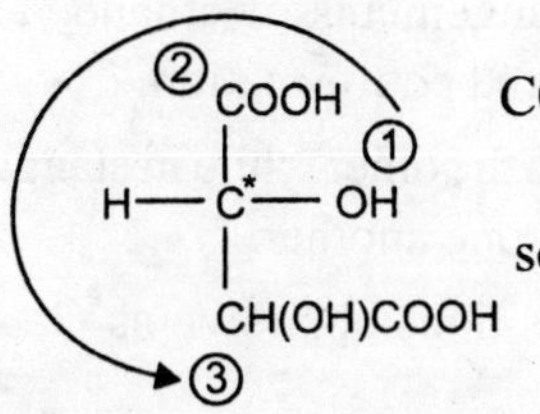

For 1st Chiral carbon, order of priority will be OH > COOH > CH(OH) COOH > H.

'H' is on horizontal plane, order of priority is in anticlockwise direction, so Carbon 1st is assigned as 'R'.

For 2nd Chiral carbon, order of priority will be same as that for 1st carbon.

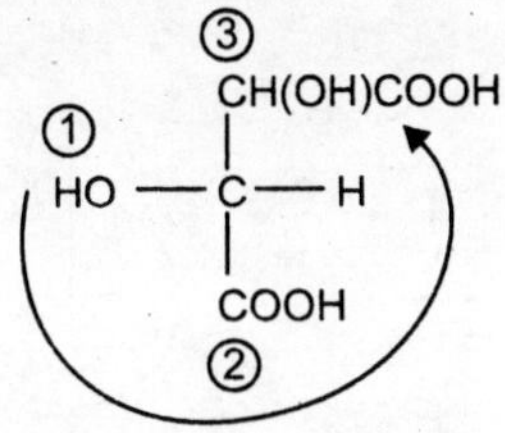

Order of priority is in anticlockwise direction, so 2nd chiral carbon is assigned as 'R'. Compound is (2R, 3R)-2, 3 dihydroxy butane-dioic acid.

(*b*) 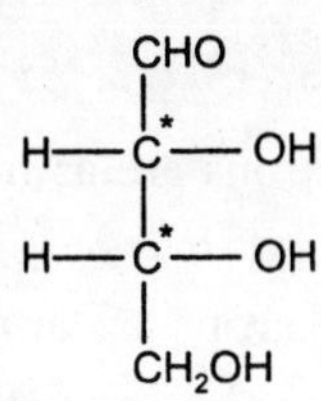

For 1st Chiral Carbon, order of priority will be -OH > -CHO > -CH(OH)CH2OH > H

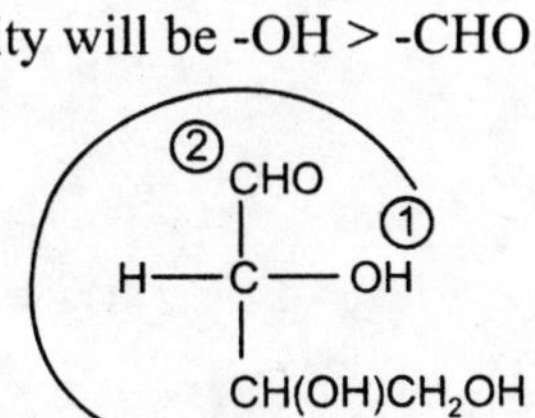

Order of priority is in anticlockwise direction so 1st Chiral Carbon is assigned as 'R'. For 2nd Chiral Carbon, order of priority will be -OH > -CH(OH)CHO > -CH2OH > H

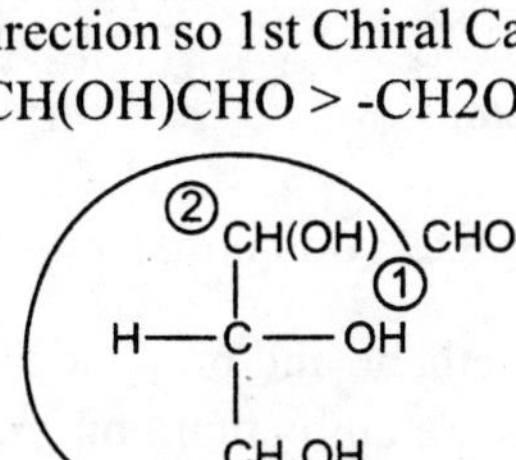

Order of priority is in anticlockwise direction so 2nd chiral carbon is also assigned as 'R'.

Compound is (2R, 3R) – 2, 3, 4 trihydroxy butanal

CONFORMATIONAL ISOMERISM

Different spatial arrangement of atoms that result from the free rotation of groups about C—C bond axis are called conformations and the phenomenon as conformational isomerism.

(a) Conformation of Ethane: Of the infinite number of possible arrangements of ethane, , two conformations represent the extremes. These are called Eclipsed and staggered conformation.

In eclipsed conformation hydrogens of one carbon atom are directly behind the other while in staggered conformation, hydrogens of two carbon atoms are staggered with respect to one another.

Any other rearrangement which will be between these two extreme positions is known as Gauche or skew.

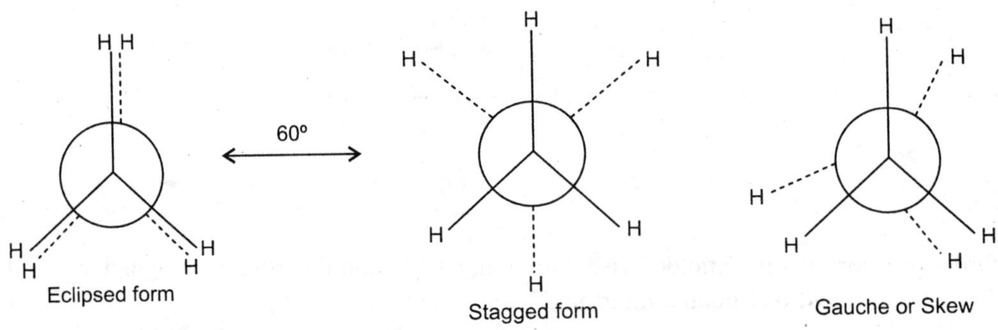

Fig.6.26: Confirmations of Ethane

Eclipsed form is least stable because hydrogens and bonding pairs of electrons are as near are as possible. Thus there is maximum repulsion and potential energy is high.

Staggered form is most stable because hydrogens and bonding pairs of electrons are at maximum distance. Thus there is minimum repulsion and potential energy is low.

(*b*) Conformations of Propane:

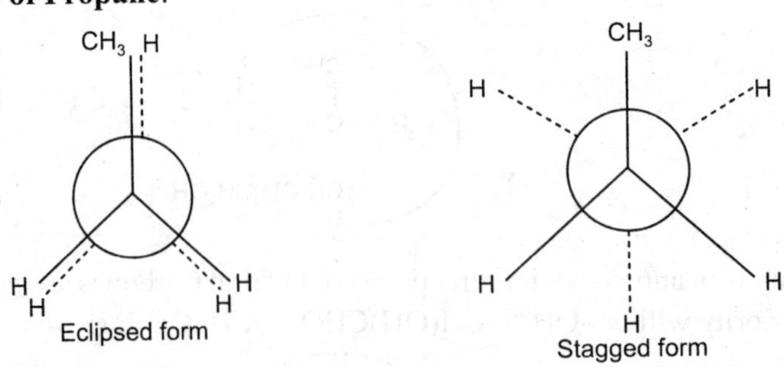

Fig.6.27: Conformations of Propane

(*c*) Conformation of butane: In ethane and propane all eclipsed conformations are equivalent and similarly all staggered conformations are equivalent, but in butane two distinct eclipsed and staggered conformations are possible.

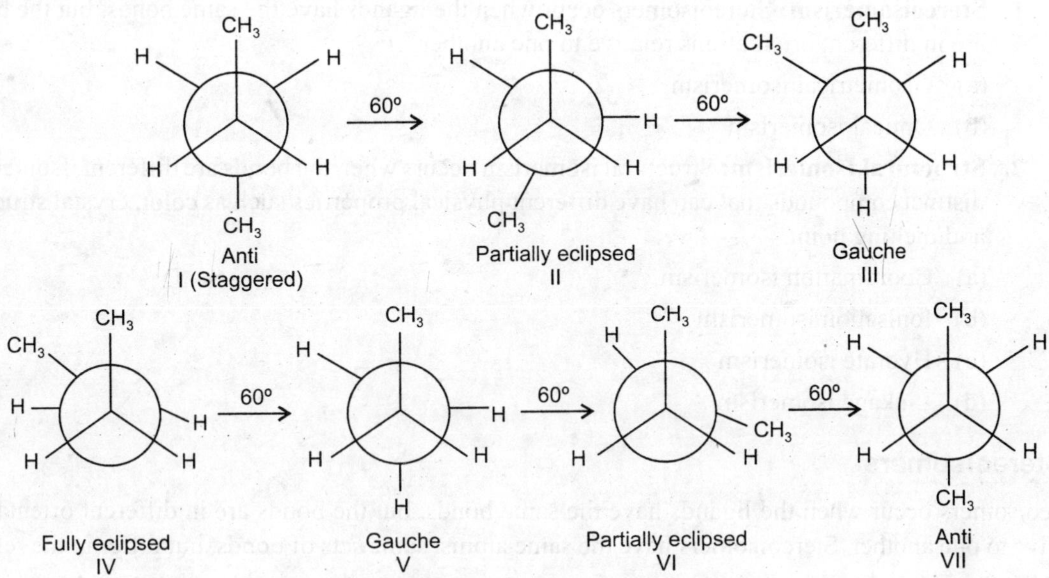

Fig.6.28: Conformation of n-butane

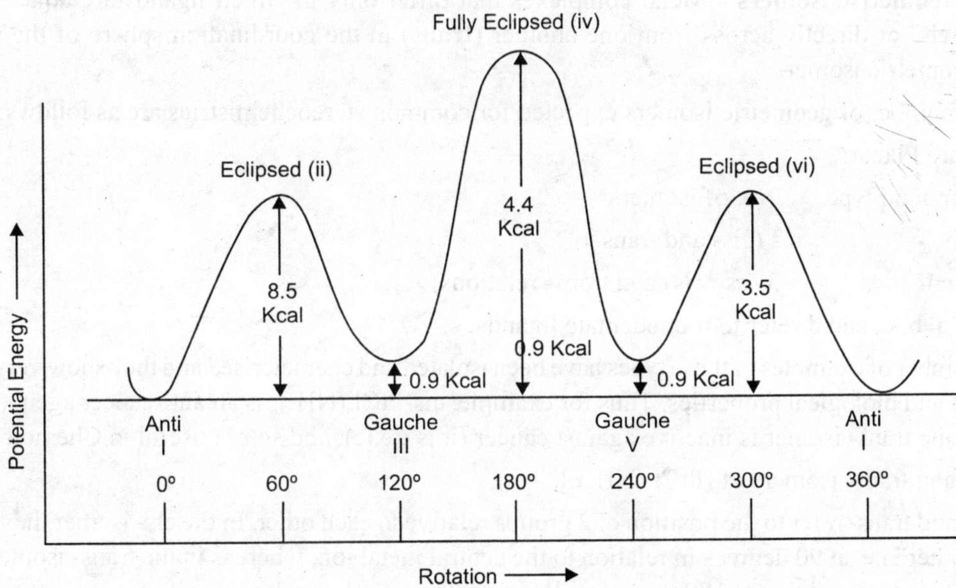

Fig.6.29: Energy chages during rotation around the central C-C bind in n-butane

ISOMERISM IN TRANSITION METAL COMPOUNDS

The existence of coordination compounds with the same formula but different arrangements of the ligands was crucial in the development of coordination chemistry. Two or more compounds with the same formula but different arrangements of the atoms are called isomers.

Two principal types of isomerism are known among coordination compounds. Each of which can be further subdivided.

1. **Stereoisomerism:** Stereoisomers occur when the ligands have the same bonds, but the bonds are in different orientations relative to one another.
   (a) Geometrical isomerism
   (b) Optical isomerism
2. **Structural Isomerism:** Structural isomerism occurs when the bonds are different. Isomers are distinct compounds that can have different physical properties such as color, crystal structure, and melting point.
   (a) Coordination isomerism
   (b) Ionisation isomerism
   (c) Hydrate isomerism
   (d) Linkage isomerism

1. Stereoisomers

Stereoisomers occur when the ligands have the same bonds, but the bonds are in different orientations relative to one another. Stereoisomers have the same atoms, same sets of bonds, but differ in the relative orientation of these bonds.

(a) Geometric isomers - Metal complexes that differ only in which ligands are adjacent to one another (**cis**) or directly across from one another (**trans**) in the coordination sphere of the metal are called geometric isomers.

The number of geometric isomers expected for common stereochemistries are as follows:

Square Planar:

Compound type	No. of isomers
Ma_2b_2	2 (cis- and trans-)
Mabcd	3 (use cis- and trans- relations)

here a, b, c, and d refer to monodentate ligands.

A number of examples of these types have been isolated and characterised and they show very different chemical and biological properties. Thus for example, cis-$PtCl_2(NH_3)_2$ is an anti-cancer agent (cisplatin) whereas the trans- isomer is inactive against cancer (it is toxic), and so not useful in Chemotherapy.

cis- and trans- isomers of $[PtCl_2(NH_3)_2]$

cis- and trans- refer to the position of 2 groups relative to each other. In the cis- isomer they are "next to each other" i.e. at 90 degrees in relation to the central metal ion, whereas in the trans- isomer they are "opposite each other", i.e. at 180 degrees relative to the central metal ion.

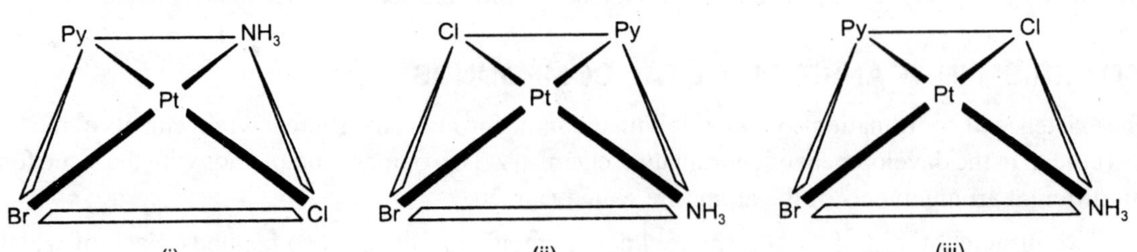

Fig.6.30: Three geometric isomers of a square planar complex $[PtBrClNH_3(pyr)]$.

Three geometric isomers of a square planar complex [$PtBrClNH_3$(pyr)].The first report of the three geometric isomers being isolated and characterised for complexes of the type [Mabcd] was by Il'yaChernyaev in 1928. The example above was reported by Anna Gel'man in 1948.

Octahedral

Compound type	No. of isomers
Ma_4b_2	2 (cis- and trans-)
Ma_3b_3	2 (fac- and mer-)
MAA_2b_2	3 (2*cis- and 1 trans-)

here a, and b, represent monodentate ligands and AA is a bidentate ligand.

In the second example, new labels are introduced to reflect the relative positions of the ligands around the octahedral structure. Thus; placing the 3 groups on one face of the octahedral gives rise to the facial isomer and placing the 3 groups around the centre gives rise to the meridional isomer. fac- and mer- isomers of [$RhCl_3(pyr)_3$].

[Mabcdef] is expected to give 15 geometric isomers. In the case of [$PtBrClI(NO_2)(NH_3)$(pyr)], several of these were isolated and characterised by Anna Gel'man and reported in 1956. Optical isomers are possible for each of these 15 forms, making a total of 30 isomers.

The cis- isomer of MAA_2b_2 may also exhibit optical isomerism although we will concentrate largely on optical isomers of the type $M(AA)_3$ (see below).

(b) Optical Isomers

Optical isomers are related as non-superimposable mirror images and differ in the direction with which they rotate plane-polarised light. These isomers are referred to as enantiomers or enantiomorphs of each other and their non-superimposable structures are described as being asymmetric.

Priorities are assigned for mononuclear coordination systems based on the standard sequence rules developed for enantiomeric carbon compounds by Cahn, Ingold and Prelog (CIP rules). These rules use the coordinating atom to arrange the ligands into a priority order such that the highest atomic number gives the highest priority number (smallest CIP number). For example the hypothetical complex [Co Cl Br I NH_3 NO_2 SCN]$^{2-}$ would assign the I- as 1, Br as 2, Cl as 3, SCN as 4, NO_2 as 5 and NH_3 as 6.

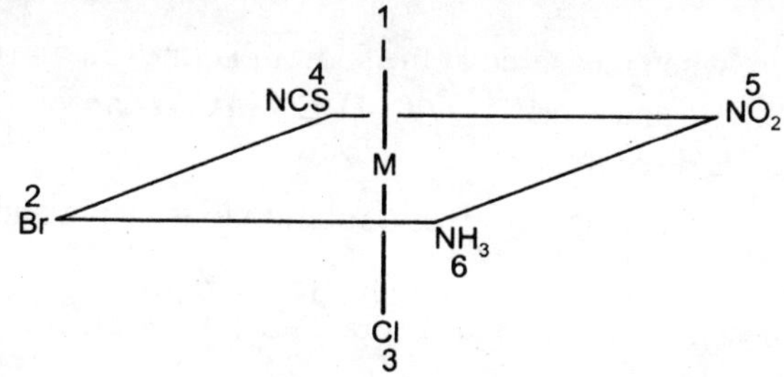

Here is one isomer where the I and Cl, and Br and NO_2 were found to be trans- to each other.

The reference axis for an octahedral centre is that axis containing the ligating atom of CIP priority 1 and the trans ligating atom of lowest possible priority (highest numerical value). The atoms in the coordination plane perpendicular to the reference axis are viewed from the ligand having that highest priority (CIP priority 1) and the clockwise and anticlockwise sequences of priority numbers are compared. The structure is assigned the symbol C or A, according to whether the clockwise (C) or anticlockwise (A) sequence is lower at the first point of difference. In the example shown above this would be C.

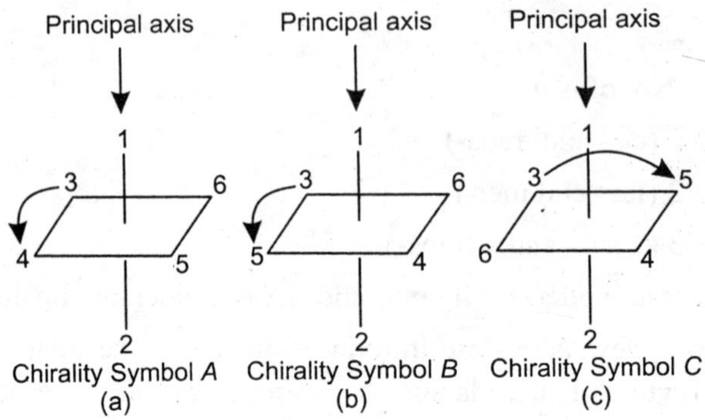

Fig.6.31: Optical isomerism in octahedral complexes

The two optical isomers of $[Co(en)_3]^{3+}$ have identical chemical properties and just denoting their absolute configuration does NOT give any information regarding the direction in which they rotate plane-polarised light. This can ONLY be determined from measurement and then the isomers are further distinguished by using the prefixes (-) and (+) depending on whether they rotate left or right.

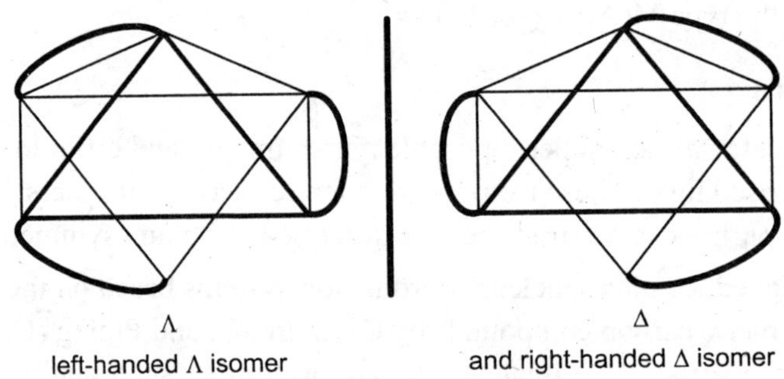

left-handed Ë isomer and right-handed Ä isomer

Fig.6.32: Optical isomerism in $[Co(en)_3]^{3+}$

To add to the confusion, when measured at the sodium D line (589nm), the tris(1,2-diaminoethane) M(III) complexes (M= Rh(III) and Co(III)) with **IDENTICAL** absolute configuration, rotate plane polarised light in **OPPOSITE** directions!

The left-handed (Ë)-$[Co(en)_3]^{3+}$ isomer gives a rotation to the right and therefore corresponds to the (+) isomer.

Optical isomers are possible for both tetrahedral and octahedral complexes, but not square planar.

2. Structural Isomers

There are several types of this isomerism frequently encountered in coordination chemistry and the following represents some of them.

(*a*) **Coordination isomerism**: In coordination isomerism, both positive and negative ions of a salt are complex ions and the two isomers differ in the distribution of ligands between the cation and the anion.

Example - $[Co(NH_3)_6][Cr(CN)_6]$ and $[Cr(NH_3)_6][Co(CN)_6]$.

where compounds containing complex anionic and cationic parts can be thought of as occurring by interchange of some ligands from the cationic part to the anionic part.

one isomer $[Co(NH_3)_6]\ [Cr(C_2O_4)_3]$

another isomer $[Co(C_2O_4)_3]\ [Cr(NH_3)_6]$

(*b*) **Ionisation isomers**: In ionisation isomerism, the isomers result in different ions in solution although they have the same composition. This type of isomerism occurs when the center ion of the complex is also a potential ligand. For example, pentaaminebromidocobalt(III) sulphate [Co(NH3)5Br]SO4 is violet and in solution gives a precipitate with barium chloride, confirming the presence of sulphate ion. Pentaaminesulphatecobalt(III) bromide [Co(NH3)5SO4]Br on the other hand, is red and tests negative for sulphate ion in solution. It gives a precipitate of silver bromide (AgBr) with silver nitrate (AgNO3).

Example - one isomer $[PtBr(NH_3)_3]NO_2 > NO_2^-$ anions in solution

another isomer $[Pt(NO_2)(NH_3)_3]Br > Br^-$ anions in solution

(*c*) **Hydrate isomerism**: In solvate or hydrate isomerism, the isomers have the same composition but differ with respect to the number of solvent ligand molecules as well as the counter ion in the crystal lattice.

Example -the best known example of this occurs for chromium chloride "$CrCl_3.6H_2O$" which may contain 4, 5, or 6 coordinated water molecules.

$[CrCl_2(H_2O)_4]Cl.2H_2O$ bright-green

$[CrCl(H_2O)_5]Cl_2.H_2O$ grey-green

$[Cr(H_2O)_6]Cl_3$ violet

These isomers have very different chemical properties and on reaction with $AgNO_3$ to test for Cl- ions, would find 1, 2, and 3 Cl- ions in solution respectively.

(*d*) **Linkage isomerism**-Linkage isomerism occurs with ambidentate ligands that can bind in more than one place.

Example- NO_2 can bind to a metal at either the N atom or an O atom. occurs with ambidentate ligands. These ligands are capable of coordinating in more than one way. In the monodentate ligands SCN^- / NCS^- and NO_2- / ONO^-.

For example:

$[Co(ONO)(NH_3)_5]Cl$ the nitrito isomer -O attached

$[Co(NO_2)(NH_3)_5]Cl$ the nitro isomer - N attached.

Uses of Coordination Compounds

I. **Dyes and Pigments:** Coordination compounds have been used from the earliest times as dyes and pigments, for example madder dye which is red, was used by the ancient Greeks and others. It is a complex of Hydroxyanthraquinone. A more modern example is the pigment copper phthalocyanine, which is blue.

2. **Analytical Chemistry:** many complex compound are being used as analytical reagent for detection and identification purpose.
   (a) **Colour Tests:** Since many complexes are highly coloured they can be used as colourimetric reagents e.g. formation of red 2,2'-bipyridyl and 1,10-phenanthroline complexes as a test for Fe(II).
   (b) **Gravimetric Analysis:** Here chelating ligands are often used to form insoluble complexes e.g. $Ni(DMG)_2$ and $Al(oxine)_3$.
   (c) **Complexometric Titrations and Masking Agents:** An example of this is the use of EDTA in the volumetric determination of a wide variety of metal ions in solution, e.g. Zn^{2+}, Pb^{2+}, Ca^{2+}, Co^{2+}, Ni^{2+}, Cu^{2+}, etc. By careful adjustment of the pH and using suitable indicators, mixtures of metals can be analysed, e.g. Bi^{3+} in the presence of Pb^{2+}(see laboratory manual). Alternatively, EDTA may be used as a masking agent to remove a metal ion which would interfere with the analysis of a second metal ion present.
3. **Sequestering Agents:** Related to their use as masking agents is the use of ligands for "sequestering" i.e. for the effective removal of objectionable ions from solution in industrial processing, e.g. EDTA is used to "soften" water. The addition of EDTA to water is used in boilers etc., to prevent "scaling" or build up of insoluble calcium salts.
4. **Extraction of Metals:** Sometimes certain metals can be leached from their ores by formation of stable complexes e.g. Ag and Au as complexes of cyanide ion.
5. **Bio-Inorganic Chemistry:** Naturally occurring complexes include haemoglobin, chlorophyll, vitamin B_{12} etc.

   EDTA and other complexing agents have been used to speed the elimination of harmful radioactive and other toxic elements from the body. (e.g. Pb^{2+}). In these cases soluble metal chelate complexes are formed.
6. **Chemo-therapy:** an example here is the use of cis-$PtCl_2(NH_3)_2$ as an anti-tumour drug.

ELECTRON DELOCALIZATION IN ORGANIC MOLECULES

Electron delocalization effects play a crucial role in stabilization of the intermediate and the product of a reaction. Dispersal of charge stabilize a species whereas localization of charge increases its reactivity. Delocalization of electron density in a species may take place by following ways:

Inductive Effect or Transmission Effect

This is an electron delocalization effect in a a-bonded chain where two a-bonded atoms differ considerably in their electronegativities.

It is a permanent effect and is always transmitted along a chain of Carbon atoms but this transmission rapidly diminishes with the distance. For example, in 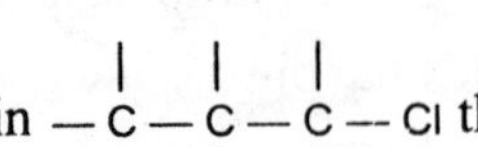 the terminal Cl atom pulls the shared pair of electron of $C_\alpha - Cl$ bond towards itself making Cl partially positive. This Carbon now is more electronegative than next C_β atom of the chain, so it pulls a-bonded electrons pair of $C_\beta - C_\alpha - Cl$ bond towards itself making C partially positive. Thus electron pulling effect of Cl atom is transmitted through the chain of C atoms but this effect diminishes with the distance and can be practically ignored beyond C_β of chain.

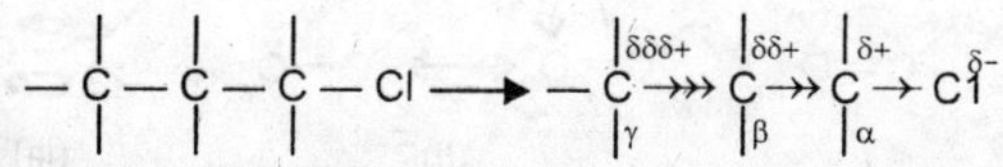

If a group at the end of Carbon chain with draws electron density towards itself, thereby decreasing electron density at C-end by inductive affect, then such groups are called –I groups and effect is termed as –I effect.

Cl, Br, CN, NO_2 are –I groups.

If a group release electrons towards Carbon chain, then such groups are termed as +1 groups and effect as +1 effect.

NH_2, CH_3, C_2H_5, OH are +1 groups.

Above concept can be applied to know the strength of Carboxylic acids. –I effect increases acidity whereas +1 effect decreases the acidity of Carboxylic acid.

For eg, halogenated Carboxylic acids are more stronger than parent acid. Moreover strength of acidity increases with increase in electronegatively of -I groups.

$$FCH_2COOH > C|CH_2COOH > BrCH_2COOH > ICH_2COOH$$

Furthermore, strength of acid also increases with the number of halogens: –COOH

As we know Inductive effect decreases with distance, thus strength of acid also decreases as distance of –I groups, increases from –COOH.

$$CH_3CH_2\underset{|}{\overset{Cl}{C}}HCOOH > CH_3CH\text{—}\overset{Cl}{C}H_2\text{—}COOH > CH_2\text{—}CH_2\text{—}\overset{Cl}{C}H_2\text{—}COOH$$

Concept of inductive can also be applied to know the basic strength of certain organic compounds especially that of amines. Since a base is electron donor, thus electron donating group which increases the electron density (showing +1 effect) increases the basic strength while electron withdrawing group which decreases the electron density of an atom (showing –I effect) decreases the basic strength.

Thus 2° aliphatic amines are more basic than 1° aliphatic amines which in turn are more basic than NH,, (due to +1 effect of alkyl group).

Order of basic strength of methyl amines is as

$$(CH_3)_2NH > CH_3NH_2 > (CH_3)_3N > NH_3.$$

Dimethyl amines is most basic, then comes methyl amine. Trimethyl amine is more basic than NH_3 due to +1 effect of 3 methyl groups but is less basic than even methyl amine due to steric effect caused by 3 methyl groups.

MESOMERIC EFFECT

Mesomeric effect refers to the polarity produced in a molecule as a result of interaction between two π bonds or a π bond and lone pair of electrons.

In case molecules have conjugated system of double bonds, the mesomeric effect is transmitted through whole of the conjugated system.

In a carbonyl group ($>C=O$) the oxygen atom is more electronegative than C atom. The π-electrons of double bond is more displaced towards Oxygem atom.

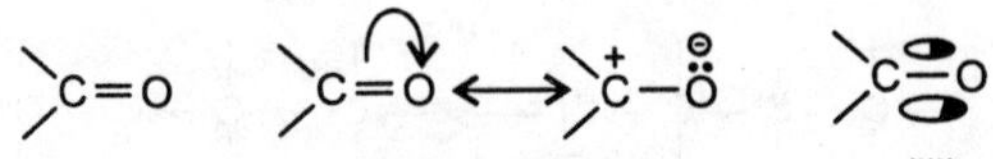

(i) (ii) (iii)

Actual structure lies between (*i*) and (*ii*) and is best represented by (*iii*).

Now if Carbonyl group is conjugated with C=C type system, above polarisation is transmitted further through π-electrons.

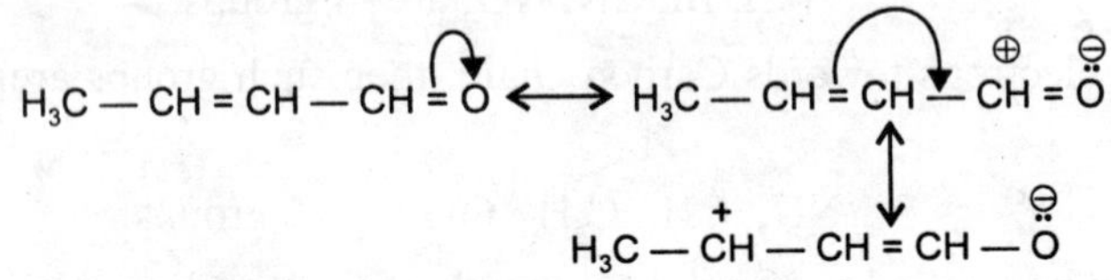

Above structure is well represented by hybrid.

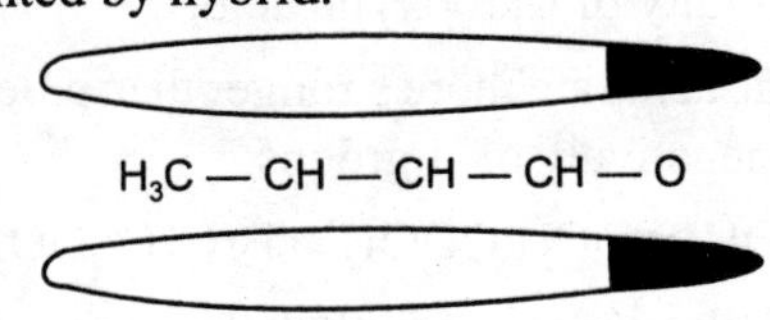

Polarisation also occurs when an atom or group having lone pair of electrons is present in conjugation with double bond.

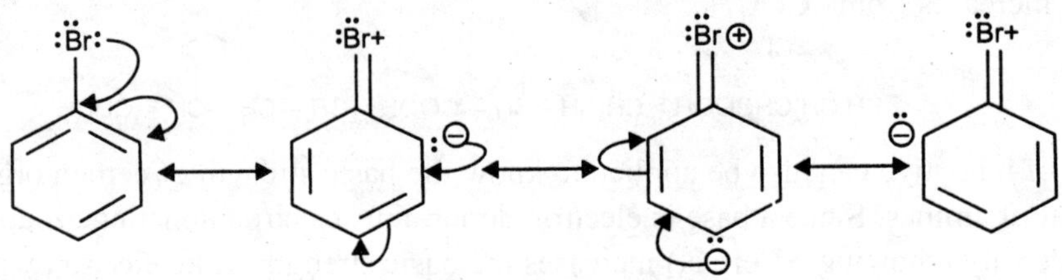

Like inductive effect, the mesomeric effect may be +M and –M.

It is +M effect when transference of electron pair is away from the atom and –M effect when the transference of electron pair is towards the atom.

$C=C-C=C-NH_2$ (+M Effect) $H_3C-C=C-C=O$ (–M Effect)

+M groups: –Cl, –Br, NH_2, OH

–M groups: $-NO_2$, $C\equiv N$, C = O

Like inductive effect, mesomeric effect is also permanent and effects physical properties of a molecule.

I and M effect, when present together, may act in same direction or opposite to each other. M effect is more powerful than I effect. For example in Vinyl Chloride due to –I effect Cl should develop a negative charge but due to +M effect it have positive charge.

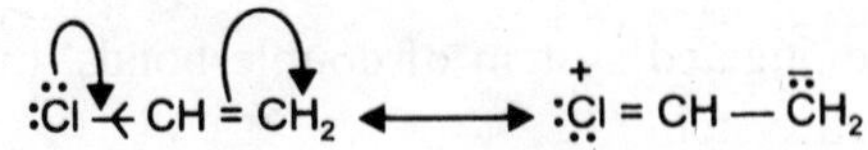

ELECTROMERIC EFFECT

The effect involving the complete transfer of a shared pair of electrons to one of the atoms formed by a multiple bond at the requirement of attacking reagent is called electromeric effect. It is represented by E.

It is temporary effect. It comes into play instantaneously at the demand of attacking reagent but as soon as reagent is removed, original condition is restored.

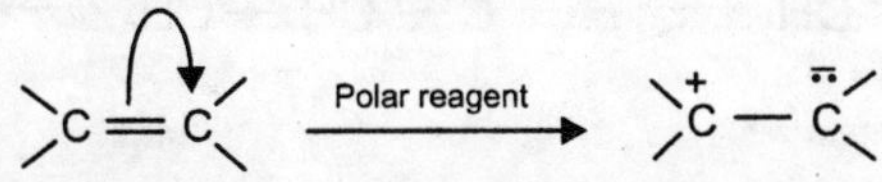

In case of carbonyl group, Cyanide group more electronegative atom O, N gets negative charge respectively.

Since electromeric effect takes place only at the time of attacking reagent, it always facilitates the reaction and never inhibits it. This effect is commonly seen during the addition of polar reagents to (>C = O) and (>C = O) bonds.

In cases where I effect and Electromeric effect simultaneously operate, usually E effect predominates.

There are two types of E effect:

(*a*) **+E effect:** When transfer of electrons take place towards the attacking reagent, the effect is called +E effect.

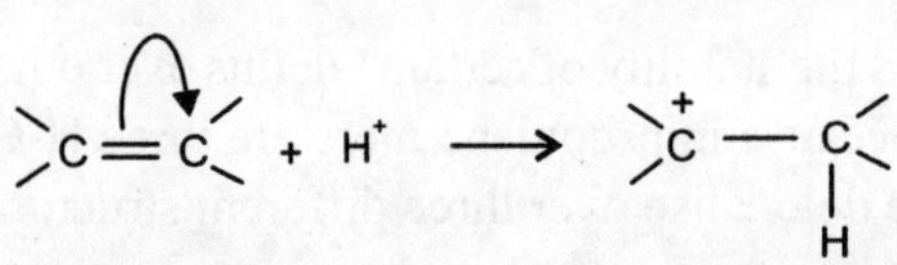

Attacking reagent attaches to atom to which electrons have been transferred.

(*b*) **–E Effect:** When transfer of electrons take place away fro the attacking reagent, the effect is called –E effect.

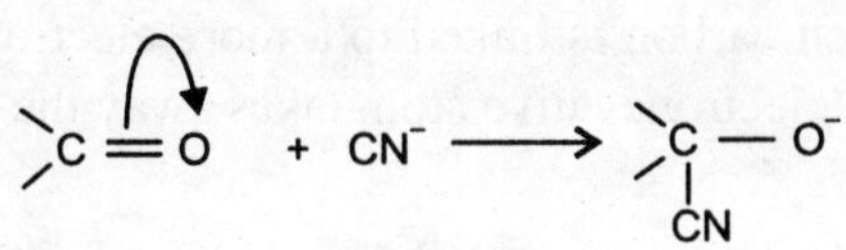

In this case attacking reagent does not attach to that atom on which electrons have been transferred.

HYPER CONJUGATIVE EFFECT

Baker and Nathan suggested that alkyl groups with at least one hydrogen atom on the α-Carbon atom attached to an unsaturated Carbon atom, are able to release electrons by a mechanism similar to that of E effect eg

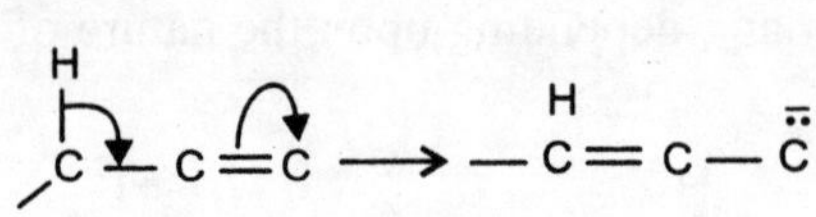

The delocalisation involves a and JT bonds orbitals, thus also known as a-n conjugation. This type of electron release due to the presence of the system H–C–C=C is known as hyper conjugaıion.

Since in resulting structure there is no bond between one of the H and one of the C so it is also known as no-bond resonance.

Hyper conjugation in propene:

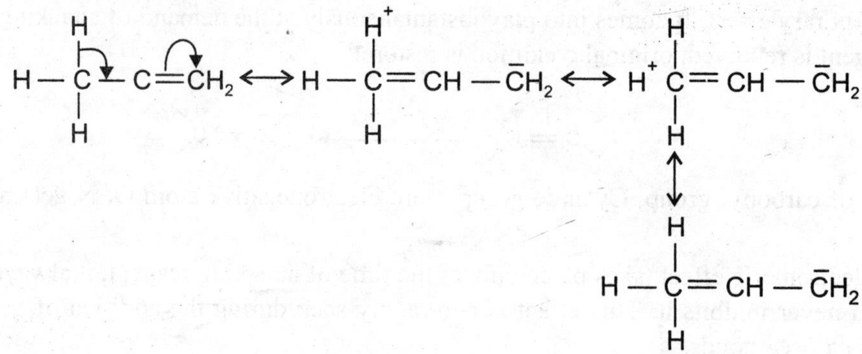

Although I effect and hyperconjugative effect causes the release of electrons magnitude of two effect changes in opposite direction along a series of alkyl groups.

$$CH_3–, CH_3CH_2–, (CH_3)_2CH–, (CH_3)_3C–$$

I effect increases, H effect decreases along above series of alkyl groups.

Hyperconjugation explains the stability of certain Olefins over other alkenes. For example, propylene is more stable than ethylene because in propylene there are three H-C hyperconjugated bonds and thus a electron of C-H bond can be delocalise over three different structures.

REACTION INTERMEDIATES

Carbocation (Carbonium ion)

When a covalent bond, in which carbon is linked to a more electronegative atom or group, breaks up by heterolytic fission, the more electronegative atom takes away the electron pair while carbon loses its electron.

Such organic species in which carbon atom has only six electrons and a positive charge is called Carbocation.

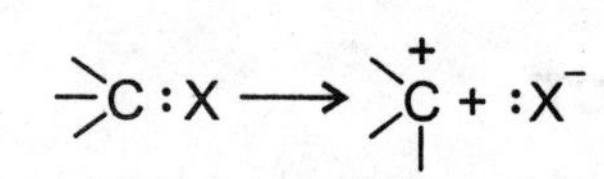

Carbocation

Carbonium ions are named by adding the words 'carbonium ion' to parent alkyl group. These are also termed as primary, secondary, tertiary, depending upon the nature of the carbon atom bearing positive charge.

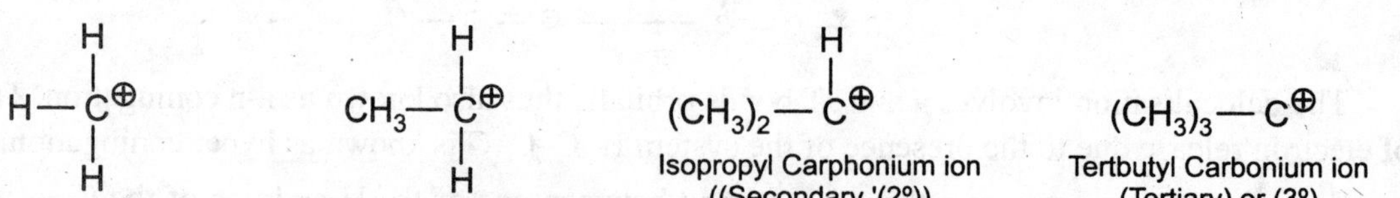

Methyl Carbonium ion

Ethyl Carbonium ion
Primary (1°) Carbocation

Isopropyl Carphonium ion
((Secondary '(2°))

Tertbutyl Carbonium ion
(Tertiary) or (3°)

Structure of Carbocation: Carbon atom of carbocation is sp^2 hybridized. The three sp^2 hybrid orbitals are utilised in forming bonds to three substituents, the remaining unused p orbital remains vacant.

Stability of Carbocation: The stability of carbocation is influenced by both resonance and inductive effects.

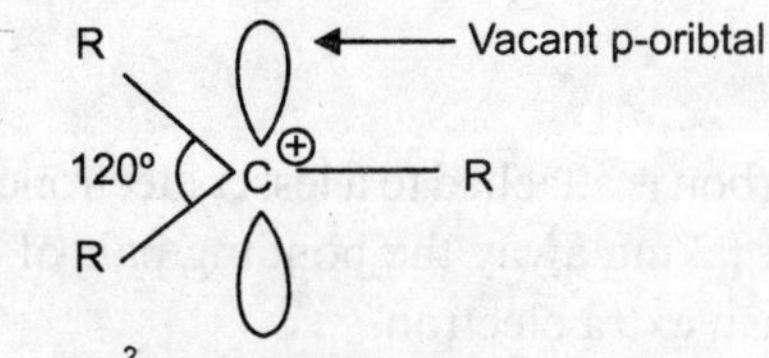

Sp[2] hybridized Carbon

An alkyl group has an electron releasing I effect. An alkyl group attached to positively charged Carbon of carbocation tends to release electrons towards that carbon, thus dispersing the positive charge of Carbon of Carbocation. This dispersal of the charge stabilizes the carbonium ion. More the number of alkyl groups, the greater the dispersal of positive charge, and therefore, will be the stability of Carbonium ion.

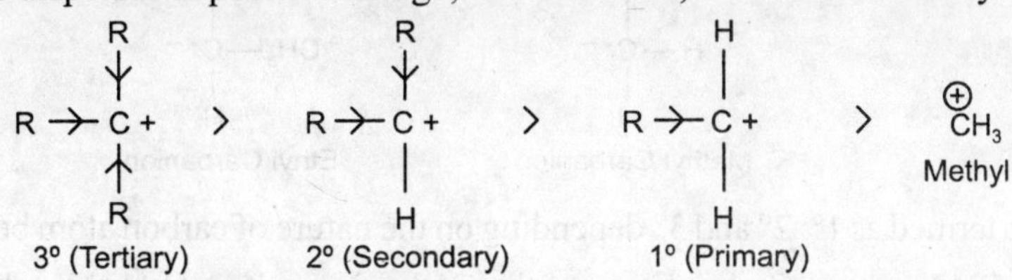

Alkyl and benzyl carbonium ions are much more stable as these are stabilized by resonance.

$$CH_2 = CH - \overset{+}{C}H_2 \longleftrightarrow \overset{+}{C}H_2 - CH = CH_2 \quad \text{(Allyl)}$$

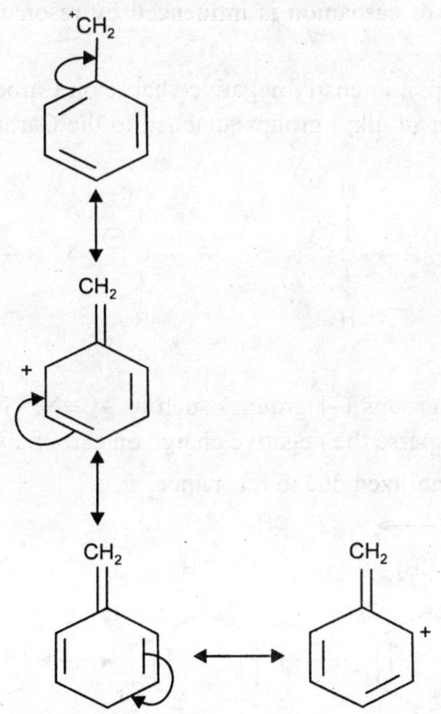

(Benzyl)

Thus relative stability of carbonium are:

$$(C_6H_5)_3C^+ > C_6\overset{+}{H}_5 - \overset{+}{C}H_2 > CH_2 - CH = CH_2 > 3° > 2° > 1° > \text{ methyl}$$

Electron withdrawing groups like –NO, and –Br which have –I effect reduce the stability of carbocation.

Carbanions

When a covalent bond, in which carbon is attached to a lesser electronegative atom, breaks up by heterolytic cleavage, the atom leaves without taking away the bonding pair of electrons and thus the Carbon atom acquires a negative charge due to an extra electron.

$$\gt C : Y \longrightarrow \gt C^- : + Y^+$$

Such organic species which contains a negatively charged carbon atom are called carbanions.

They are named after parent alkyl group and adding the word carbanion.

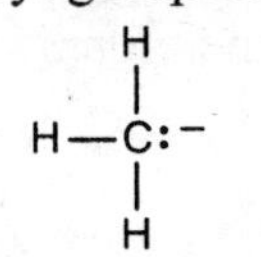

Methyl Carbanion

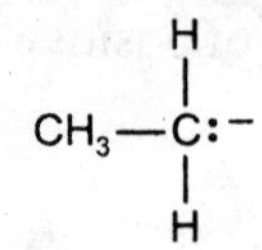

Ethyl Carbanion

These are also termed as 1°, 2° and 3° depending on the nature of carbon atom bearing negative charge.

Structure of Carbanion : Carbon atom of the Carbanion is in sp^3 hybridised state, three sp^3 hybrid orbitals form covalent bonds with three atoms or groups while the fourth sp^3 hybrid orbital has a non-bonding pair of electrons. Thus carbanion has a pyramidal shape just like NH_3.

Stability of Carbanion: Stability of carbanion is influenced by resonance, inductive effect and *s*-character of orbitals.

Electron releasing groups (+I groups), intensify negative charge on Carbon of Carbanion, making it less stable. Hence greater the number of alkyl groups attached to the Carbon atom having negative charge, lesser is its stability.

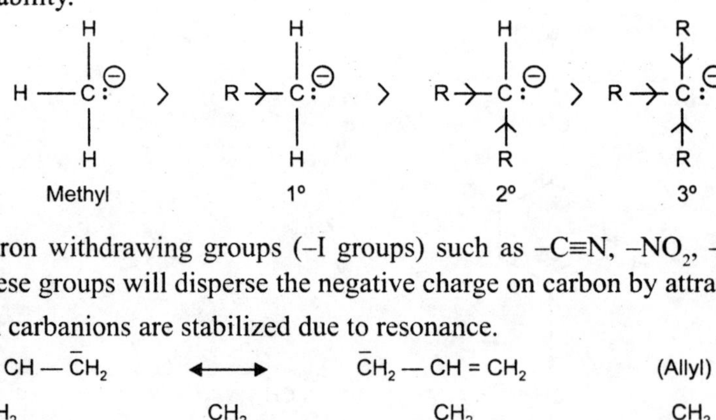

Methyl 1° 2° 3°

Conversely electron withdrawing groups (–I groups) such as –C≡N, $-NO_2$, –Br will stabilise a carbanion because these groups will disperse the negative charge on carbon by attracting electrons.

Aikyl and benzyl carbanions are stabilized due to resonance.

$$CH_2 = CH - \bar{C}H_2 \longleftrightarrow \bar{C}H_2 - CH = CH_2 \quad \text{(Allyl)}$$

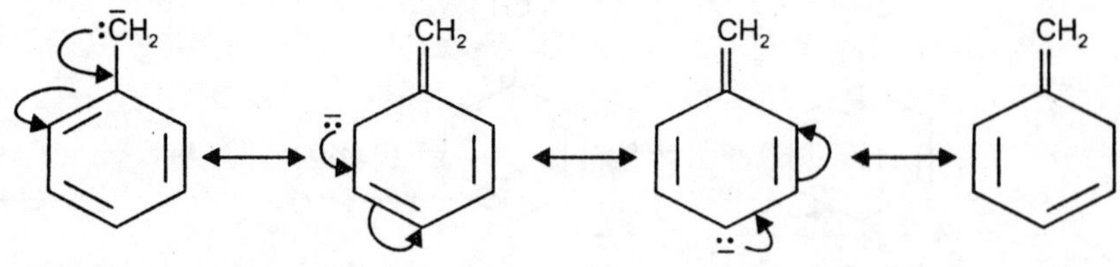

(Benzyl)

Stability of carbanion increases with increases in s-character of orbitals

$$\underset{50\%\ \text{s-character}}{R-C\equiv\bar{C}} > \underset{33\%}{R_2C=\bar{C}H} > \underset{25\%}{R_3C-\bar{C}H_2}$$

Since s electrons are more close to nucleus.

Among different carbanions stability order is

$$(C_6H_5)\,C > C_6H_5CH_2 > \text{Allyl} > \text{Methyl} > 1° > 2° > 3°$$

Carbenes

Carbenes are reactive, short lived, neutral species in which carbon atom has six electrons in outer shell, out of which two constitute a lone pair and two are shared. So they are divalent carbon species containing two unpaired electrons and possess no charge.

They can be formed as : .

$$CH_2N_2 \xrightarrow{h\nu} CH_2 + N_2$$

Carbenes are of two types

***(a)* Singlet:** In singlet carbenes, the two electrons have opposite spins and are paired in one orbital (sp^2 hybridization).

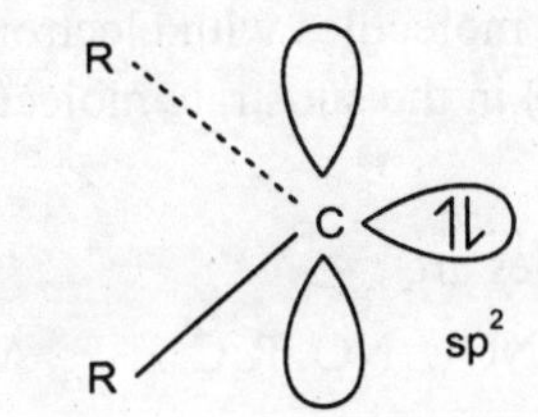

***(b)* Triplet:** In triplet carbenes, the electrons have the same spin and are in different orbitals (*sp* hybridized).

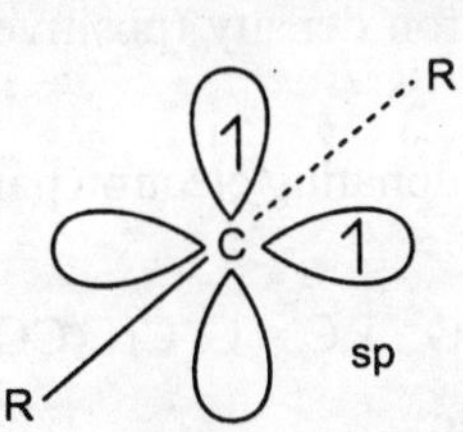

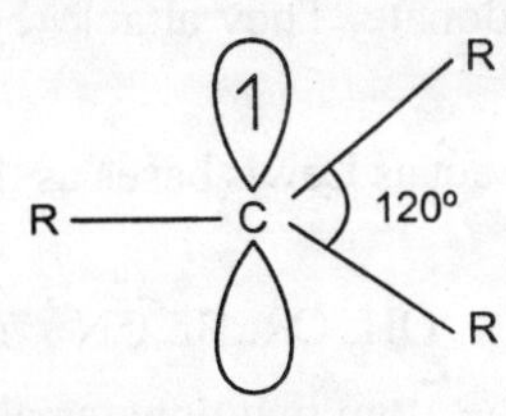

Triplet Carbene is more stable than singlet carbine

Free radicals

A free radical may be defined as an atom or group of atom having an unpaired electron. They are generated as a result of homolytic fission of a covalent bond. Due to unpaired electron they are highly reactive.

Structure of free radicals: The carbon atom of alkyl free radicals is sp^2 hybridised in which odd electron remains in the p orbital. Thus free radical has planar structure.

Stability of free radicals: Stability of carbon free radicals is not influenced by inductive effect because they have no charge. However, they are stabilised by hyperconjugation.

Relative stability of free radicals is as such

$$(C_6H_5)_3C° > C_6H_5CH_2° > \text{allyl} > 3° > 2° > 1° > \text{methyl} > \text{vinyl}$$

Extra stability of the aromatic and alkyl radicals is due to resonance. The relative stability order of 3°, 2° and 1° free radicals is explained on the basis of hyperconjugation.

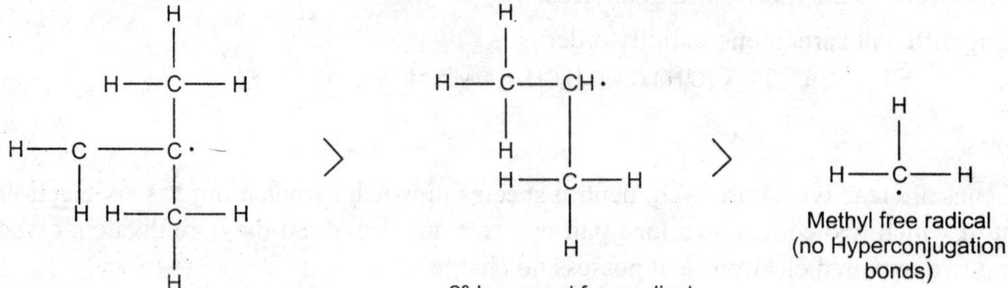

2° butyl free radical (9 H-C Hyperconjugation bonds)

2° Iso propyl free radical (6 H-C Hyperconjugation bonds)

Methyl free radical (no Hyperconjugation bonds)

Free radical, formed easily has greatest stability.

CLASSIFICATION OF REAGENTS

1. Electrophile: Electrophiles or electrophilic reagents are electron loving species. These species carry either positive charge or neutral molecules with electron deficient centre. These attack regions of high electron density (negative centre) in the substrate molecule. .

Electrophiles act as Lewis acids.

Some of the important electrophiles are :

$$H_3\overset{+}{O},\ Br^+,\ H^+,\ \overset{+}{N}H_4,\ NO_2^+,\ \overset{+}{N}O,\ R_3C^+,\ BF_3,\ AlCl_3,\ ZnCl_2,\ FeCl_3,\ SO_3 \text{ etc.}$$

2. Nucleophile: Nucleophiles or nucleophilic reagents are electron rich or electron donating species. Nucleophile means nucleus loving. They are either negatively charged or neutral molecules with free electron pairs to donate, They attack regions of low electron density (positive centre) in the substrate molecule.

Nucleophiles act as Lewis bases as they are capable of donating electron pair. Some of the important nucleophiles are:

$$^-OH,\ \bar{O}R,\ Br,\ CN^-,\ ^-NH_2,\ HSO^-_3,\ RCOO^-,\ RC = \bar{C},\ \bar{C}H\,(COOC_2H_5)_2$$

(* indicates the atom that donates electrons to substrate).

TYPES OF ORGANIC REACTIONS

1. Substitution reaction
2. Addition reaction
3. Elimination reaction
4. Rearrangement reaction

Substitution reaction: Substitution reaction are those reaction in which an atom or group of atoms attached to a Carbon atom is replaced by another atom or group of atoms. Substitution reaction is further classified into

(a) Free radical substitution reaction.

(b) Electrophilic substitution reaction

(c) Nucleophilic substitution reaction

Nucleophilic Substitution Reaction : Nucleophilic substitution reaction are those substitution reaction which involve the attack of nucleophile. They are represented by SN (S stands for substitution and N stands for Nucleophilic). They are more common in aliphatic compounds. For eg →

$$R—X + aqs\ KOH \longrightarrow ROH + KX$$

SN reaction may follow two mechanism.

1. Unimolecular mechanism: S_N reaction which proceed via unimolecular mechanism are represented by S_N1

The rate of such reaction depends only on concentration of substrate and are independent of nucleophile.

$$\text{Rate} \propto [\text{substrate}]$$

S_N1 reactions are completed in 2 steps

In first step substrate ionises to give carbocation. This step is slow step and hence rate determining step (r.d.s.)

$$R — X \xrightarrow{\text{slow}} R^+ + :X^{\ominus}$$

In second step nucleophile attack on carbocation to give final product.

$$R^+ + :OH^- \xrightarrow{\text{fast}} ROH$$

Since first step is r.d.s. involves only one molecule, the mechanism is terned as Unimolecular and represented as S_N1).

Stereochemistry: Carbocation generated is planar because it has sp^2 hybridized C. Planar carbocation can be attacked by nucleophile on either side to give two isomers (*d* and *l*).

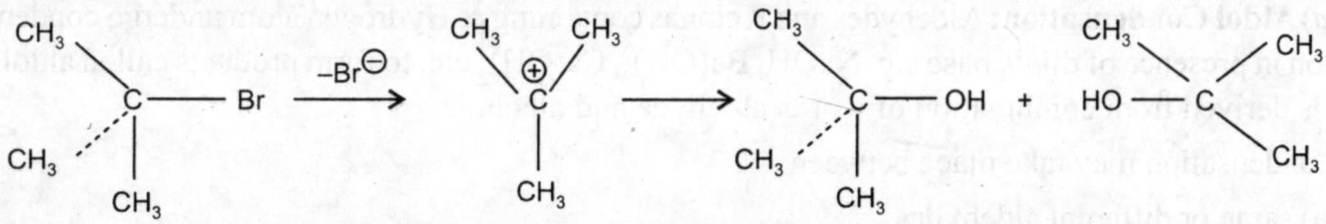

Thus in S_N1 reaction retention as well as inversion of configuration is observed.

2. Bimolecular Mechanism: Nucleophilic substitution reactions which proceed through bimolecular mechanism are represented as S_N2 reactions where '2' stands for bimolecular.

The rate of S_N2 reactions depends on the concentration of substrate as well as the nucleophile.

$$\text{Rate} \propto [\text{Substrate}]\ [\text{Nucleophile}]$$

S_N2 reactions are completed in one step.

Nucleophile attacks the substrate from side opposite to that of leaving group, a transition state (TS) results in which both leaving group and nucleophile are partially bonded to Carbon atom.

In transition state leaving group starts taking hold of electron pair via which it is bonded to Carbon and nucleophile offers a pair of electrons for the formation of bond with Carbon Transition state has maximum potential energy and hence minimum stability.

Finally leaving group leaves the molecule and nucleophile forms a covalent bond with molecule to give product.

Thus S_N2 reaction is one step process in which two processes, that is, formation of new bond and breaking of older bond take place simultaneously. Since nucleophile attacks from back side hence configuration of product is opposite to that of reactant i.e. complete inversion takes place. This change in configuration is called Walden Inversion.

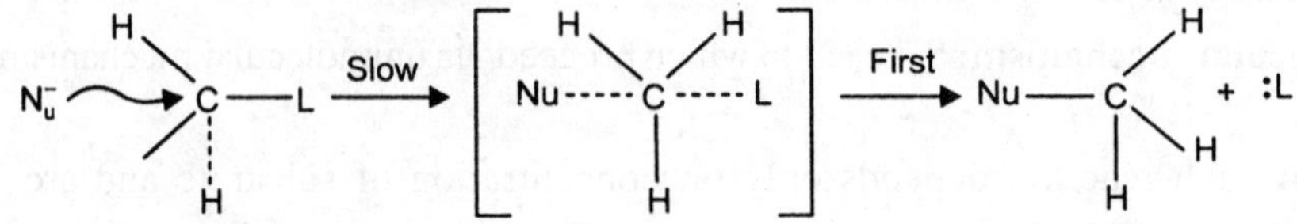

Transition State

[Nu → nucleophile like OH^-]

[L → Leaving group like halide ion]

Since formation of T.S is rate determining step (as it is slow) and it involves both nucleophile as well as substrate, the mechanism is termed as bimolecular and represented as S_N2

S_N2 reactions are favoured by strong nucleophile and presence of simple and fewer alkyl groups. Thus presence of bulky groups hinder S_N2 reaction and facilitates the S_N1 reaction.

Thus hydrolysis of tertiary alkyl halides take place by SN, mechanism while that of primary alkyl halides take place by SNZ mechanism.

The hydrolysis of 2° alkyl halides may take place by any one of S_N1 and S_N2 mechanism.

MECHANISM OF SOME NAMED REACTIONS

***(a)* Aldol Condensation:** Aldehydes and Ketones containing α-Hydrogen atom undergo condensation reaction in presence of dilute base e.g. NaOH, $Ba(OH)_2$, $Ca(OH)_2$ etc. to form products called aldol. Term aldol is derived from combination of words aldehyde and alcohol.

Condensation may take place between

(a) same or different aldehydes

(b) same or different ketones

(c) an aldehyde and a ketone.

The carbonyl compound must have an α-H.

Mechanism: Take example of condensation between acetaldehyde molecules.

First step: Base OH^- abstracts a hydrogen ion from Cl-carbon of one of the aldehyde molecules to form resonance stabilized carbanion.

$$OH^- + H-CH_2CHO \rightleftharpoons H_2O + \overset{-}{C}H_2-CH=O$$

$$\updownarrow \text{ (resonance stabilised Carbanion)}$$

$$CH_2=CH-O^-$$

Second Step: Carbanion formed in first step acts as nucleophile attacks on carbonyl carbon of another acetaldehyde molecule.

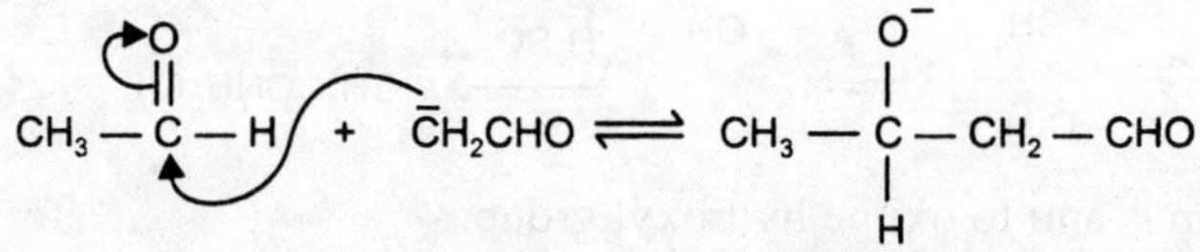

(Aldol anion)

Third Step: Aldol anion takes proton from H_2O to form aldol.

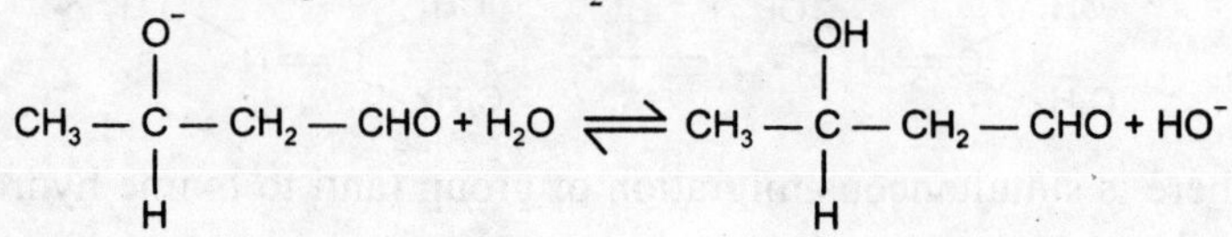

Aldol

Catalyst OH is regenerated. Aldol

(*b*) Canizarro Reaction: Formaldehyde and other aldehydes having no α-Hydrogen atoms undergo canizarro reaction when treated with alcoholic or aqueous alkali. In this reaction one molecule is oxidised at the expense of other which is reduced to alcohol.

$$\underset{\text{formaldehyde}}{2HCHO} + NaOH \longrightarrow HCOONa + CH_3OH$$

Canizarro reaction between two different aldehydes is known as crossed Canizarro reaction.

$$C_6H_5\,CHO + HCHO \xrightarrow{NaOH} HCOONa + C_6H_5CH_2$$

Mechanism:

First Step : There is nucleophilic attack of OH^- on carbonyl Carbon of formaldehyde to give hydro alkoxide ion.

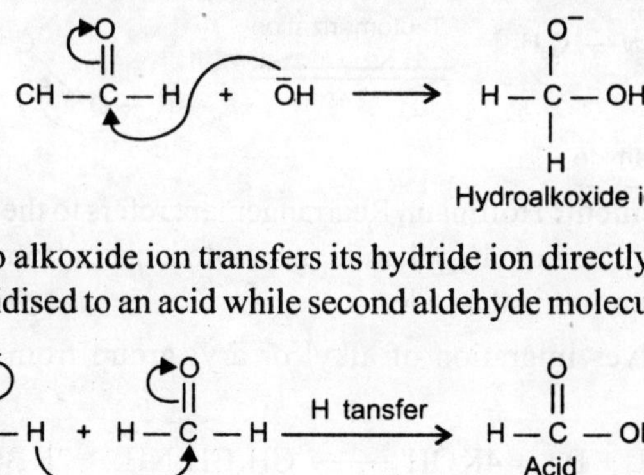

Hydroalkoxide ion

Second step: The hydro alkoxide ion transfers its hydride ion directly to another aldehyde molecule, thus hydroalkoxide ion is oxidised to an acid while second aldehyde molecule is reduced to an alkoxide ion.

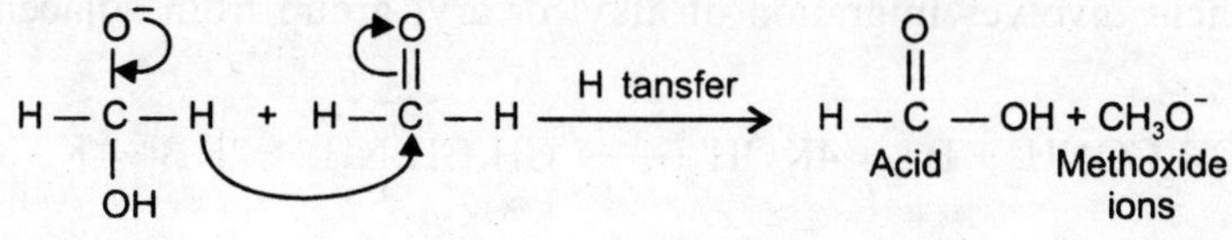

Third step: Acid and Methoxide ion exchanges their proton to give acid anion and alcohol.

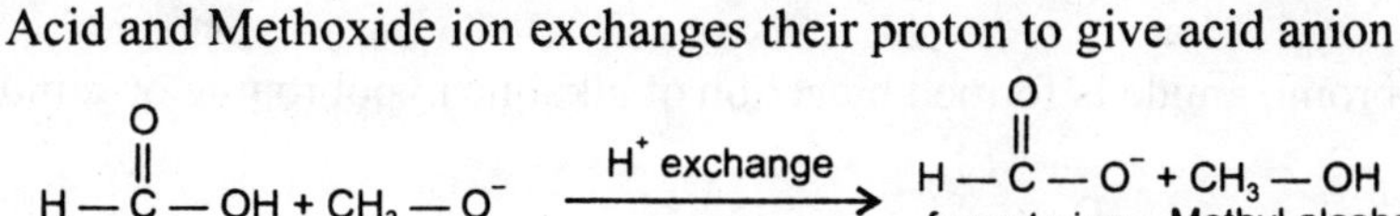

(*c*) Backmann Rearrangement: Ketoximes on treatment with an acid catalyst (PCl_5, P_2O_5, $SOCl_2$, $POCl_3$, H_2SO_4) undergo a rearrangement to form N-substituted amides. This reaction is known as Beckmann Rearrangement.

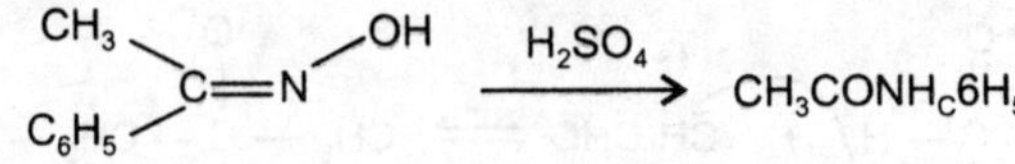

The migrating group is anti to oxime hydroxyl group.

First Step: Oxime reacts with acid catalyst to give an oxonium ion.

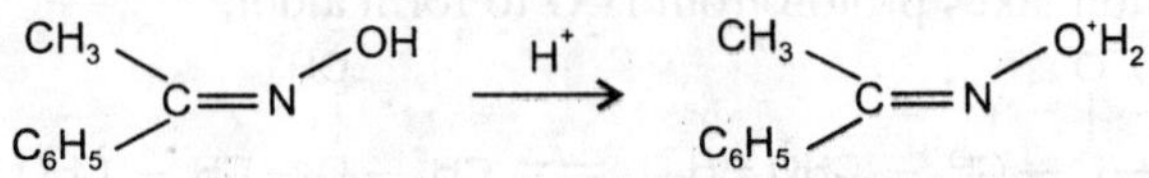

Second Step: There is simultaneous migration of group (anti to oxime hydroxyl) and removal of H_2O as neutral molecule.

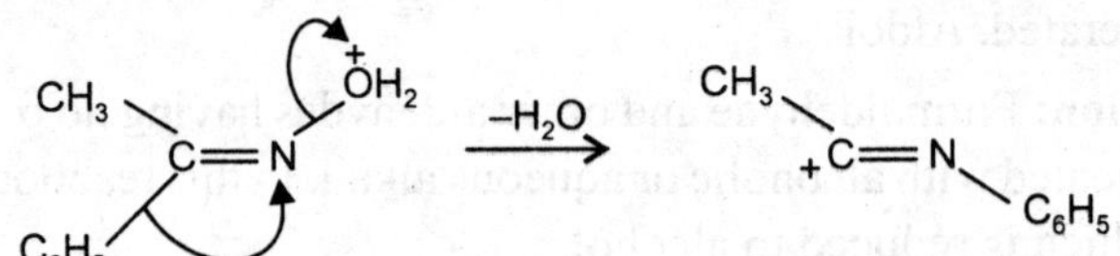

Imino Carbonium ion

Third Step: Imino carbonium ion reacts with water to give N substituted amide.

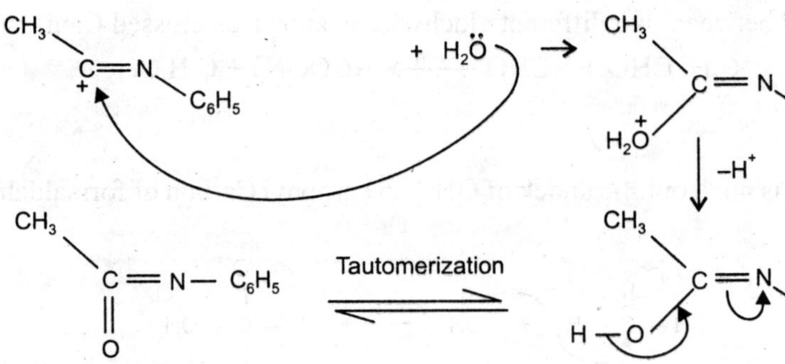

N-Substituted amide

(*d*) Hoffmann Rearrangement: Hoffmann Rearrangement refers to the conversion of primary amide into primary amines by action of sodium hypohalide (generated in situ from halogen and NaOH). Amine formed has one less carbon than original amide.

This rearrangement involves migration of alkyl or aryl group from adjacent carbon to electron deficient nitrogen.

$$CH_3CH_2CONH_2 + Br_2 + 4KOH \longrightarrow CH_3CH_2NH_2 + 2KBr + K_2CO_3 + 2H_2O$$

Mechanism :

First Step: N-bromo amide is formed by action of alkaline hypobromite on amide:

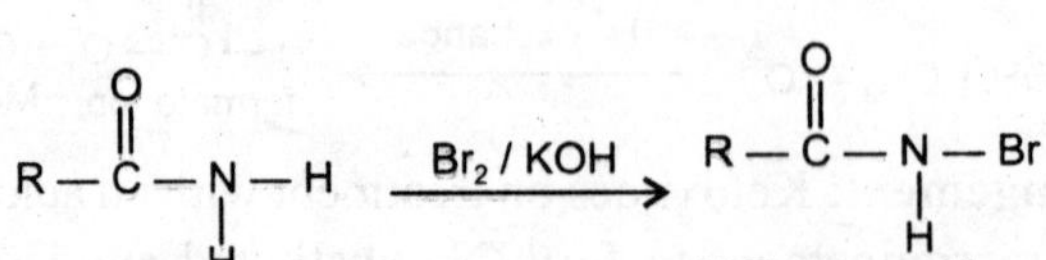

N-Bromo amide

Second Step: Base abstracts acidic Hydrogen from N-Bromo amide. Resultant anionic nitrogen species is instable and loses Br^- with simultaneous migration of alkyl or aryl group from adjacent Carbon to nitrogen.

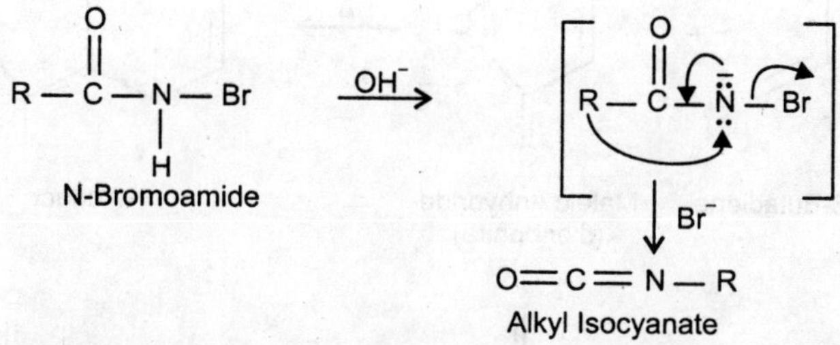

Third Step: Alkyl isocyanate is hydrolyzed to give N-substituted Carbamic acid which is unstable and decarboxylates into primary amine.

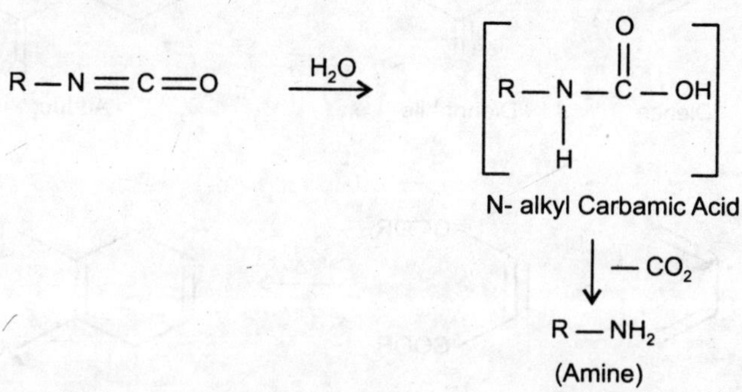

(*e*) Diels Alder reaction: [4 + 2] cycloaddition reaction between a conjugated diene [4π e^- system] and a compound containing a double or triple bond called dienophile [2π e^- system] to form an adduct is known as Diels Alder reaction. Simple alkenes or alkynes are only poor dienophiles. (Dienophiles means love for dienes). One or more electron withdrawing substituents enhance their reactivity as dienophile. Dienophiles containing electron withdrawing substituents such –COOH, NO_2 etc. are more reactive in this reaction where as electron releasing substituents like OH, OCH_3, NH_2 groups enhance the reactivity of dienes. Thus this reaction may also be regarded as nucleophile - electrophile reaction, (diene being electron rich behaves as nucleophile while dienophile being electron poor behaves as electrophile).

Mechanism

1,3 diene (conjugated diene) reacts with a dienophile to form a 6 membered ring with a double bond.

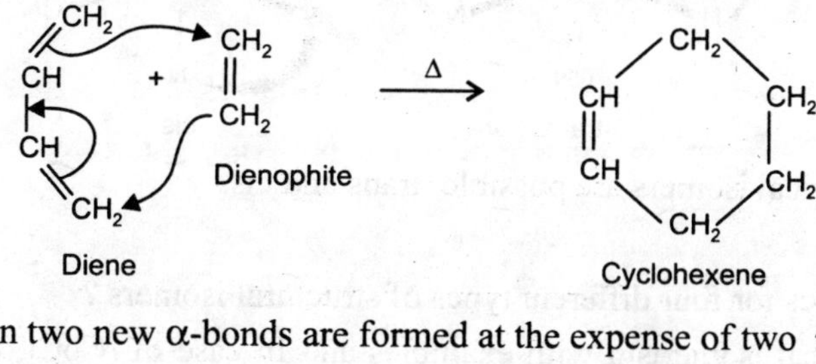

In Diels-Alder reaction two new α-bonds are formed at the expense of two π-bonds and formation of a ring.

Examples:

(a) 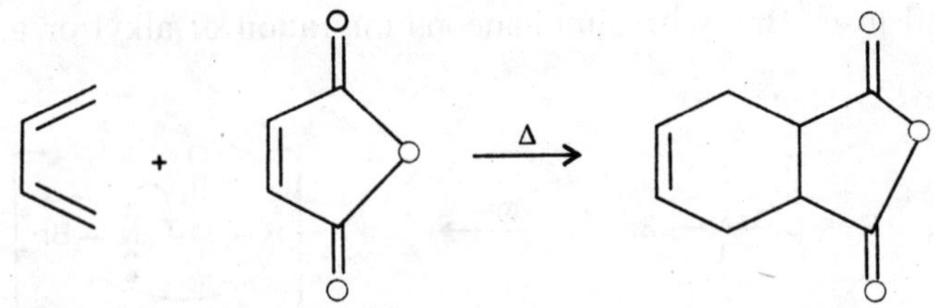

1, 3-Butadiene Maleic Anhydride (dienophite) Adduct

(b) 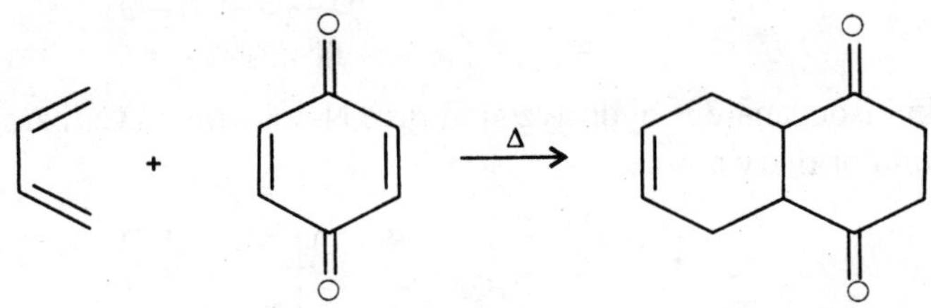

Dience Dienophile Adduct

(c) 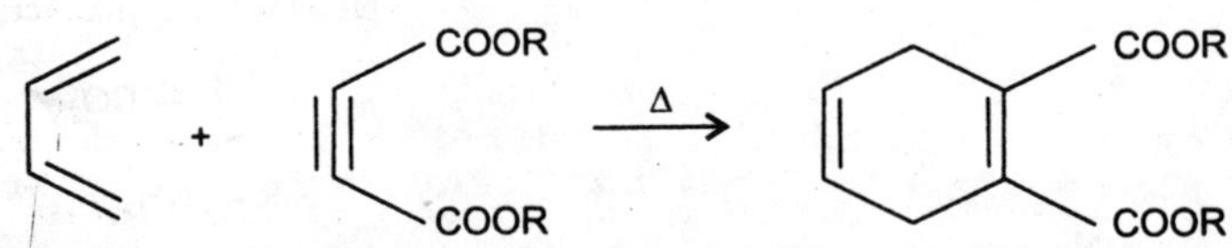

Diene Dienophile

QUESTIONS

1. What is a plane-polarized light? How is it produced?
   What are optically active compounds? What are the necessary conditions for the compound to be optically active?
2. Draw all the possible geometrical isomers for the complex $[Cr(en)_2(CN)_2]^+$.

   Answer:

   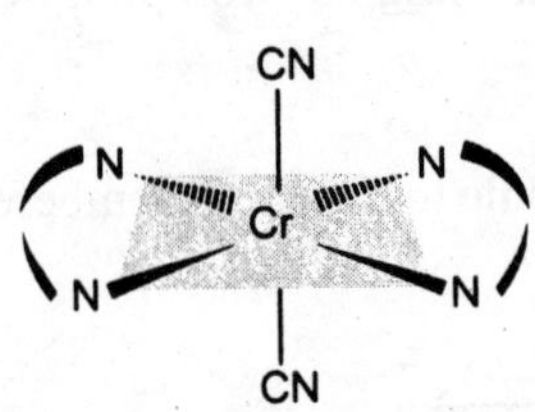

   trans

   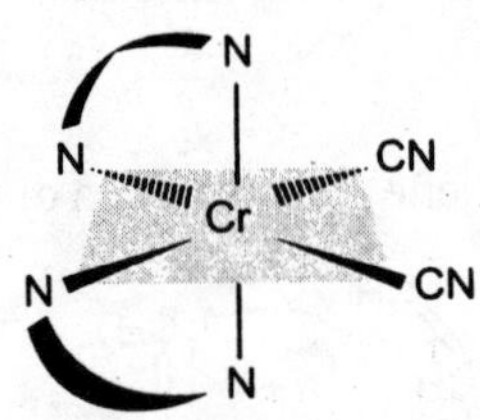

   cis

   Two geometrical isomers are possible: trans and cis.

3.
- Give examples for four different types of structural isomers ?
- Explain optical isomerism with examples and In case of $[CoCl_2(en)_2]^+$ why Trans form is optically inactive than cis form ?

- Read the following bit of text, and then answer the questions after it. You should expect to find it difficult to read. I have deliberately included as many of the unfamiliar words as possible! 2-aminopropanoic acid (alanine) has two enantiomers (optical isomers) because it has a chiral molecule containing an asymmetric carbon atom. One enantiomer is a non-superimposable mirror image of the other. The two enantiomers rotate the plane of polarisation of plane polarised light in opposite directions, but 2-aminopropanoic acid can also be found as a racemic mixture which has no effect on the plane of polarisation. 2-aminopropanoic acid has the structure:
  a) Draw the structures of the two enantiomers. Use your diagram to explain what is meant by the term non-superimposable mirror image.
  b) Explain what is meant by a chiral molecule and say how you would recognise an asymmetric carbon atom.
  c) Why doesn't a racemic mixture have any effect on the plane of polarisation of plane polarised light?
  d) Some, but not all, of the following molecules have optical isomers. For those that do, draw the structures for the two isomers.

4. Draw the structure of the smallest alkane (with a general molecular formula of C_nH_{2n+2}) which has optical isomers.

5. What do you understand by optical activity? The optical activity is associated with the presence of an assymetric carbon atom in it. Justify this statement with suitable example.

6. What are enentiomers? Describe properties of enentiomers.

7. What you meant by conformational analysis? Describe the confirmation of cyclohexane.

8. What is plane polarised light? Describe optical activity by schematic diagram of polarimeter.

9. Describe with the help of suitable example of relative and absolute configuration methods for representing the configuration of optically active compounds.

10. Explain the difference between optical isomerism and geometrical isomerism

11. Which of the following compounds will exhibit cis-trans isomerism?

a) 2-butene b) 2-butyne c) 2-butanol d) butanal

Logic:

The compounds with each doubly bonded carbon attached to two different groups (like Cab=Cab, Cab=Ccd) exhibit geometrical isomerism i.e., cis and trans forms. The geometrical isomerism arises due to restricted rotation of double bond.

However, even though there is restricted rotation for triple bond, alkynes do not exhibit geometrical isomerism, since the triply bonded carbons are attached to one group each only.

Conclusion: Option 'a' is correct

12. How many geometrical isomer is possible for $[PtCl_2(NH_3)_2]$?

a) 1 b) 2 c) 3 d) 4

Logic: It is Ma_2b_2 type molecule.

13. How many geometrical isomer is possible for $[Pt(NO_2)(NH_3)(py)(NH_2OH)]NO_2$?

a) 1 b) 2 c) 3 d) 4

REFERENCES

1. Petrucci, Harwood, Herring, and Madura. General Chemistry: Principles and Modern Applications. 9th ed. Upper Saddle River, New Jersey: Pearson Education, 2007.
2. Unified chemistry by Dr. M.M.N Tandon . D.S. Rathore, S.C. Agrawal and G.C. Saxena, published by Shivlal & Co. Indore.
3. Selected Topics in Inorganic Chemsirty by Malik ,Tuli, Madan published by S. Chand Co., New Delhi.
4. Physical Chemistry,Peter Atkins oxford University Press.
5. Fundamentals of. Molecular Spectroscopy. C. N. Banwell. ,Mcgraw-Hill Book CompanyLondon.
6. Vollhardt, Peter, and Neil E. Schore. Organic Chemistry: Structure and Function. 5th ed. New York: W. H. Freeman & Company, 2007.
7. Advanced Physical chemistry by V.K. Gupta and R. G. Sharma K. Nath & Co. Meerut.
8. Berson, Jerome. Chemical Creativity: Ideas from the Work of Woodward, Hückel, Meerwein, and Others. New York: Wiley-VCH, 1999.
9. Badger, G.M. Aromatic Character and Aromaticity. London, England: Cambridge University Press, 1969.
10. Lewis, David and David Peters. Facts and Theories of Aromaticity. London, England: Macmillan Press, 1975.
11. Organic Chemistry Reaction and Reagents by O P Agrawal, Goel Publishing House.
12. hyperphysics.phy-astr.gsu.edu/hbase/Solids/band.html
13. https://chem.libretexts.org
14. https://onlinecourses.nptel.ac.in
15. http://infofavour.blogspot.com/2016/10/valence-bond-vb-theory-and-its.html
16. https://en.wikibooks.org/wiki/Introduction_to_Inorganic_Chemistry/Coordination_Chemistry
17. https://courses.lumenlearning.com/boundless-chemistry/chapter/coordination-compounds/
18. http://electron6.phys.utk.edu/phys250/modules/module%203/hydrogen_atom.htm
19. http://www.umich.edu/~chem461/QMChap7.pdf
20. http://eguruchela.com/chemistry/learning/Introduction_of_Molecular_orbital.php
    https://www.allaboutcircuits.com/textbook/semiconductors/chpt-2/band-theory-of-solids/

Objectives

Unit 1 – ATOMIC AND MOLECULAR STRUCTURE

Learning Objective

Model curriculum has been framed to meet the expectations of an academically challenging environment with current standards and to enrich the students. Technology is being increasingly based on the electronic, atomic and molecular level modifications. Quantum theory is more than 100 years old and to understand phenomena at nano - meter levels. To fulfil above standards, objectives of this chapter are–

- Students will be able to learn Schrodinger equation through conceptual interpretation of quantum chemistry.
- Be able to understand and draw atomic and molecular orbital energy level diagrams for diatomic molecules, poly atomic molecules and unsaturated systems.
- Be able to identify, differentiate and calculate aromaticity of organic compounds.
- Differentiate and Shapes and geometry of transition metal ions as well as their magnetic properties.
- The students will learn band structure of solids and the role of doping in elements.

Summary

Schrodinger Wave equation

$$\frac{\partial^2 \Psi}{\partial X^2} + \frac{8p^2 m}{h^2}(E - V)\Psi = 0$$

Function for Hydrogen like Atoms - Integrals in spherical-polar coordinates over a spherically-symmetrical integrand can be significantly simplified. And after reduction written as

$$\int_0^{\infty}\int_0^{\pi}\int_0^{2\pi} f(r) r^2 \sin\theta dr\, d\theta\, d\varphi = \int_0^{\infty} f(r) 4\pi r^2 dr$$

since integration over θ and φ gives 4π, the total solid angle of a sphere. The normalization of the 1s wavefunction can thus be written as

$$\int_0^{\infty} [\psi 1s(r)]^2 4\pi r^2 dr$$

Bonding Molecular Orbitals -When addition of wave function takes place, the type of molecular orbitals formed are called Bonding Molecular orbitals and is represented by - $\Psi_{MO} = \Psi_A + \Psi_B$

Anti-Bonding Molecular Orbitals -When molecular orbital is formed by subtraction of wave function, the type of molecular orbitals formed are called **Antibonding Molecular Orbitals** and is represented by $-\Psi_{MO} = \Psi_A - \Psi_B$

Increasing order of energy level for MO - $\sigma 1s, \sigma^*1s, \sigma 2s, \sigma^*2s, [\pi 2p_x = \pi 2p_y], \sigma 2p_z [\pi^*2p_x = \pi^*2p_y], \sigma^*2p_z$

Bond Order- defined as the half of difference between the number of electrons present in the bonding orbitals and the antibonding orbitals that is,

Bond order(B.O.) = (No. of electrons in BMO - No. of electrons in ABMO)/ 2

Aromaticity -When deciding if a compound is aromatic, go through the following checklist. If the compound does not meet all the following criteria, it is likely not aromatic.

1. The molecule is cyclic (a ring of atoms)
2. The molecule is planar (all atoms in the molecule lie in the same plane)
3. The molecule is fully conjugated (p orbitals at every atom in the ring)
4. The molecule has $4n+2\pi$ electrons (n=0 or any positive integer)

Postulates of CFT

- Ligand are either negative ions *e.g.*, F^-, Cl^-, Br^-,CN^-, etc. or neutral molecule consisting dipole *e.g.* NH_3, H_2O etc.
- Complex is a group of central metal ion and ligand surrounding the central metal ion.
- At the time of formation of complex by metal ion and ligands, ligends as point charge move towards the metal ions with their negative charge facing the metal ion.
- Metal ion orbitals get splitted in the set of different energy due to non- uniform increase in the energy of orbitals of metal ion. This splitting of energy depends on the geometry of the complex.

Spectrochemical Series: Increasing value of **crystal field splitting (Δ)**

I- < Br- < S^{2-}< SCN^- <Cl^- < N_3^- < F^- < urea, OH^- < ox, O^{2-} < H_2O < NCS^- < py, NH_3 < en < bipy, phen < NO_2^- < CH_3^-, C_6H_5-< CN^-

CFSE of octahedral complex = $- 0.4 \times n(t_2g) + 0.6 \times n(e_g) \Delta_0$

For a tetrahedral complex = $0.4 \times n(t_{2g}) - 0.6 \times n(e_g) \Delta_t$

Magnetic properties

a. **Diamagnetism:** Due to motion and spinning paired electrons

b. **Paramagnetism:** Due to motion and spinning of unpaired electrons, $\mu = \sqrt{n(n+2)}$

c. **Ferromagnetism:** Ferromagnetic compounds retain magnetism ever after removing from magnetic field. Ex. alloys of Fe, Co. Ni, Mn.

d. **Antiferromangetism:** Substance which do not show paramagnetism even after the presence of unpaired electrons are called ferromagnetic substances eg. MnO, MnF_2

Conductors- No band gap between their valence and conduction bands. Examples – Metals.

Insulators - The band gap between the valence band the conduction band is so large that electrons cannot make the energy jump from the valence band to the conduction band. Example- wood, non metals, plastics.

Semiconductors -Semiconductors have a small energy gap between the valence band and the conduction band. Electrons can make the jump up to the conduction band, but not with the same ease as they do in conductors.

Unit 2 – SPECTROSCOPIC TECHNIQUES AND APPLICATIONS

Learning Objective

- Students will be able to know Principles of UV-Visible, IR spectroscopy and Its applications in daily life.
- They may understand selection rules of all kind of spectroscopic analysis.
- Able to differentiate between vibrational and rotational spectroscopy and formula related of diatomic molecules.
- Learn about NMR and MRI and its application in medical and other fields.
- Students may have clear concept of techniques related to surface characterization.
- Get knowledge of diffraction and scattering techniques.

Summary

Spectroscopy is a technique that uses the interaction of energy with a sample to perform an analysis.
Electromagnetic waves : EM waves are composed of oscillating magnetic and electric fields.

Born Oppenheimer Approximation -the total energy of a molecule is given by ,

$$E = E_{translational} + E_{rotational} + E_{vibrational} + E_{electronic}$$

$$E_{electronic} << E_{vibrational} << E_{rotational} << E_{translational}$$

The Franck–Condon principle is a rule in spectroscopy and quantum chemistry that explains the intensity of vibronic transitions. Vibronic transitions are the simultaneous changes in electronic and vibrational energy levels of a molecule due to the absorption or emission of a photon of the appropriate energy.

Jablonski Diagram Once a molecule has absorbed energy in the form of electromagnetic radiation, there are a number of routes by which it can return to ground state. Graphical representation of this is termed as Jablonski diagram.

Types of Electronic Transitions: $\sigma - \sigma^*$ transition, $n - \sigma^*$ transition, $\pi - \pi^*$ transition, $n - \pi^*$ transition.

Fluorescence: If the photon emission occurs between states of the same spin state (*e.g.* $S_1 \longrightarrow S_0$) this is termed fluorescence.

Phosphorescence: If the spin state of the initial and final energy levels are different (*e.g.* $T_1 \longrightarrow S_0$), the emission (loss of energy) is called phosphorescence.

Fluorescence spectroscopy: Fluorescence spectrometry is a fast, simple and inexpensive method to determine the concentration of an analyte in solution based on its fluorescent properties.

Infrared Spectroscopy: The infrared absorption spectrum of a substance is sometimes called its molecular fingerprint. Although frequently used to identify materials, infrared spectroscopy also may be used to quantify the number of absorbing molecules.

Normal modes of vibrations: nonlinear: 3N – 6 independent vibrational modes. linear: 3N – 5 independent vibrational modes

Raman Spectroscopy: Raman scattering of light by molecules may be used to provide information on a sample's chemical composition and molecular structure.

Nuclear Magnetic Resonance: The principle behind NMR is that many nuclei have spin and all nuclei are electrically charged. The energy transfer takes place at a wavelength that corresponds to radio frequencies and when the spin returns to its base level, energy is emitted at the same frequency. The signal is NMR spectrum.

Chemical shift is a function of the nucleus and its environment. It is measured relative to a reference compound. For ^1H NMR, the reference is usually tetramethylsilane, $Si(CH_3)_4$TMS.

$$\delta = \frac{H_0\,(\text{reference}) - H_0\,(\text{sample})}{H_0\,(\text{reference})} \times 10^6 \text{ppm}$$

Magnetic resonance imaging (MRI) Magnetic resonance imaging is a medical application of nuclear magnetic resonance (NMR) develop during period of 1970s and 1980s. MRI scanners use strong magnetic fields, electric field gradients, and radio waves to generate images of the organs in the body.

X-Ray Diffraction - XRD provides detailed information on the crystallographic structure and physical properties of materials and thin films.

Unit 3 – Intermolecular forces and potential energy surfaces

Learning Objective

- To learn concept and types of bonding in chemical compounds due to molecular interactions.
- To know reason, requirements, types and utility of hydrogen bond.
- Can able to understand idea of Vander Waals Forces in molecule sand chemical compounds.
- Gain complete knowledge of equations of state of real gases and critical phenomena and related equations after application of all corrections.
- Students may gain information of potential energy surfaces of various chemical compounds.

Summary

Molecular Interactions may be two types: Intramolecular interaction and Intermolecular interaction

Three types of intermolecular forces - Ionic interaction,Dipole-dipole interaction and Van der waals interactions.

Ionic Forces or Ionic: Interaction are interactions between charged atoms or molecules. This interactions arise from electrostatic attraction between two groups of opposite charge.

Dipole-Dipole Interaction: Dipole-dipole interaction occurs whenever two polar molecules get near each other. The positively charged portion of one molecule is attracted to the negatively charged portion of another molecule.

Hydrogen Bonding: A dipole-dipole interaction where the H atom is noncovalently attracted to an electronegative atom like F, O, N.

Van der Waals Forces -Van der Waals forces are the interaction between uncharged atoms or molecules.

Ionic Bonds > H Bonding > Van der Waals Dipole-Dipole Interactions > Van Der Waals Dispersion Forces.

Ideal gases-A gas which obeys Boyle's law, Charles' law etc. under all condition of temperature and pressure is known as an ideal gas or a perfect gas.

Real gas – A real gas is one which obeys the gas law fairly well under low pressure or high temperature.

Critical temperature - Critical temperature of a gas may be defined as that temperature above which it cannot be liquefied however high pressure may be applied on the gas.

Critical Pressure - The pressure required to liquefy the gas at the critical temperature is called critical pressure .

Critical volume - The volume occupied by 1 mole of the gas at critical temperature and critical pressure is called critical volume.

Andrews isotherms of carbondioxide is a pressure - volume isotherms of carbon dioxide gas at a series of temperature.

Thomson's experiment

Joule-Thomson Effect Joule-Thomson showed that when a compressed gas is forced through a porous plug into a region of low pressure, there is appreciable cooling.

Potential energy surfaces of H_3, H_2F, HCN - Potential energy surface is a mathematical function that gives the **energy** of a molecule as a function of its geometry.

A potential energy surface (PES) describes the potential energy of a system, especially a collection of atoms, in terms of certain parameters, normally the positions of the atoms.

Unit 4 – USE OF FREE ENERGY IN CHEMICAL EQUILIBRIA

Learning Objective

- Students will able to learn concept of thermodynamic functions: energy, entropy and free energy and related equations and formulas.
- Can be able to understand emf of cell, cell potentials, Nernst equation its applications.
- Definition, explanations and examples of various theories of acid and base can be learnt by all readers.
- Introductory idea and explanation of oxidation- reduction and solubility equilibrium.
- Types of water, hardness, softening methods are explained under the portion of water chemistry.
- Students may be learning what is corrosion? How it is formed and types of it and uses of free energy considerations in metallurgy through Ellingham diagrams.

Summary

Free energy (ΔG) Free energy has the dimensions of energy, and its value is determined by the state of the system and not by its history. It is expressed in two forms:

1. Helmholtz free energy or work function **F** **2.** Gibbs free energy **G.**

$$G = H - TS$$

Free energy is used to calculate Work Done $-\Delta F = -\Delta U + T\Delta S$, $-\Delta G = -\Delta H + T\Delta S$

ΔG is negative for $T > 273.16$ K, indicating that the direction of spontaneous change is from ice to water.

ΔG is positive for $T < 273.16$ K, where the reverse reaction of freezing takes place.

Enthalpy -The enthalpy (ΔH) is a measure of the actual energy that is liberated when the reaction occurs (the "heat of reaction"). $dH = dQ + VdP$, $dU = dQ - PdV$

Sign of dH – When heat absorb during chemical reaction dH will have positive sign and when heat is evolved , then will have negative sign.

Entropy (ΔS) -**Entropy** S, which is a measure of the order or disorder. $\Delta S = \frac{Q_{Rev}}{T}$

Entropy change in adiabatic process in an adiabatic change heat is neither given to the system nor taken out of the system. Therefore, q=0, and the change in entropy during the process, will be

$$\Delta S = \frac{Q_{Rev}}{T} = \frac{q}{T} = 0$$

change in entropy -

$$\Delta S = 2.303 n\, C_v \log \frac{T_2}{T_1} + nR\ 2.303 \log \frac{V_2}{V_1}$$

Entropy changes for other process

- Isochoric Process – $\Delta S_v = nC_v \ln \frac{T_2}{T_1}$
- Isobaric Process – $\Delta S_p = nC_p \ln \frac{T_2}{T_1}$
- Isothermal Process – $\Delta S_T = \Delta S_T = nR \ln \frac{P_2}{P_1}$
- Adiabatic Process – $\Delta S = q/T = 0$; $nC_p \ln$
- Entropy Change in Physical Change $\Delta S = \frac{q}{T} = \frac{\Delta H}{T}$
- Entropy of fusion $\Delta S_f = \frac{\Delta H_f}{\Delta T_f}$ ΔH_f become negative, process of freezing the decrease in entropy occurs.
- Entropy of evaporation $\Rightarrow \Delta S_v = \frac{\Delta H_v}{T_b}$ ΔH_v = negative the process of condensation of vapours

- Change from one crystalline form to other crystalline

$$\Delta S_t = \frac{\Delta H_t}{T_t}$$

Gibbs Free Energy

The **Gibbs free energy** G is the thermodynamic potential. The two measures of energy (enthalpy H and entropic energy TS) are brought together in the Gibbs free energy equation: (the chemical potential is the equivalent for a component). $G = H - TS$

Relationship Among G, S, and V $dG = VdP - SdT$

It means that changes in Gibbs free energy are produced by changes in pressure and temperature acting on the volume and entropy of a phase.

Cell potentials-The cell potential, E_{cell}, is the measure of the potential difference between two half cells in an electrochemical cell..

Components of Electrochemical Cell - Two half cells

1. Two metal electrodes
2. One voltmeter
3. One salt bridge
4. Two aqueous solutions for each half cell

Standard Cell Potential-The standard cell potential (E°cell) is the difference of the two electrodes, which forms the voltage of that cell. To find the difference of the two half cells, the following equation is used:

$$E^o_{Cell} = E^o_{Red,Cathode} + E^o_{Oxd,Anode}$$

Nernst Equation – When the cell potential equals zero, the reaction is at equilibrium.

$$E = E^\circ - \frac{RT}{nF} \ln Q_c$$

Theories of acis and bases

- **The Arrhenius Theory Acid** - Acids are substances which produce hydrogen ions (H^+) in solution.**Base** - Bases are substances which produce hydroxide ions (OH^-) in solution.
- **The Bronsted-Lowry Theory -Acid** - An acid is a proton (hydrogen ion) donor.**Base** -A base is a proton (hydrogen ion) acceptor.
- **The Lewis Theory** -An acid is an electron pair acceptor. A base is an electron pair donor.

Oxidation and Reduction

Oxidation and reduction in terms of oxygen transfer

- Oxidation is gain of oxygen.
- Reduction is loss of oxygen.

Oxidation and reduction in terms of hydrogen transfer

- Oxidation is loss of hydrogen.
- Reduction is gain of hydrogen.

Oxidation and reduction in terms of electron transfer

- Oxidation is loss of electrons.
- Reduction is gain of electrons.

Hardness of water- Water which does not produces lather with soap are termed as hard water. The hardness is mostly due to the presence of Ca and Mg metallic ions which react with soap and form precipitates.

Types of Hardness is of two types -(A) Temporary Hardness (B) Permanent Hardness

Methods for removal of hardness are-Lime –Soda method, Zeolite Method, Ion-Exchange method

Corrosion -Any process of deterioration (or destruction)and consequent loss of a solid metal through an unwanted chemical or electrochemical attack of its environment , starting at its surface is called corrosion.

"Corrosion is a reverse process of extraction"

Diagram -Thermodynamics of Metallurgy - H.G.T Ellingham proposed the Ellingham diagram to predict the spontaneity of reduction of various metal oxides. diagram is basically a curve which related the Gibbs energy value with the temperature.

Unit 5 – PERIODIC PROPERTIES

Learning Objective

Students will be able-

- To learn and cherish various characters of elements based on periodic classification like Electronic Configurations, Atomic and Ionic sizes, Ionization Energies, Electron affinity, Electro negativity, Polarizability and Oxidation states.
- Determine the number of valence electrons for elements from their electron configuration or position on the periodic table.
- They may able to get clear idea of effective nuclear charges and penetration of orbitals.
- Can correlates various elements and its properties according to variations of s, p, d and f orbitals.
- Able to locate an element on the periodic table and determine which group and which period it is in.
- May able to calculate coordination numbers and find out Geometries of various molecules.

Summary

EFFECTIVE NUCLEAR CHARGES

The effective nuclear charge is the net charge an electron experiences in an atom with multiple electrons.
$Z_{eff} = Z - S$

Penetration refers to how effectively electrons can get close to the nucleus. Different orbitals have greater nuclear **penetration** than others.

Order for filling the electrons in orbitals-**1s, 2s, 2p, 3s, 3p, 4s, 3d, 4p, 5s, 4d, 5p, 6s, 4f, 5d, 6p, 7s, 5f, 6d, 7p**

- **Pauli Exclusion Principle:** No two electrons in an atom can have all the four quantum numbers exactly the same.
- **Hund's Rule:** When assigning electrons in orbitals, each electron will first fill all the orbitals with similar energy (also referred to as degenerate) before pairing with another electron in a half-filled orbital.
- **Aufbau principle:** According to this principle, the electrons will first occupy the orbitals that have the lowest energy.

Elements can be classified according to their electronic configuration are s,p,d and f-block elements

Atomic Radii: The atomic size is, therefore, generally expressed as the radius of the sphere or the distance between the centre of the nucleus and the outermost shell of electrons.

Types of atomic radii -(i) Covalent radius (ii) Metallic radius (iii) Vander Waal's radius (iv) Ionic radius

Ionisation Energy: The electrons in an atom are attracted by the positively charged nucleus. In order to remove an electron from an atom, energy has to be supplied.

$IE_3 > IE_2 > IE_1$

Electron Affinity: The amount of energy released when an electron is added to a neutral gaseous atom of an element to form an anion is called first electron affinity. For example,

Electronegativity: Electronegativity is defined as a measure of the ability of an atom in a covalent molecule to attract the shared electron pair towards itself.

POLARIZABILITY

The power of an ion to destroy the other ion is known as its polarizing power and the tendency of the ion to the ion to distort is known as its polarizability.

Fajans' Rules: Rules are used to predict whether a chemical bond will be covalent or ionic, and depend on the charge on the cation and the relative sizes of the cation and anion.

Oxidation State: An oxidation state is a number that is assigned to an element in a chemical combination. This number represents the number of electrons that an atom can gain, lose, or share when chemically bonding with an atom of another element.

Coordination Number -The coordination number is the sum of the atoms in a given molecule bonded to the central metal atom of a compound. Coordination numbers typically ranges from 2 to 16 but is usually 6.

Coordination Number Geometry-Coordination Number 2 – linear, Coordination Number 4 - tetrahedral, square planar, Coordination Number 6 - hexagonal planar, trigonal prism, octahedral.

Hard and soft acids and bases -According to HSAB concept, hard acids prefer binding to the hard bases to give ionic complexes, whereas the soft acids prefer binding to soft bases to give covalent complexes.

Applications of HSAB principle

- Application of in hydrogen bonding
- Linkage of ambidentate ligands to metal atoms
- Site preference in organic reactions
- Inorganic reactions
- Precipitation reactions

Molecular geometries

Molecular geometry is the three-dimensional arrangement of the atoms that constitute a molecule. It influences several properties of a substance including its reactivity, polarity, phase of matter, color, magnetism and biological activity.

Unit 6 – STEREOCHEMISTRY AND ORGANIC REACTIONS

Learning Objective

Students will be able:

- To get complete idea about representations of 3 dimensional structures of various organic compounds and its theories.
- To learn and draw the formula and concepts of structural isomers and stereoisomerisms and optical isomerisms.
- To learn, identify, differentiate and calculate configurations, chirality, enantiomers, diastereomers and conformational analysis of organic compounds .
- Isomerism in transitional metal , effect of unpaired electrons in transition metals to arrange it geometry can be learnt by students through this chapter.

Summary

Three-Dimensional Representations: Representation of three dimensional structures of Organic molecules are

Wedge-Dash Notation ,Newman Projections,Sawhorse Projections,Fischer Projections and Haworth Projections

Isomerism -isomers are molecules with the same molecular formula (i.e. the same number of atoms of each element), but different structural or spatial arrangements of the atoms within the molecule.

- **Structural isomerism:** Isomers can be split into two broad groups – structural (or constitutional) isomers, and stereoisomers.
- **Chain Isomers:** Chain isomers are molecules with the same molecular formula, but different arrangements of the carbon 'skeleton'.
- **Position Isomers:** Position isomers are based on the movement of a 'functional group' in the molecule. A functional group in organic chemistry is the part of a molecule that gives it its reactivity.
- **Functional Isomers:** Also referred to as functional group isomers, these are isomers where the molecular formula remains the same, but the type of functional group in the atom is changed.
- **Metamerism:** This form of isomerism is rare and is limited to molecules having a divalent atom like O or S and alkyl groups around it. The main examples come from ethers and thioethers.
- **Tautomerism:** Due to spontaneous inter conversion of two isomeric forms with different functional groups. The prerequisites for this is the presence of the C=O, C==N or N=O in the usual cases and an alpha H atom. The most usual is the 'keto-enol' tautomerism, but there can be others like nitro-aci and amine-imine forms.

- **Ring-Chain isomerism:** Here one isomer is an open chain molecule and the other a cyclic molecule.

Stereoisomerism

Molecules with the same connectivity but different arrangements of the atoms in space are called stereo isomers. Two main types of stereoisomerism – Geometric isomerism and Optical isomerism.

Isomerism in Transition Metal Compounds

The existence of coordination compounds with the same formula but different arrangements of the ligands was crucial in the development of coordination chemistry. These are of two types

1. **Stereoisomerism:** Stereoisomers occur when the ligands have the same bonds, but the bonds are in different orientations relative to one another. It is of two types-

   a) Geometrical isomerism

   b) Optical isomerism

2. **Structural Isomerism:** Structural isomerism occurs when the bonds are different. Isomers are distinct compounds that can have different physical properties such as color, crystal structure, and melting point.

   a) Coordination isomerism

   b) Ionisation isomerism

   c) Hydrate isomerism

   d) Linkage isomerism

INDEX

ENGINEERING CHEMISTRY

DR. MANISHA AGRAWAL AND DR. SATYA PRAKASH

FEEDBACK FORM

1. Which topics from your examination point of view are inadequately or not covered in this book?

____

____

2. Is there any misprints/ mistakes/ factual inaccuracy in this book? If so, please specify.

____

____

3. What is your assessment of this book as regards the presentation of the subject-matter, expression, precision and price in relation to the other recommended books available on this subject?

____

____

4. Which competing books you regard as better than this? Please mention their author and publishers.

____

____

5. Any further suggestions/comments you would like to make for the improvement of the book?

____

____

PS: If needed, please attach a separate sheet.

Name : ______________

College : ______________

Address : ______________

Email : ______________

Mobile : ______________

Please mail the feedback form at:

KHANNA BOOK PUBLISHING CO. (P) LTD.

4C/4344, Ansari Road, Daryaganj, New Delhi-110002

You can also email this feedback form at contact@khannabooks.com or Whatsapp at +91 – 99109 09320

BEST FEEDBACK WILL BE REWARDED SUITABLY